Innovative Processing Technologies for Healthy Grains

Innovative Processing Technologies for Healthy Grains

Edited by

Milica Pojić
Institute of Food Technology
University of Novi Sad
Novi Sad
Serbia

Uma Tiwari
School of Food Science and Environmental Health
Technological University Dublin
City Campus
Dublin
Ireland

WILEY Blackwell

This edition first published 2021

Registered Offices
John Wiley & Sons, Inc., 111 River Street, Hoboken, NJ 07030, USA
John Wiley & Sons Ltd, The Atrium, Southern Gate, Chichester, West Sussex, PO19 8SQ, UK

Editorial Office
The Atrium, Southern Gate, Chichester, West Sussex, PO19 8SQ, UK

For details of our global editorial offices, customer services, and more information about Wiley products visit us at www.wiley.com.

Wiley also publishes its books in a variety of electronic formats and by print-on-demand. Some content that appears in standard print versions of this book may not be available in other formats.

Library of Congress Cataloging-in-Publication Data

Names: Pojić, Milica, editor. | Tiwari, Uma, editor.
Title: Innovative processing technologies for healthy grains / edited by Milica Pojić, Uma Tiwari.
Description: Hoboken, NJ : Wiley-Blackwell, [2020] | Includes bibliographical references and index.
Identifiers: LCCN 2020028487 (print) | LCCN 2020028488 (ebook) | ISBN 9781119470168 (cloth) | ISBN 9781119470212 (adobe pdf) | ISBN 9781119470199 (epub)
Subjects: LCSH: Cereal products. | Grain–Processing–Technological innovations.
Classification: LCC TP434 .I55 2020 (print) | LCC TP434 (ebook) | DDC 664/.7–dc23
LC record available at https://lccn.loc.gov/2020028487
LC ebook record available at https://lccn.loc.gov/2020028488

Cover Design: Wiley
Cover Images: courtesy of Milica Pojić

Set in 9.5/11.5pt Times Ten by SPi Global, Pondicherry, India

10 9 8 7 6 5 4 3 2 1

Contents

About the *IFST Advances in Food Science* Book Series

The Institute of Food Science and Technology (IFST) is the leading qualifying body for food professionals in Europe and the only professional organization in the UK concerned with all aspects of food science and technology. Its qualifications are internationally recognized as a sign of proficiency and integrity in the industry. Competence, integrity, and serving the public benefit lie at the heart of the IFST philosophy. IFST values the many elements that contribute to the efficient and responsible supply, manufacture, and distribution of safe, wholesome, nutritious, and affordable foods, with due regard for the environment, animal welfare, and the rights of consumers. *IFST Advances in Food Science* is a series of books dedicated to the most important and popular topics in food science and technology, highlighting major developments across all sectors of the global food industry. Each volume is a detailed and in-depth edited work, featuring contributions by recognized international experts, and which focuses on new developments in the field. Taken together, the series forms a comprehensive library of the latest food science research and practice, and provides valuable insights into the food processing techniques that are essential to the understanding and development of this rapidly evolving industry. The *IFST Advances* series is edited by Dr. Brijesh Tiwari, who is Senior Research Officer at Teagasc Food Research Centre in Ireland.

Forthcoming Titles in the IFST Series

Food Formulation: Novel Ingredients and Processing Techniques edited by Shivani Pathania and Brijesh K. Tiwari.

Recent Advances in Micro- and Macroalgal Processing: Food and Health Perspectives edited by Gaurav Rajauria and Yvonne V. Yuan.

Oil and Oilseed Processing: Opportunities and Challenges edited by Tomás Lafarga, Gloria Bobo, Ingrid Aguiló.

List of Figures and Tables

List of Figures

List of Tables

List of Contributors

Mehran Aalami Department of Food Science & Technology, Gorgan University of Agricultural Sciences and Natural Resources, Gorgan, I.R. Iran

Caleb Acquah School of Nutrition Sciences, University of Ottawa, Ottawa, Ontario, Canada

Dominic Agyei Department of Food Science, University of Otago, Dunedin, New Zealand

Catherine Barry-Ryan School of Food Science and Environmental Health, Technological University Dublin, Dublin, Ireland

Gaetano Cardone Department of Food, Environmental and Nutritional Sciences (DeFENS), Università degli Studi di Milano, Milan, Italy

Michael Kobina Danquah Department of Chemical Engineering, University of Tennessee, Chattanooga, TN, USA

Tamara Dapčević Hadnađev Institute of Food Technology, University of Novi Sad, Novi Sad, Serbia

Christian Kwesi Ofotsu Dzuvor Bioengineering Laboratory, Department of Chemical Engineering, Monash University, Clayton, Victoria, Australia

Bojana Filipčev Institute of Food Technology, University of Novi Sad, Novi Sad, Serbia

Abirami R. Ganesan School of Applied Sciences, College of Engineering, Science and Technology, Fiji National University, Nabua Campus, Suva, Nabua, Fiji Islands

Muriel Henrion Nestlé Research Center Orbe, Société des Produits Nestlé SA, Orbe, Switzerland

Nicholas M. Holden UCD School of Biosystems and Food Engineering, University College Dublin, Dublin, Ireland

Jaison Jeevanandam CQM – Centro de Química da Madeira, Universidade da Madeira, Funchal, Portugal

Nivedha Krishnakumar Flour Milling, Baking and Confectionery Technology Department, CSIR-Central Food Technological Research Institute, Mysore, Karnataka, India

Emilie Labat Nestlé Research Center Orbe, Société des Produits Nestlé SA, Orbe, Switzerland

Lisa Lamothe Nestlé Research, Société des Produits Nestlé SA, Lausanne, Switzerland

Sahar Akhavan Mahdavi Department of Food Science & Technology, Gorgan University of Agricultural Sciences and Natural Resources, Gorgan, I.R. Iran

Anamarija Mandić Institute of Food Technology, University of Novi Sad, Novi Sad, Serbia

Alessandra Marti Department of Food, Environmental and Nutritional Sciences (DeFENS), Università degli Studi di Milano, Milan, Italy

Aleksandra Mišan Institute of Food Technology, University of Novi Sad, Novi Sad, Serbia

Maria Ambrogina Pagani Department of Food, Environmental and Nutritional Sciences (DeFENS), Università degli Studi di Milano, Milan, Italy

Sharadwata Pan School of Life Sciences, Weihenstephan, Technical University of Munich, Freising, Germany

Milica Pojić Institute of Food Technology, University of Novi Sad, Novi Sad, Serbia

Pichan Prabhasankar Flour Milling, Baking and Confectionery Technology Department, CSIR-Central Food Technological Research Institute, Mysore, Karnataka, India

Gaurav Rajauria School of Agriculture and Food Science, University College Dublin, Lyons Research Farm, Celbridge, Co. Kildare, Ireland

Cristina M. Rosell Institute of Agrochemistry and Food Technology (IATA-CSIC), Paterna, Spain

Vijaykrishnaraj Muthugopal Sasthri Flour Milling, Baking and Confectionery Technology Department, CSIR-Central Food Technological Research Institute, Mysore, Karnataka, India

Uma Tiwari School of Food Science and Environmental Health, Technological University Dublin, City Campus, Dublin, Ireland

Chibuike C. Udenigwe School of Nutrition Sciences, University of Ottawa, Ottawa, Ontario, Canada

Marco Vassallo CREA - Research Centre for Agricultural Policies and Bio-Economy, Rome, Italy

Mingjia Yan UCD School of Biosystems and Food Engineering, University College Dublin, Dublin, Ireland

Preface

Cereals and other "healthy" grains have captured the consumer's interest in recent times, with growing health awareness and an increasing motivation for plant-based eating, thus challenging the cereal processing industry to deliver products without compromising their healthiness, tastiness, and overall likability. Although cereal processing is one of the oldest forms of food processing, a holistic approach to cereal processing is nowadays needed more than ever before, not only to preserve the health benefits of cereal grains, but also to increase safety, assure sustainability, and decrease the carbon footprint, which is possible by coupling alternative processing techniques with conventional ones.

The concept of healthy grains is based on both the major and minor cereals, and pseudocereals (also known as gluten-free grains), being important sources of energy and macro- and micronutrients in the human and animal diets. Healthy grains are utilized in many food products with high nutritional and biological values, which are required more and more by consumers with high levels of nutrition knowledge and healthy food behaviors.

Throughout its 11 chapters, this book provides an overview of recent advances and innovations, not only those limited to cereal and pseudocereal product development, but also in processing. Hence, topics such as advances in traditional and innovative cereal and pseudocereal processing techniques and innovative products thereof and their functionality, cereal-based animal feed, trends that are driving market demands, and the consumption of healthy grains, as well as the environmental impact of healthy grain processing are represented. The contents of this book provide useful information not only for researchers, academia and students, but for all stakeholders along the cereal and pseudocereal value chain – industry, policy makers, civil society, and retailers – to understand the need for innovation in the cereal and pseudocereal processing sector. Once the challenge of innovation is accepted, whether it is continuously incremental or radical, improvements in terms of productivity, cost, speed, quality, and/or flexibility of production and products are made possible for the benefit of all.

Both editors wish to thank all the contributors for their valuable expertise and their invaluable time in contributing to this book. Also, we thank the editorial support and assistance from series editor Prof. Brijesh Tiwari for helpful suggestions at all stages of the book's development.

Dr. Milica Pojić
Dr. Uma Tiwari

1
Processing Technologies for Healthy Grains: Introduction

Milica Pojić[1] and Uma Tiwari[2]

[1] *Institute of Food Technology, University of Novi Sad, Novi Sad, Serbia*
[2] *School of Food Science and Environmental Health, Technological University Dublin, City Campus, Dublin, Ireland*

1.1 Healthy Grains: What Are They?

Cereal grains have been a principal part of humans' daily diet, consumed in different forms and/or products, for many years. Cereals are traditionally utilized as a breakfast meal or as a main meal of the day, not only to provide carbohydrates, but also to increase the level of dietary fiber. Nowadays, the increase in awareness of health and demand for healthy products by consumers are becoming a challenge for the food industry to develop new and nutritious cereal products. However, when it comes to nutrition, health, and wellbeing, one might think cereal grains inadequate foodstuffs, considering that they have been attributed as a major contributor to obesity due to their high content of easily digestible carbohydrates. Thus, in the early 2000s a decline in wheat consumption was observed in the USA, attributed to the "low carb" diet craze. Moreover, protein, iron, zinc, and vitamin A deficiencies are observed in developing countries with the highest per capita consumption of refined cereal grains, which are low in micronutrients. On the other hand, a vast number of scientific studies that have been emerging demonstrate protective positive effects of whole grains against cardiovascular disease, cancer, diabetes, obesity, and other chronic noncommunicable diseases, which have resulted in a growing consumption of whole-grain products (Awika 2011). The consumption of gluten (and "gluten-like") proteins from major cereals – wheat (including khorasan and spelt), barley (including malts), rye, and triticale – as well as gluten-containing food additives (in the form of flavoring, stabilizing, or thickening agents) and foods contaminated with gluten-containing products (such as oat) causes gastrointestinal problems and malabsorption syndrome in approximately 0.5–1.0% of the world's population, i.e. those diagnosed with celiac disease (El-Chammas and Danner 2011). Gluten-free diets, although predominantly

Innovative Processing Technologies for Healthy Grains, First Edition.
Edited by Milica Pojić and Uma Tiwari.

designed for patients with celiac disease and nonceliac gluten sensitivity, have been gaining increasing popularity in recent years. A growing demand for gluten-free food is not only due to the increasing number of diagnosed patients, but also due to the higher availability of different gluten-free foods in the market (e.g. salty snacks, crackers, fresh bread, pasta, ready-to-eat cereals, baking mixes, cookies, flour, frozen bread/dough, etc.) and due to the advertising campaigns, press coverage and promotion of this type of diet (Newberry et al. 2017). As a result, in the period 2004–2011 the sale of gluten-free products had an annual growth of nearly 28% and in 2012 was close to US$2.6 billion (Asbran 2017; Remes-Troche et al. 2020). Moreover, a survey conducted in 2015 in the US, whose results were published in the report *Gluten-Free Foods in the US* (5th Edition), showed varying attitudes of the population toward these products. The survey indicated that 36% of respondents consumed gluten-free products for reasons other than gluten sensitivity: 65% because they thought it was healthier, 27% because they thought it helped in weight loss, 7% to reduce inflammation, and 4% to fight depression, whilst only 5.7% of respondents claimed the consumption of gluten-free products due to formal medical conditions (Békés et al. 2017). Therefore, in recent years the utilization of pseudocereals, being gluten-free, has captured consumers' interest and more research is now focused on partial or full utilization with cereals to produce "healthy" grain products. Furthermore, health and wellness retail showed growth in healthy products of 3.3% in Asia and the Pacific and 4.2% in the Middle East and Africa (Mascaraque 2018). Similarly, in Europe, the sales of healthy grain products reached €12.8 billion in 2018, with a projection for the market value to increase about 6% from the previous year (CBI 2019). Globally, over the last few years an increase in market demand was observed for products perceived as more natural and "healthier" – a product group consisting of organic, "free-from", and naturally "healthy" products (Mascaraque 2018).

1.2 Cereals and Pseudocereals: Production, Nutritional Value, and Utilization

Cereals (monocotyledonous) and pseudocereals (dicotyledonous) are species that are taxonomically not closely related to each other, but share certain characteristics, such as the structure and composition of their kernels, especially in terms of starch and protein content in approximately the same relative proportions. Moreover, they are cultivated, harvested, processed, and used in the same manner as cereals (Rosentrater and Evers 2018).

Although the increasing worldwide demand for pseudocereals in recent years caused their increased production, they are still considered underutilized feedstock. Their significance is increasing due to high-quality allergy-free proteins and large amounts of micronutrients and bioactive compounds, which increases their market price. Although the worldwide interest in pseudocereals is a relatively recent phenomenon, some of the species were cultivated as traditional crops in certain part of the world for centuries (Rosentrater and Evers 2018). Among pseudocereals, amaranth, quinoa, and buckwheat are of the highest commercial potential. On the other hand, traditional cereals are considered *major* and *minor* based on the volume of their production and utilization. Wheat, maize, rice, and barley are classified as major cereals, while sorghum, millet, oats, rye, spelt, and primitive and wild wheat species are minor cereals. The differences between major and minor cereals are not only in the quantity of production, but also in the nutritional profile, with higher levels of certain antioxidant substances, which makes minor cereals useful in preventing a wide range of diseases linked with oxidative damage (Akkoc et al. 2019).

According to the Food and Agriculture Organization of the United Nations (FAO) Statistics Database (FAOSTAT), the total production level of cereal crops worldwide significantly increased from the year 2000 to 2018. For example, crop production increased by 80% for wheat, 52% for maize, and 77% for rice, followed by barley, sorghum, millet, oats, and rye. Similarly, FAOSTAT also estimated pseudocereal production of buckwheat decreased slightly, while quinoa production increased over the 18 years to 2018 (Figure 1.1). Additionally, FAOSTAT also showed that production increased by 67% for quinoa, but reported a significant decrease by 30% for buckwheat (FAO 2020). However, due to the growing demand to feed the growing world population, the estimated world buckwheat utilization is expected to increase to 7 million tonnes by 2020 (FAO 2020). They predicted that wheat consumption will increase by 12 million tonnes, while world rice utilization will increase to 514 million tonnes in the year 2019–2020.

The major nutritional components of cereals are starch and nonstarch carbohydrates accounting for approximately 87%, while their protein content ranges from 6 to 15% (Goldberg 2003). The major storage proteins present in the cereal grains are gliadins and glutenins for wheat, oryzenin for rice, zeins for maize, kafirins for sorghum and millet, and hordeins and glutelins for barley, while in oats the main proteins are albumins and globulins (Kulp and Ponte 2000). Pseudocereal grains mainly consist of starch and proteins accounting for 55–75% (Venskutonis and Kraujalis 2013) and 12–16% (Mota et al. 2016), respectively. Unlike true cereals, pseudocereals contain high amounts of essential amino acids, particularly methionine, lysine, arginine, tryptophan, and sulfur-containing amino acids (Schoenlechner et al. 2008). Additionally, cereals and pseudocereals also contain good amounts of bioactive compounds including dietary fibers, phenolic acids, carotenoids, β-glucans, as well as other phytochemicals such as tocopherols, alkylresorcinols, and flavonoids associated with the prevention of diseases (Akkoc et al. 2019).

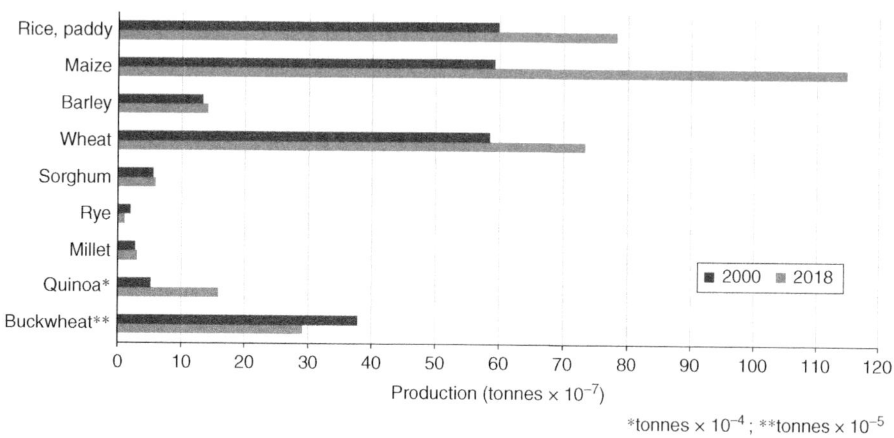

Figure 1.1 Global production of cereals and pseudocereals from 2000 to 2018. *Source*: FAO (2020).

Based on the healthy and nutritive value of cereal and pseudocereal grains, consumers are attracted toward increasing their consumption of the combination of these grains. For this reason, the popularity of healthy grains in many countries has gained importance and researchers are focused on creating new and innovative products.

1.3 Cereal Byproducts for Food and Feed Utilization

The increased demands for sustainability of food production, climate change, and limited natural resources for food for an increasing global population reaching 10 billion by 2050 impose the need to improve the efficiency of food systems and find alternative food solutions (Fasolin et al. 2019; Galanakis 2020). One of them is valorization of byproducts and side streams, and when it comes to cereals and pseudocereals, they are generated in dry milling, pearling, and malting processes. These processes generate byproducts in different forms composed of highly valuable compounds, which are most commonly utilized directly as animal feed livestock with no additional processing costs. Numerous recent studies have shown that cereal byproducts can be also redirected from animal to human consumption and used directly, as in the case of cereal brans and germs which can be used as food ingredients in a wide range of food products as natural sources of fibers and other bioactive compounds. Moreover, cereal byproducts can be further subjected to fractionation, extraction, and purification to obtain high added value compounds for food, feed, and nonfood uses (pharmaceutical, biomedical, cosmetic, etc.) (Dapčević-Hadnađev et al. 2018; Galanakis 2020). However, whether used for food or animal feed purposes, certain challenges in the valorization of cereal byproducts have arisen related to safety – the presence of toxic compounds (mycotoxins, heavy metals, and pesticides) and the presence of high amounts of antinutritional factors (Pojić et al. 2018).

A further increase of the efficacy of cereal material utilization can be achieved within the biorefinery concept of processing which enables the integral valorization of byproducts to obtain antioxidants, biofuels, bioenergy, bioproducts, and biofertilizers, as well as improve the technological and nutritional functionality of byproducts for their further use (Galanakis 2020).

1.4 Challenges in Healthy Grain Processing: Traditional vs Innovative Processing

The most common cereal and pseudoceral processing operations – dry milling, wet milling, pearling, malting, and baking – are confined to traditional technologies characterized by a small pace of innovation. In an era in which innovation is considered a key driver of economic growth, the innovation of cereal and healthy grain processing needs to be boosted. Innovation in the cereal processing sector is not only driven by increasing consumer demands for sustainable, safe, and nutritious high-value cereal and gluten-free products, but also the need to decrease the environmental impact of processing by minimizing energy demands and reducing food losses and waste. For example, traditional milling and baking processes are characterized by the implementation of incremental innovation, which improved the efficacy of processing, reduced energy consumption and decreased the need for manpower.

On the other hand, we are witnessing increasing research dynamics in the field of innovative process technologies, mainly applied to increase the extraction of bioactive compounds by means of cell rupture and disrupting or damaging the cellular membrane (Hernández-Hernández et al. 2019). Their implementation in the cereal processing sector can be perceived through the improvement of product quality, enhancement of (techno-) functionality, alteration of allergenicity, enzyme deactivation, microbial and chemical decontamination (removal of pesticides, mycotoxins, and antinutritive factors), acceleration of heat and mass transfer, control of Maillard reactions, and extension of shelf-life (Hernández-Hernández et al. 2019). Therefore, if they are combined with traditional cereal processing methods they can provide benefits to consumers, while companies that have implemented them can maintain or increase their market share and profitability (Albertsen et al. 2020). However, especially in the food sector, scientific or technological innovations often encounter mistrust and rejective reactions from consumers, resulting in decreasing acceptance of those innovations. It was found that consumer acceptance of innovative food products is conditioned by relative advantage, naturalness, and novelty, but also by discomfort described by insecurity and uneasiness. Therefore, in order to increase consumer acceptance of food innovations, effective communication strategies must be applied to reduce existing mistrust (Albertsen et al. 2020). It must be noted that the majority of innovative processing technologies are still in the research and development stage, while those already commercialized are barely applied in the food industry and only on a small scale. Another condition for their higher commercial exploitation is the development of high-capacity industrial-scale equipment (Pojić et al. 2018).

1.5 Relevance of this Book

This book, *Innovative Processing Technologies for Healthy Grains*, aims to address innovative cereal science and technology and create a knowledge base relevant for students, educators, researchers, food processors, and product developers by bringing together essential information on the nutritional and techno-functional properties of cereals and pseudocereals and processing techniques utilized to deliver final products in line with consumer expectations. Innovative cereal processing is associated with the addition of value to raw materials and final products – increasing safety, modification of technological properties, and better utilization of functional ingredients and byproducts. Therefore, innovative cereal processing has a huge potential, but also represents a real challenge for science, industry, and policymakers, and to a certain extent for consumers, too. The acceptability of novel foods by consumers is a complex and challenging issue influenced by many factors, including sensory preferences and personal factors that need to be perceived and overcome as a prerequisite for the increased acceptance of food innovations.

This book comes at a time when food and nutrition are intertwined with a number of trends: the trend toward healthful eating patterns, the increasing adoption of plant-based diets, and the consumption of high-protein foods, as well as "clean" and "free-from" labelling – all of them being mostly favorable for grain and cereal-based food. Moreover, this book comes at a time when efforts are made to ensure the sustainability of production and the utilization of byproducts, when legislative restrictions limit the number of fumigants and storage insecticides, and when the safety and technological properties of grains are compromised by incidents of extreme weather conditions as a result of climate change (e.g. mycotoxin contamination).

Acknowledgment

M. Pojić would like to acknowledge the financial support of the Ministry of Education,Science and Technological Development of the Republic of Serbia (No. 451-03-68/2020-14/200222).

References

Akkoc, Y., Lyubenova, L., Grausgruber, H. et al. (2019). Minor cereals exhibit superior antioxidant effects on human epithelial cells compared to common wheat cultivars. *Journal of Cereal Science* 85: 143–152.

Albertsen, L., Wiedmann, K.-P., and Schmidt, S. (2020). The impact of innovation-related perception on consumer acceptance of food innovations – development of an integrated framework of the consumer acceptance process. *Food Quality and Preference* 84: 103958.

Asbran (Associação Brasileira de Nutrição) (2017). Celíacos: faltam estatísticas e protocolos no Brasil. https://www.asbran.org.br/noticias/celiacos-faltam-estatisticas-e-protocolos-no-brasil (accessed 17 June 2020).

Awika, J.M. (2011). Major cereals grains production and use and around the world. In: *Advances in Cereal Science: Implications to Food Processing and Health Promotion* (eds. J.M. Awika, V. Piironen and S. Bean), 1–13. American Chemical Society.

Békés, F., Schoenlechner, R., and Tömösközi, S. (2017). Ancient wheats and pseudocereals for possible use in cereal-grain dietary intolerances. In: *Cereal Grains*, 2e (eds. C. Wrigley, I. Batey and D. Miskelly), 353–389. Woodhead Publishing.

CBI (2019). The European market potential for healthy snacks with grains, pulses and oilseeds. https://www.cbi.eu/market-information/grains-pulses-oilseeds/healthy-snacks/market-potential (accessed 6 July 2020).

Dapčević-Hadnađev, T., Hadnađev, M., and Pojić, M. (2018). The healthy components of cereal by-products and their functional properties. In: *Sustainable Recovery and Reutilization of Cereal Processing by-Products* (ed. C.M. Galanakis), 27–61. Woodhead Publishing.

El-Chammas, K., and Danner, E. (2011). Gluten-free diet in nonceliac disease. *Nutrition in Clinical Practice* 26: 294–299.

FAO (2020). FAOSTAT Database: Data on crop production for the year 2000–2018. http://www.fao.org/faostat/en/#data/QC (accessed 6 April 2020).

Fasolin, L.H., Pereira, R.N., Pinheiro, A.C. et al. (2019). Emergent food proteins – towards sustainability, health and innovation. *Food Research International* 125: 108586.

Galanakis, C. (2020). Food waste valorization opportunities for different food industries. In: *The Interaction of Food Industry and Environment* (ed. C. Galanakis), 341–422. Academic Press.

Goldberg, G. (ed.) (2003). *Plants: Diet and Health. The Report of the British Nutrition Foundation Task Force*. Oxford: Blackwell.

Hernández-Hernández, H.M., Moreno-Vilet, L., and Villanueva-Rodríguez, S.J. (2019). Current status of emerging food processing technologies in Latin America: novel non-thermal processing. *Innovative Food Science & Emerging Technologies* 58: 102233.

Kulp, K. and Ponte, J.G. (eds.) (2000). *Handbook of Cereal Science and Technology*, 2e. New York: Marcel Dekker.

Mascaraque, M. (2018). What the new health and wellness data is telling us: a look into latest trends. https://blog.euromonitor.com/new-health-wellness-data-look-latest-trends (accessed 24 April 2020).

Mota, C., Santos, M., Mauro, R. et al. (2016). Protein content and amino acids profile of pseudocereals. *Food Chemistry* 193: 55–61.

Newberry, C., McKnight, L., Sarav, M. et al. (2017). Going gluten free: the history and nutritional implications of today's most popular diet. *Current Gastroenterology Reports* 19: 54–61.

Pojić, M., Mišan, A., and Tiwari, B. (2018). Eco-innovative technologies for extraction of proteins for human consumption from renewable protein sources of plant origin. *Trends in Food Science & Technology* 75: 93–104.

Remes-Troche, J.M., Cobos-Quevedo, O.D.J., Rivera-Gutiérrez, X. et al. (2020). Metabolic effects in patients with celiac disease, patients with nonceliac gluten sensitivity, and asymptomatic controls, after six months of a gluten-free diet. *Revista de Gastroenterología de México (English Edition)* 85 (2): 109–117.

Rosentrater, K.A. and Evers, A.D. (eds.) (2018). Introduction to cereals and pseudocereals and their production. In: *Kent's Technology of Cereals*, 5e, 1–76. Woodhead Publishing.

Schoenlechner, R., Siebenhandl, S., and Berghofer, E. (2008). Pseudocereals. In: *Gluten-Free Cereal Products and Beverages* (eds. E.K. Arendt and F.D. Bello), 149–190. Academic Press.

Venskutonis, P. and Kraujalis, P. (2013). Nutritional components of amaranth seeds and vegetables: a review on composition, properties, and uses. *Comprehensive Reviews in Food Science and Food Safety* 12: 381–411.

2
Introduction to Cereal Processing: Innovative Processing Techniques

Uma Tiwari[1] and Milica Pojić[2]

[1] School of Food Science and Environmental Health, Technological University Dublin, City Campus, Dublin, Ireland
[2] Institute of Food Technology, University of Novi Sad, Novi Sad, Serbia

2.1 Introduction

Globally, cereals and their products have become a part of one's daily diet. Cereals are edible grains of the family of Poaceae or Gramineae. The largest group within this grass family, cereals consist of more than 10000 species and are commonly consumed around the world. Shewry and Tatham (1999) studied the taxonomy of cultivated cereals and classified them under different subfamilies: Bambusoideae, Festucoideae, Panicoideae and Chloridoideae (Figure 2.1).

Many cereals belong to the subfamily Pooideae (also known as Festucoideae), such as wheat, barley, and rye, belonging to the tribe Triticeae, while oats belong to the tribe Aveneae. Rice belongs to the subfamily Bambusoideae, while the minor grains such as finger millet (also known as ragi) and teff are classified under the subfamily Chloridoideae (Shewry and Tatham 1999). According to the Food and Agriculture Organisation of the United Nations (FAO) Statistics Database (FAOSTAT) on crop statistics (FAO 2016), rice, wheat, maize (corn) and sorghum occupy vast harvested areas compared with other cereal crops. Therefore, they are classified as major cereals, as opposed to minor cereals based on their production and utilization levels. For instance, according to Healthy Minor Cereals (2016), wheat and barley are the most important cereals while spelt, einkorn, rye, and oats are minor cereals.

The FAOSTAT database shows that the overall production of cereal grains increased from the year 2018 to 2000 for maize (48.4%), wheat (20%), rice (23%), sorghum (6%), and barley (6%). The OCED-FAO Agricultural Outlook (2018–2027) indicates that global cereal (e.g. wheat, rice, and maize) production will increase by 17.6 Mha between 2017 and 2027. Thus, cereal production is indispensable to feeding the growing

Innovative Processing Technologies for Healthy Grains, First Edition.
Edited by Milica Pojić and Uma Tiwari.

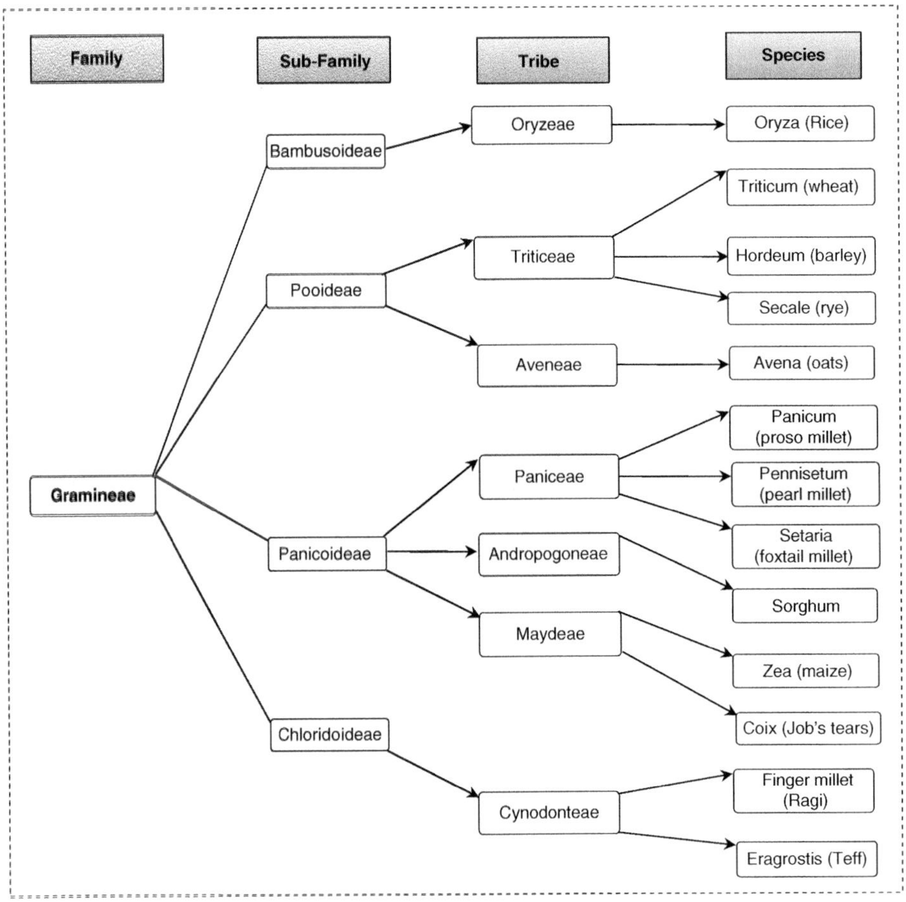

Figure 2.1 Taxonomic relationships of cereals. *Source*: adapted from Shewry et al. (1992).

population (i.e. 6–8.3 billion by 2030) and world consumption is forecast to increase from 2.6 to 2.9 billion tonnes (OECD-FAO 2018). Due to the high content of starch, cereal foods provide high amounts of energy in a diet; this is followed by other major nutrients such as dietary fiber, nonstarch carbohydrates, and proteins, and minor nutrients. Cereals are utilized in various forms (e.g. bread and bakery products, breakfast cereals, cookies, porridges, extruded snacks, etc.) around the world and consumed either partially or fully processed. However, cereals require appropriate post-harvest management followed by primary and/or secondary processing to produce suitable end products. Nowadays, researchers working on cereal and cereal products are focussed on the implementation of innovative processing methods in combination with traditional methods to achieve healthy and beneficial cereal-based products. Cereal scientists are moving toward the trend of sustainable production of end or final products with more nutrients, that are high in functional properties and low in allergenicity, and increase the safety of the products with processing techniques. Therefore, this chapter provides a detailed overview of cereal characteristics, grain structure and composition, and processing methods with special emphasis on innovative processing techniques.

2.2 Characteristics of Cereals

The characteristics of cereals vary in terms of the specification of inflorescences, roots, stem, types of leaves, and kernel structure. The kernel structure is the main characteristic that determines the mode of processing.

2.2.1 Cereal's Inflorescences

Inflorescence structure directly affects the yield of grains and it varies with species, diversity of branching architecture, size, and number of kernels (Kyozuka et al. 2014; Bommert and Whipple 2018). Cereals such as wheat, rye, barley, and oats have a spiral arrangement of leaves on the stem while some species consist of alternate leaf arrangements (Kellogg et al. 2013). In cereals, each flower produces one seed, depending on the design of the inflorescence (panicle or spike), which controls the yield of the cereal grains – the panicle inflorescence (rice and sorghum), spike inflorescence (wheat, barley, and rye), panicle attached to the central axis (oats and millet), etc. Inflorescences also differ in their arrangements of branches, i.e. short or long branches. For example, rice has many long branches bearing single spikelets whereas sorghum consists of short branches bearing two spikelets (Doust 2007; McSteen et al. 2000; Vollbrecht et al. 2005).

2.2.2 Cereal's Roots

Gramineae possess two distinct root systems, mainly consisting of primary or seminal roots and coronal roots. Most cereals, such as rice, wheat, oats, millet, and sorghum, have both primary and secondary root systems. The root system of rice is generally shallow and suitable for flooded conditions, while maize has a more complex root system with an embryogenic primary root (first root) followed by seminal roots, crown roots and aerial nodal roots (Hetz et al. 1996). The primary roots or seminal roots are generally intact until the time of harvest, while the coronal root provides stronger anchorage and prevent the plant falling over. These root systems absorb and secure uptake of water and nutrients including nitrogen (Aiken and Smucker 1996).

2.2.3 Cereal's Stems and Leaves

The stems are usually hollow, divided into series of nodes and internodes, elongated in shape, and grow up to approximately 30–40 m long, but this varies with different species. In rice, the lower internodes are shorter than the upper internodes, providing greater plant resistance against waterlogging (Weaver and Zink 1945). As is well known, the stem connects the roots and other parts of the plant, and thus helps the transport of water, minerals, and sugars. In some rice varieties, aerenchyma tissue formation enables oxygen to be supplied to the roots and root nodules, thus reducing waterlogging. In some cases, the stem diameter and height also influence the resistance of the plant to waterlogging. The leaves of the grass family consist of a sheath that encloses the culm and opens out into the leaf blade, which is long and narrow with blunt tip. For example, sorghum leaf blades are smaller, flat, and pointed in structure while maize leaves are broader in shape. Moreover, leaves are arranged alternately on the stem with one leaf per node, allowing the production of sufficient carbohydrates during photosynthesis (Weaver and Zink 1945).

2.3 Kernel Structures

All cereals share the common basic anatomy of a kernel (Alldrick 2017). The anatomy of wheat and rice kernels is given in Figure 2.2.

Following the fertilization stage, seeds are developed from the ovule and contain the embryo surrounded by outer layers such as the husk, seed coat (pericarp and testa), aleurone layers and endosperm. Monocotyledon grains (rice, wheat, barley, oats, etc.) consist of seeds that contain one cotyledon or one embryonic seed leaf (Hoseney 1994). The kernel tissue of greatest nutritional significance is the endosperm, composed of cells filled with nutrients to sustain the embryo during the germination process. It is well known that cereals are major source of carbohydrates, protein, certain vitamins, minerals, and phytochemicals, which satisfy energy needs and provide health benefits for humans (Goldberg 2003). In particular, cereals mainly consist of carbohydrates in the form of polysaccharides – primarily starch located in the endosperm (56–74%) and fiber, primarily arabinoxylans, β-glucans, and cellulose located in the bran layer – followed by protein ranging from 8 to 12% (Koehler and Wieser 2013). Table 2.1 gives an overview of the major nutrients present in cereals.

2.3.1 Rice

Rice or paddy is covered by a hull or husk, which constitutes about 18–28% of the weight of the grain, and the caryopsis (also known as brown rice), which constitutes about 72–82% of the weight of the grain. The caryopsis consists of the outer pericarp

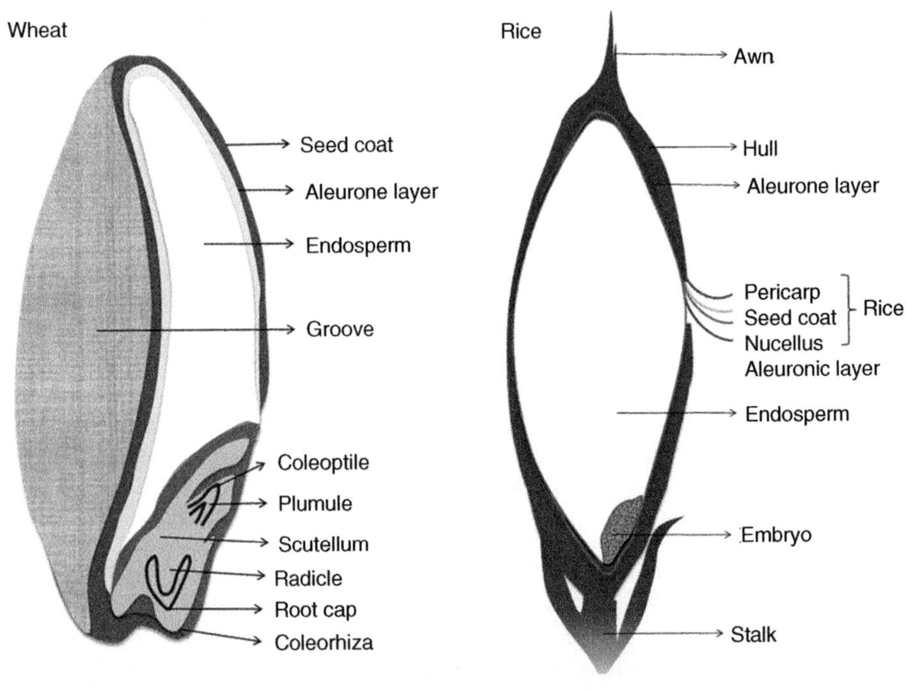

Figure 2.2 Structure of wheat (left) and rice kernels (right).

Table 2.1 Composition of cereals.

Cereals	Crude protein (%)	Crude fat (%)	Ash (%)	Crude fiber (%)	Digestible CHO (%)	Starch (%)	Total dietary fiber (%)	Total phenolics (mg/100 g)
Wheat	10.6	1.9	1.4	1	69.7	64	12.1	20.5
Maize	9.8	4.9	1.4	2	63.6	62.3	12.8	2.91
Brown Rice/ Paddy	7.3	2.2	1.4	0.8	64.3	77.2	3.7	2.51
Barley	11	3.4	1.9	3.7	55.8	58.5	15.4	16.4
Sorghum	8.3	3.9	2.6	4.1	62.9	73.8	11.8	43.1
Pearl Millet	11.5	4.7	1.5	1.5	63.4	60.5	7	51.4
Oats	9.3	5.9	2.3	2.3	62.9	52.8	15.4	16.4
Rye	8.7	1.5	1.8	2.2	71.8	68.3	16.1	13.2

(*Source*: FAO 1999; Saldivar 2003)

layer (1–2%), the aleurone layer or bran (5%), the germ or embryo (2–3%), and the starchy endosperm (89–91%). The aleurone layer, which encloses the embryo, has one to five cell layers, being thicker at the dorsal surface than the ventral surface. Generally, the aleurone layers are thicker in short-grain than in long-grain rice, and the starchy endosperm is the whitest portion of the rice caryopsis (Juliano and Tuaño 2019). Milling of rice removes the outer cover by dehusking (removal of husk) and polishing (removal of bran), producing the edible endosperm (white polished rice) for human consumption. The dehusking process also removes the different layers of rice and thereby removes quantities of fat, carbohydrates, protein, and fiber, influencing the nutritional value of rice (Fernando 2013). The more that polishing given to the rice grains, the more that fats, proteins, thiamine, and other vitamin-rich compounds are removed.

2.3.2 Wheat

The wheat kernel consists of three main anatomical parts: the bran (seed coat), the endosperm and the embryo (germ). Generally, the germ comprises about 2–3% of the kernel, the bran 13–17%, and the starchy endosperm makes up about 83–85% of the kernel's weight (Pomeranz 1982). The aleurone layer is the outermost layer of the endosperm, generally attached to the outer coat, which in successive grinding and sieving operations in an industrial roller mill ends up in tail-end break and reduction flour mill streams or attached to bran particles (Pojić et al. 2014). Milling separates these layers from the wheat kernel prior to the production of refined flour. Generally, the inner bran layer is high in protein, fat, and minerals, and the outer layer of the bran is high in cellulose and hemicelluloses. Wheat germs are also good sources of vitamins B and E, minerals, lysine, and unsaturated fatty acids.

2.3.3 Maize

The maize kernel consists of the pericarp, the hull or bran, the germ or embryo, the endosperm, and the tip cap, a conical structure of dead tissue where the kernel joins the cob. The maize kernel has a relatively larger germ than other cereals, placed in the lower portion of the endosperm. The endosperm represents approximately 70–86% of the kernel and contains mainly starch (87.6%) and protein (8%). Maize germ ranges from 7 to 22% of the kernel and consists of high levels of lipids (18–41%), protein (12–21%), and starch (6–21%), but it is also rich in unsaturated fatty acids, tocopherols, tocotrienols, and carotenoids (FAO 1992; Navarro et al. 2016). Moreover, the maize kernel also contains phytate, acting as an endogenous toxic compound and antinutritive factor in monogastric species that are not able to utilize a large amount of minerals (Humer et al. 2015). The corn kernel is flattened, wedge-shaped, and broader at the apex end than at the point of attachment to the cob. Maize kernels are processed by dry milling to produce primary products such as brewers' grains, snack food grits, and flour, and wet milling to obtain corn starch and a wide assortment of byproducts such as corn bran, germ meal, and corn protein meal (FAO 1992; Papageorgiou and Skendi 2018).

2.3.4 Barley

Barley kernels are spindle-shaped, comprising the caryopsis (one-seeded fruit) covered by the hull or husk. The hull or husk represents 10–13% of the dry weight of the kernel, but this might vary with the dehulling process, which may remove up to 20% of the kernel weight. The endosperm cell walls are mostly composed of β-glucan (70%) (Tiwari and Cummins 2009). The aleurone layer contains cells in two or three layers, depending on cultivars. The caryopsis consists of the pericarp, the seed coat, the germ or embryo, and the starchy endosperm, accounting for 80% of the total grain weight. The barley embryo is generally located at the end of the caryopsis on its dorsal side. "Hull-less" barley has a loosely attached hull that falls off during harvesting and threshing (the removal of grains from the chaff) (Evers and Millar 2002).

2.3.5 Oats

Oat caryopses (groats or kernels) are similar to those of wheat and barley, and composed of the bran, the endosperm and the germ. The caryopsis and the hull account for 65–75% and 25–35% of the whole kernel respectively. The oat germ is located on the dorsal side of the caryopsis so that it is partly covered by the lemma, which comprises about 2–3 leaf shoots of the plumule and about 2–3 rudimentary roots of the radicle (Welch 2012). The bran comprises layers of tissue and aleurone cells located in the outer layers of the groat, whereas the endosperm (55–80%) is located inside the wall layers of the groat and composed of starch, protein, lipids, and the major concentration of β-glucans (Tiwari and Cummins 2009).

2.3.6 Rye

Rye grains are arranged in a zigzag fashion on the rachis and are covered with a lemma, a palea, and a glume. On maturity the grains fall off easily during threshing. The grains are usually grayish-yellow in color with a shrivelled and rough surface. Rye kernels are composed of 86.5% starchy endosperm, followed by the bran (10%) and the germ (3.5%). During the milling process, the bran and germ of the rye kernel are separated from the endosperm and milled into flour (Bushuk 2004).

2.3.7 Sorghum

The three principal anatomical components of the basic sorghum kernel are the pericarp, the germ, and the endosperm, which account for 6, 10, and 84% of kernel weight, respectively. However, these proportions vary with different sorghum cultivars. The endosperm is the largest part of the kernel and has a comparatively poor mineral and oil content. The endosperm contributes mainly to the kernel's protein (80%), starch (94%) and B-complex vitamin (50–75%) compositions, whereas the germ contains 68% of the minerals, 75% of the oil and 15% of the protein of the whole kernel (FAO 1995). Therefore, processing leads to removal of the outer pericarp, increases the relative protein level, and reduces the cellulose, lipid, and mineral content in the grain. For example, Alvarenga et al. (2018) demonstrated the effects of milling sorghum into various fractions to produce animal feed with a good protein content. They concluded that mill-feed fractions contained a higher level of crude protein (13.4%) compared with flour (9.68%), indicating the potential benefits of utilizing the milling fraction for human and animal feed.

2.3.8 Millet

Millet kernels comprise about 7–10% pericarp, 15–21% germ, and 70–76% endosperm. Four major millet species are pearl millet, foxtail millet, proso millet, and finger millet. The pericarp of pearl millet is strongly attached to the seed (caryopsis), whereas in proso, finger, and foxtail millets the pericarp is attached to one point on the seed. The endosperm of millet is divided into the peripheral, outer, hard endosperm and the inner, floury endosperm, while the germ constitutes up to one third of the pearl millet caryopsis. The relative proportions of the endosperm and germ in millet are about 4.5 : 1, i.e. the germ constitutes ~20% of the weight of the whole kernel (FAO 1995). The distribution of the total amount of protein within the pearl millet grain is 60% in the endosperm, followed by 31% in the germ, and 9% in the pericarp. The protein content in pearl millet is in the range 8–23%, while proso millet contains 11–13% protein (Lestienne et al. 2005; Serna-Saldivar and Rooney 1995).

2.4 Processing of Cereals

In order to derive the nutritional benefit from cereal grains and increase their digestibility and palatability they must be subjected to a certain type of processing involving one or a combination of different mechanical treatments – threshing during which the outer seed coats are removed, milling during which the particle size is reduced and grain converted into a flour of some type, and/or thermal processing (e.g. cooking, roasting, or baking).

The anatomy of the cereal kernel also affects the types and routes of contaminants (e.g. mycotoxins, pesticides, or heavy metals) and endogenous toxic compounds presenting a potential hazard when consumed. Although the level of endogenous toxic compounds in cereals is low, two compounds of interest are phytate and tannins. Their anatomical distribution depends on the type of cereal grain: phytate is found to be predominantly located in the germ of maize, the aleurone layer of wheat, and uniformly distributed through millet (Alldrick 2017). The need to make grain digestible and palatable, but also safe for consumption, conditioned the development of novel process technologies as one of the mitigation strategies to reduce the risk of contamination (Alldrick 2017).

Although cereal grains are structurally different and are harvested in different seasons, the general post-harvest treatments and processing regimes are applicable to most of them. Between harvest and consumption, cereals are subjected to a number of processing stages, which can be divided into:

1. *Cereal grain storage*, which includes preparative operations on harvested grain for safe storage such as threshing, pre-cleaning the grain mass, drying, and segregation;
2. *Primary processing*, which includes further cleaning, grading, removing the husk or reduction of the size, milling and sieving, tempering, parboiling, and soaking; and
3. *Secondary processing*, which includes all processing operations that transform the grains into edible products, such as fermentation, baking, puffing, flaking, frying, and extrusion.

The principal primary cereal processing operation is milling, which is classified in two categories: dry and wet milling. Dry milling separates the outer fibrous materials and germ from the starchy endosperm to obtain semolina, grits and flour (in the processing of wheat, rye, and maize), but it also refers to pearling in which the seed coat (testa and pericarp) and aleurone and subaleurone layers are removed by abrasion to obtain the polished grain (in the processing of rice, oats, and barley). Wet milling is applied for the production of starch and gluten (in the processing of wheat and maize) (Papageorgiou and Skendi 2018). All of these processing steps are characterized by a slow pace of innovation due to the utilization of traditional technologies. Today, innovation is considered a key driver of economic growth; the innovation performance of cereal and healthy grain processing needs to be boosted. Several innovative processing technologies have emerged which can be coupled with traditional cereal processing technologies to increase the quality and safety of the grains, to minimize the changes to or loss of nutritional composition and sensory attributes, to increase the sustainability of production and to decrease the amount of byproducts (Figure 2.3). Given that traditional cereal processing has already been extensively represented in the literature, this chapter will address only emerging innovative processing solutions for cereals and other healthy grains and their products.

2.5 Innovations in Post-harvest Processing

2.5.1 Irradiation of Cereal Grains

The utilization of ionizing radiation (e.g. x-rays or γ-rays) in the post-harvest treatment of cereals mainly refers to the disinfection of grains with pathogens such as *Salmonella*, *Campylobacter*, *Escherichia coli*, and viruses, and the disinfestation of grain mass (Bhattacharjee and Singhal 2016). The application of irradiation as an alternative technology to fumigation with ethylene oxide should be noted, due to the remaining toxic residues (Arvanitoyannis and Stratakos 2010).

2.5.2 Ozone Technology in Post-harvest Cereal Processing

Due to strong the oxidant properties of ozone, ozone treatment is considered an eco-friendly and cost-effective food processing technique applicable in the cereal industry as an effective agent for (i) the fumigation of stored grains, (ii) microbiological

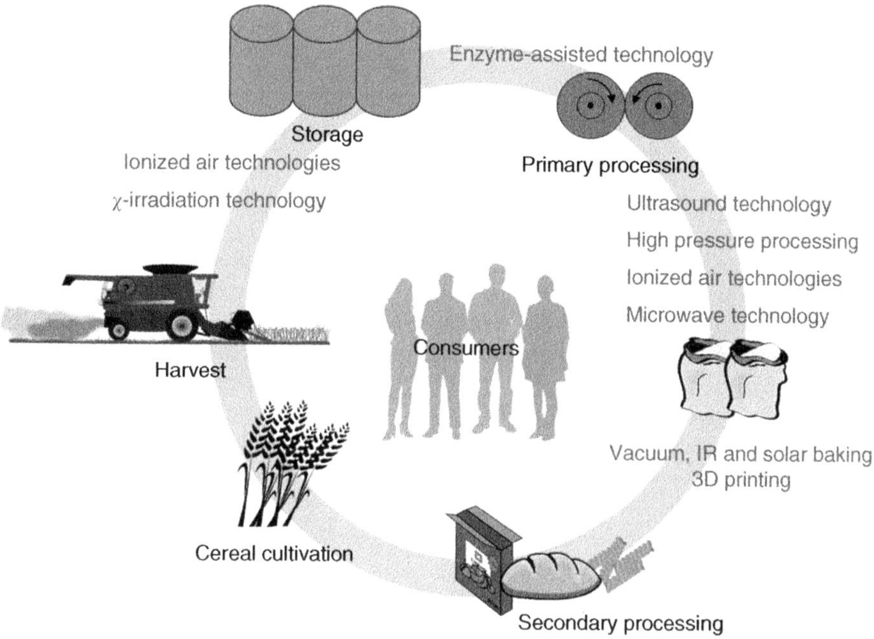

Figure 2.3 Innovative processing technologies in the cereal value chain.

disinfection, (iii) the treatment of mycotoxin-contaminated cereals, and (iv) the modification of the physico-chemical properties of the major components of cereals (e.g. starch and protein). However, ozone treatment may influence the color, storage, and germination capacity of the grains, as well as the rheological and textural properties of the products thereof (Tiwari et al. 2010; Zhu 2018). Granella et al. (2018) reported the application of ozone technology along with drying in the post-harvest processing of naturally contaminated wheat seeds. The beneficial effect of ozone treatment (ozone exposure for 45 minutes and drying at an air temperature of 50 °C) was reflected in a total reduction of the fungal count of 92.86%, i.e. a reduction from 1.87 to 0.13 cfu/g, during which the physiological qualities of the wheat grains – germination, vigor, and electrical conductivity – were not affected. Since fungal contamination of grains is often associated with mycotoxin production, it was shown that ozonization was an effective treatment for the inactivation of *Fusarium graminearum* and the reduction of deoxynivalenol contamination when grains were exposed to 60 mmol/mol of ozone for 120 minutes (Savi et al. 2014), as well as the reduction of zearalenone and ochratoxin A in contaminated corn (Qi et al. 2016).

The authors noted small changes in color (whiteness increased while yellowness decreased) and fatty acid content following ozone treatment of 180 minutes (Qi et al. 2016). The effective applicability of ozone technology for the degradation of pesticide residues in stored grains was demonstrated by several authors, where the degradation efficiency was directly proportional to the period of exposure: pirimiphos-methyl residues (Freitas et al. 2017), deltamethrin and fenitrothion (Savi et al. 2015), and bifenthrin and pirimiphos-methyl residues (Savi et al. 2016). The potential of ozonation in controlling the enzymatic activity of wheat flour in its fluidized state was demonstrated by Piechowiak et al. (2018).

2.5.3 Cold Plasma Technology in Post-harvest Cereal Processing

The utilization of cold plasma technology within the cereal production chain has a twofold character: for effective bio-decontamination and for eco-innovative transformation of the techno-functional properties of grain and grain-based products (Los et al. 2017). The bio-decontamination of microorganisms, mycotoxins and pesticides is based on the generation of reactive oxygen and nitrogen species that can react with the contaminants on the surface of the cereal (Los et al. 2017; Wielogorska et al. 2019). So far, highly variable efficacy has been reported in terms of the decontamination of mycotoxins, being dependent on a number of factors, such as the nature of gases, plasma exposure time, and the formation of byproducts and their toxicity. However, a detrimental influence of cold plasma treatment on cereal matrixes (especially those rich in antioxidants and lipids) is possible, affecting their nutritional composition. Therefore, more research is needed to determine the balance between the possible detrimental effects on the nutritional composition of cereals and the beneficial effect of their detoxification (Wielogorska et al. 2019).

2.6 Innovations in Primary Cereal Processing

2.6.1 Dry Milling of Cereals

Dry milling (roller or abrasive milling), as a traditional process, could be considered as having low innovation potential because the basic principle of dry milling – sequential processes of particle size reduction and separation – has not changed in over decades. For instance, by successive grinding and sieving operations, the gradual fragmentation of wheat kernels occurs, resulting in a partial separation of bran, germ, and starchy endosperm. However, roller mills are usually applied to wheat and corn, while abrasive milling techniques are applied to barley, rice, and other cereals due to strong adherence of their hull to the pericarp (Pojić et al. 2014).

2.6.1.1 Roller Milling Traditional milling industries have been developing with incremental innovations that imply the application of more efficient roller mills and automation (Bock and Sweley 2018). Innovation in the dry milling sector is not frequent, but it still exists. A patented process for the production of ultrafine-milled whole-grain wheat flour (with particle sizes less than or equal to 150 μm) by a combined process of roller and gap milling enables flour to be obtained with the full nutritional value of wheat kernels, while retaining the texture of refined wheat flour (Arndt and Korolchuk 2014; Korolchuk 2008).

Another innovative solution applicable in roller milling is the integration of accelerating-breaking rollers, which have a larger diameter than those within the grain guiding hopper, to enable high-speed milling (Kértesz 2009).

2.6.1.2 Other Milling Methods Conventional grain processing by roller milling is often associated with the loss of the nutritional value of the whole cereal kernel. Therefore, certain novel processing techniques have been proposed to improve the quality of the milling products while preventing nutritional losses in an eco-friendly manner (Singh et al. 2015). Milling procedures that involve the utilization of different types of mills result in flour of the desired quality for the desired end purpose. The application of burr and blade mills is suitable for obtaining coarse flours; pin, hammer, and turbo mills are suitable for obtaining fine flours; while jet mills are suitable for obtaining ultrafine flours (Lee et al. 2019). These grinding methods are more suitable for grinding cereal kernels other than wheat, and novel cereal-based raw materials, such as germinated cereals. Jet milling is a high-air-pressure milling technique in which powder microparticles are obtained with a relatively narrow particle size distribution (Lazaridou et al. 2018). In a jet milling process it is possible to obtain fine and ultrafine flour fractions with particle sizes of $d_{50} < 21\,\mu m$ and $d_{50} < 12\,\mu m$ respectively. The reduction of flour particle size in a jet milling process affects the level of damaged starch and the content of water-soluble arabinoxylans, but does not affect their molecular structure and apparent peak molecular weight. Moreover, the reduction of flour particle size by jet milling affects the rheological behavior of dough and dough-handling properties, enabling the production of a specific end-use flour. The change in rheological behavior of dough was reflected in an increase in the dough's resistance to deformation, elasticity, and zero shear viscosity. With the decrease in flour particle size, the dough became harder with higher consistency, more sticky and gummy, and exhibited longer half relaxation time, lower relaxation rate, and higher elongational viscosity (Lazaridou et al. 2018).

2.6.1.3 Micronization Ultrafine grinding, also known as micronization, is performed to increase the accessibility of the bioactive compounds from the ground material. The novelty of the micronization process in contrast to traditional dry milling is that the entire cereal kernel is processed without any byproducts. A similar process was utilized for the enrichment of barley flours with β-glucans, alkylresorcinols and phenolic compounds (Ferrari et al. 2009; Gómez-Caravaca et al. 2015).

2.6.2 Novel Fractionation Methods

2.6.2.1 Air Classification Micronization is sometimes combined with an air classification process, in which the resulting flour particles are separated into different fractions according to the size and density. Obtaining a fraction with a higher content of the compounds of interest is possible, by varying some of the air classification parameters, such as air flow rate. Ficco et al. (2018) demonstrated the utilization of air classification to obtain anthocyanin-rich fractions of durum and soft pigmented wheats. They reported a significant reduction in the estimated glycaemic index following the incorporation of the durum fractions in bread making. Likewise, a similar technique was utilized to obtain protein-enriched barley ingredients, as demonstrated by Silventoinen et al. (2018).

2.6.2.2 Electrostatic Separation Electrostatic separation appears to be a superior dry classification method to air classification, and can also be coupled with ultrafine grinding (Sibakov et al. 2014). The principle of electrostatic separation is based on charging the material to be separated by tribo-electrification, then introducing it into an electric field, where the material is separated depending on the acquired charge. Sibakov et al. (2014) demonstrated obtaining the β-glucan-enriched fractions from oat bran, which were electrostatically separated from particles rich in arabinoxylan.

Although ultrafine grinding and electrostatic separation represent energy-demanding technologies, they do not require any liquids or solvents to obtain enriched components, which is considered environmentally friendly. Moreover, they are more efficient techniques than conventional fractionation methods such as sieving and air classification, which allow β-glucan fractions up to 20–25% concentration to be obtained, compared with 42.2–48.4% after one or two successive electrostatic separations (Sibakov et al. 2014). Hemery et al. (2011) demonstrated the utilization of electrostatic separation to obtain purified fractions from wheat bran, where particles rich in highly branched and cross-linked arabinoxylans from the pericarp were separated from particles rich in β-glucan, ferulic acid, and para-coumaric acid from the aleurone cell walls. The most positively charged fraction represented was 34% of the initial bran and contained 62% of the ferulic acid present in the initial bran.

2.6.3 Alteration of the Techno-functional Properties of Cereals and Flours

2.6.3.1 Irradiation Technology Apart from microbiological effects, irradiation causes side effects on nutritional, rheological, and textural properties, by affecting starch, proteins, and other biomolecules (pectins, cellulose, or added hydrocolloids) (Arvanitoyannis and Stratakos 2010; Bhattacharjee and Singhal 2016). Thus, gamma irradiation appeared to be a useful technology applicable to the modification of the physico-chemical and functional properties of cereal flours in order to obtain targeted raw materials for the production of specific bakery products like breads, biscuits, and cookies (Bashir et al. 2017; Bhat et al. 2016).

Irradiation causes the physical modification of starch molecules and the breaking of hydrogen bonds within starch molecules. Thereby, the starch molecules are cleaved into smaller polysaccharide units and the viscosity of the irradiated starches is reduced. The extent of starch damage depends on irradiation doses, being visually undamaged at low doses of irradiation, but severely damaged at higher doses of irradiation (100 kGy). Doses of radiation typically applied in cereal industries for phytosanitation purposes are within the range of 0.15–0.50 kGy (Ravindran and Jaiswal 2019). However, changing physico-chemical properties requires higher irradiation doses. Bashir et al. (2017) demonstrated changes to the physico-chemical, thermal, and functional properties of whole wheat flour induced by γ-irradiation, as well as the properties of starch extracted after irradiation. The changes to physico-chemical properties from γ-irradiation were reflected in decreases in pasting parameters (peak viscosity, final viscosity, setback, and breakdown values) and increased freeze–thaw stability, water solubility, and water absorption capacity, while a decreasing trend was observed in case of syneresis. Similar changes to the pasting properties of oat starch extracted from oat seed irradiated by 5, 10, 15, and 20 kGy were reported (Mukhtar et al. 2017). Irradiation also influenced changes to syneresis, solubility index, swelling index, and light transmittance values in comparison with nonirradiated counterparts. Irradiation influenced the formation of ridges on the surface of starch granules, as well as causing a significant improvement in antioxidant activity in the extracted starches (Mukhtar et al. 2017). Improvements in antioxidant activity in pigmented brown rice flour were observed by Sultan et al. (2018) at irradiation doses of 2, 5 and 5 kGy. By applying γ-irradiation doses of 2.5 and 5 kGy to whole wheat flour, Bhat et al. (2016) determined decreases in water and oil absorption, swelling power, and emulsion capacity, and increases in the water solubility index, emulsion stability, foaming capacity, and stability. Lee et al. (2013) demonstrated the formation of resistant starch in corn starch with different amylose content (e.g. normal, waxy, and high-amylose corn starch) when irradiated at 5, 10, 25, and 50 kGy. The increase in resistant starch content was noticed with the lowest radiation dose of 5 kGy, being most pronounced at doses of 50 kGy. It was found that the irradiation-induced resistant starch content was most evident in waxy corn starch, followed by high-amylose corn starch and normal corn starch.

2.6.3.2 Ozone Technology The potential of ozonation for controlling the enzymatic activity of wheat flour in its fluidized state was demonstrated by Piechowiak et al. (2018). Ozonization decreased the total activity of amylases, proteases, and lipases, while increasing the activity of lipoxygenases. This can serve as a starting point for the development of solutions in milling and baking plants where ozonization can be coupled with pneumatic transport and/or raw material dosing. Moreover, to eliminate the usage of chemicals with an oxidizing effect commonly used for flour quality standardization, ozonization may be applied as an alternative to the chlorine or potassium bromate treatment of soft wheat flour for cake- and/or bread-making (Chittrakorn et al. 2014; Sandhu et al. 2011). Ozonization of flour resulted in a higher number of crumb cells and larger bread loaf volumes when optimal treatment conditions were selected (36 minutes; and 2, 4.5, and 9 minutes of ozone exposure, respectively) (Sandhu et al. 2011). Obadi et al. (2018) reported improvements in properties such as the specific volume, color, and crumb cell numbers of bread produced from ozonized wheat flour. Gozé et al. (2017) demonstrated the effects of ozone on the molecular properties of wheat grain proteins and consequently on the bread-making quality of the flours thereof. Due to the action of ozone, a significant reduction in the sodium dodecyl sulfate (SDS)-solubility of the wheat prolamins was observed, due to the formation of new intermolecular S S bonds and other types of intermolecular covalent crosslinks, and changes in secondary structure. They observed changes in the rheological properties of dough, such as an increase in the tenacity of the dough and a decrease in the extensibility of the dough. Likewise, the ozone treatment of wheat seeds showed no physico-chemical modification of starch or changes in its molecular structure, except that a slight increase of carboxyl groups was reported with increasing ozonation (Gozé et al. 2016). On the other hand, ozonation caused a decrease in the pasting temperature, decreasing the retrogradation tendency and increasing the gelatinization percentage of wheat starch (Çatal and İbanoğlu 2014). Bai et al. (2017) indicated the potential of ozonated water for processing semi-dried buckwheat noodles: when used as an ingredient of the noodles, the initial total plate count was reduced by 47%, thus extending the shelf life with acceptable sensorial properties. Additionally, study by Obadi et al. (2018) reported that bread made from wheat flour ozonized for 15 minutes was fresher than a control sample of bread stored under the same conditions, due to the lower relative starch crystallinity.

2.6.3.3 Cold Plasma Technology One of the eco-innovative transformations of the techno-functional properties of cereals and cereal products is based on the alteration of the secondary structure of proteins induced by exposure to cold plasma treatment. It was observed that dielectric-barrier-discharge atmospheric-pressure cold plasma (DBD-ACP) induced changes in the structural and functional properties of strong and weak wheat flours, reflected in changes in the rheological properties of dough. DBD-ACP treatments induced increases in the viscoelasticity of the dough depending on the applied voltage and treatment time, while an improvement in the dough strength and optimum mixing time for both weak and strong wheat flours was observed (Misra et al. 2015). Likewise, the inactivation of destructive endogenous enzymes was also reported (Tolouie et al. 2018). Furthermore, they observed changes in the secondary structure and the release of amino-acids, with increases in the glutamic acid, asparagine, serine, histidine, threonine, γ-aminobutyric acid, tryptophan, isoleucine, phenylalanine, and proline content, with the cold plasma treatment of short and long grain rice flour (Pal et al. 2016). The effects of low pressure plasma treatment on parboiled rice flour were reflected in the improvement of the flour hydration and gel hydration properties, and the inducement of crosslinking in the flour and depolymerization of the starch (Sarangapani et al. 2016). They also observed that plasma treatment induced a decrease in the amylose content, a change in the amylose to amylopectin ratio and an increase in the gelatinization temperature. When plasma was applied to the cereal germination process, it decreased the germination time, while maximizing the content of bioactive phytochemicals (Yodpitak et al. 2019).

2.6.3.4 Ultrasound Technology Ultrasound technology uses sound waves at a frequency of 20 kHz, inducing a cavitation phenomenon, which increases the porosity of the treated material by inducing microstructural changes and the formation of microfissures (Pojić et al. 2018). Durak et al. (2016) demonstrated the utilization of ultrasonication to inactivate the proteolytic enzymes in suni-bug-damaged wheat, which appeared to be a successful treatment for this purpose. Along with decreasing the proteolytic activity of the bug-damaged wheat, sonication did not affect the quality of the sound wheat. The study noted that a significant increase in the sedimentation values and the wet and dry gluten content was obtained, depending on the sonication time. Additionally, they also reported that the total free amino acid and free proline content in the samples decreased with sonication time. Kaur and Gill (2019) demonstrated the utilization of high-intensity ultrasonic treatment for the physical modification of starch from different cereals. It was shown that ultrasonication increased the swelling power and solubility of the starches, increased the levels of rapidly digestible starch and resistant starch, and caused surface and microstructural changes without compromising the overall integrity of the starch granule. The potential of ultrasonication to loosen the treated matrix and form micropores was utilized to treat brewing rice, affecting more rapid water absorption and hydration and decreasing hardness (Li et al. 2019). Ultrasonic-treated rice exhibited a higher water binding capability, shorter cooking time and better degree of gelatinization in comparison to untreated rice. Ultrasound technology appeared to be very useful in upcycling cereal byproducts and the extraction of bioactives, proteins, and arabinoxylans from brans and germs (Jiang et al. 2019; Roth et al. 2019; Wen et al. 2019). While treating the matrix to increase the extraction yield of the compound of interest, the ultrasonic treatment is able to alter the secondary and tertiary structures of proteins and enhance the exposure rate of hydrophobic amino acids (Wen et al. 2019). Furthermore, the potential of high-intensity ultrasound (20–100 kHz) to alter the allergenicity of several foods has been demonstrated (Li et al. 2016; Shriver and Yang 2011), but its potential to alter the allergenicity of cereals has yet to be demonstrated.

2.6.3.5 High-Pressure Technology High-pressure processing (HPP) is one of the emerging processing technologies utilized in the cereal industry for the inhibition of the growth of foodborne pathogens, using pressures from 400 to 900 MPa

(Alcázar-Alay and Meireles 2015). Although the main initial utilization of high pressure was for food preservation, it was recently identified for its potential to change the functional properties of food biopolymers – particularly proteins and starches. HPP can induce the denaturation, aggregation, or gelation of proteins, depending on the protein system, the treatment temperature, the duration and amount of applied pressure, and the protein solution conditions (e.g. pH and ionic strength) (Ahmed 2016; Pei-Ling et al. 2010; Tattiyakul and Rao 2016). HPP can induce the gelatinization of starch depending on the source of the starch, the crystallinity type (A-, B-, or C-), and the starch concentration and suspending media, as well as the pressurization temperature and time (Hu et al. 2017; Yang et al. 2017). At an applied pressure of <800 MPa, it was found that A-type starches (e.g. wheat and normal maize), as well as those with a lower amylose content (e.g. waxy maize), are more susceptible to HPP than B-type starches (e.g. potato) and those with a higher amylose content (e.g. amylomaize) (Yang et al. 2017). HPP can be employed in the starch industry for physical modifications and modification of the gelatinizing properties of starch. Zhu and Li (2019) demonstrated the utilization of HPP to modify the physico-chemical properties of quinoa flour. With HPP in the range of 500–600 MPa the peak viscosity, gel hardness, gelatinization enthalpy, and *in vitro* starch digestibility of quinoa flour decreased, while water solubility increased. -Cappa et al. (2016) showed the applicability of HPP in slowing down the staling process of gluten-free breads when the main ingredients, corn starch and rice flour, were treated with pressure of 600 MPa for 5 minutes at 40 °C. The potential of high-pressure processing to promote the formation of structure in gluten-free products and improve the functional properties of gluten-free breads was demonstrated. By applying 200, 400 or 600 MPa treatments on buckwheat, white rice, and teff batters (40 g/100 g) for 10 minutes, their rheological properties were altered and affected by HPP-induced starch gelatinization and protein polymerization due to thiol/disulfide-interchange reactions relevant to white rice and teff batters, while buckwheat batter, due to the absence of free sulfhydryl groups, did not exhibit the crosslinking mechanism (Vallons et al. 2011). Kalagatur et al. (2018) demonstrated the utilization of HPP to control the growth and level of deoxynivalenol and zearalenone in maize grains. They observed a complete reduction in colony-forming units, deoxynivalenol, and zearalenone when maize grains were subjected to a pressure of 550 MPa for 20 minutes at a temperature of 45 °C.

2.6.3.6 Microwave Technology Microwave technology has been successfully utilized on a commercial scale in the meat and fruit and vegetable processing industries, as well as in the production of ready-to-eat meals (Sumnu and Sahin 2005). However, the utilization of microwave technology in cereal processing, although reported in some studies, is still developing. Mahroug et al. (2019) applied microwave technology on wheat kernels and flour to remove gluten celiac immunotoxicity. However, the authors noted that microwave treatment triggered the disaggregation of gluten and the secondary structure of gluten, affecting the extractable gliadins, but not the extractable glutenin content. Hence further research may be required to understand the process of microwave treatment on gluten and its effect on celiac patients. In another study, Gianfrani et al. (2017) applied the R5-antibody-based Enzyme-Linked Immunoabsorbant Assay (ELISA test) after microwave treatment on wet wheat kernels, and found that the reduction of gluten was up to 20 ppm. They also confirmed that the conformational modifications, reducing the alcohol solubility of gliadins and altering the access of the R5 antibody to the gluten epitopes, was induced by microwave treatment. Padalino et al. (2019) utilized the microwave treatment of hydrated durum wheat kernels for pasta production, which in turn induced the denaturation of gluten proteins, mainly gliadins. This increased the exposure of further free sulfhydryl groups and hence caused weak protein network organization during pasta processing. These changes negatively affected the rheological behavior of the dough and consequently may have influenced the quality of the final product and modified the sensorial properties to a small extent.

2.7 Innovations in Secondary Cereal Processing

2.7.1 Innovations in Bioprocessing

2.7.1.1 Fermentation Fermentation belongs to a group of traditional technologies used to improve the shelf life and sensory properties of cereal products. The utilization of certain fermentative microorganisms increased the ability to manage the fermentation process and enhance the nutritional and health properties of the final product (Gobbetti et al. 2010; Lamsal and Faubion 2009) The fermentation of cereals (e.g. wheat and rice) improves their nutritional value, reflected in an increase in essential amino acids (e.g. lysine, methionine, and tryptophan), an increase in the availability of B vitamins, a decrease in the carbohydrate content, some nondigestible poly- and oligosaccharides and antinutrients (e.g. phytates, tannins, and polyphenols). Due to the formation of the several volatile compounds that contribute to a complex blend of flavors during fermentation, the flavor of fermented cereal products is enhanced (Karovičová and Kohajdova 2007). Although this is a highly traditional process, due to increasing consumer demands for nondairy probiotic, prebiotic, and symbiotic products, innovative nondairy fermented functional food products are emerging. Innovation in this sector implies the utilization of nanoscience and nanotechnological techniques to create specific bioactive nanoparticles for the creation of fermented cereal beverages (Salmerón 2017).

Recently, fermentation biotechnology has been applied to cereal byproducts derived from the dry and wet milling industries (e.g. germ and bran), the brewing industry (e.g. brewers' spent grain), the baking industry, and the starch industry (Verni et al. 2019). When cereal bran and germ, brewers' spent grain, and other byproducts of the cereal industry are subjected to fermentation processes, an increase in mineral content, vitamin bioavailability, protein content and digestibility, peptide and free amino acid content (especially lysine), and fiber and phenolic content (especially hydroxycinnamic and ferulic acid) are noticeable. The fermentation process can be coupled with enzymatic treatment for more extensive breakdown of the cell walls. In that sense, the fermentation of wheat with sourdough lactobacilli and fungal proteases has proven effective for the elimination of gluten celiac immunotoxicity (Gobbetti et al. 2019; Stoven et al. 2012). The complete hydrolysis of gluten in wheat flour can be achieved by specific combinations of lactobacilli and fungal proteases. The lactobacillus-treated wheat sourdough can be mixed with gluten-free flours (e.g. oat, millet, and buckwheat flours) to produce a bread of acceptable texture similar to that of wheat sourdough breads that does not increase intestinal permeability when consumed by coeliac patients (Gobbetti et al. 2019).

When the fermentation process of wheat bran with selected microbial strains (*Lactobacillus brevis* and *Kazachstania exigua*) is combined with enzymatic treatment (xylanase, endoglucanase, and β-glucanase), the microstructure of the bran is altered, reflected in the increased solubility of arabinoxylans. Moreover, the fermentation of wheat bran by lactic acid bacteria and endogenous proteases increases the concentration of peptides and free amino acids, together with increasing the *in vitro* digestibility of proteins and the bioaccessibility of phenolic acids, especially hydroxycinnamic acid and ferulic acid, which is esterified with arabinoxylans. When wheat germ is subjected to fermentation there is a decrease in the aldehydes responsible for the perception of rancidity, as well as in alcohols, ketones, furanones, lactones, and other volatile compounds occurring in lipid oxidation. During the fermentation of wheat bran a decrease in lipase activity and increase in free amino acids, especially lysine, is noticeable (Verni et al. 2019).

2.7.1.2 Enzyme-assisted Processing of Cereals The application of enzymes in cereal processing has increased in the last few years, mainly for improving the processing behavior or properties of cereal foods. For example, microbial enzymes are commonly used – hemicellulases and cellulases, which hydrolyze the complex polysaccharides of cereal cell walls (e.g. arabinoxylans, β-glucans, and cellulose). The exogenous application of enzymes on whole cereals for easier transformation of insoluble cell wall polysaccharides without nutrient loss was proposed as an alternative to the mechanical and chemical polishing of cereal grains, which intensifies the research in this area. However, due to the high costs associated with the production, storage, and transportation of enzymes, the scaling-up of this solution is still not realized (Singh et al. 2015). Biotechnological concepts have also been applied to grain processing to modulate the properties of the grains such as taste, texture, and shelf life, with limited detriment to nutritional value (Singh et al. 2015). The combination of enzymatic treatment and wet milling can be utilized for the fractionation of valuable fractions from cereal brans (e.g. wheat, barley, and oat brans, and rice polish), which maximizes the extraction rate of valuable cell wall components and aleurone cells from bran (Coimbra et al. 2012). In the first phase of this process, physical separation of the main bran fractions occurs, i.e. the insoluble phase (the pericarp and the aleurone layer), the germ-rich fraction, the residual endosperm fraction and the soluble sugars. In the second phase of this process, the previously cleaned bran is physically separated from the main fractions, i.e. the insoluble phase (the remaining cell wall components), the protein-rich fraction, the soluble hemicellulose, and the oligosaccharides.

Enzyme technology can be used for the modification of the constituents to produce health-related compounds such as soluble high-molecular-weight dietary fiber (e.g. endoxylanase), prebiotic arabinoxylo-oligosaccharides (e.g. endoxylanase), resistant starch (e.g. amylases, debranching enzymes, and/or transferases), and bioactive peptides (e.g. endoproteases). Moreover, the utilization of enzymes makes it possible to obtain high-quality gluten-free food products by inducing crosslinking (e.g. transglutaminases and oxidases) (Goesaert et al. 2008), and the modification of the immunogenic sequences of gluten to avoid recognition by the immune systems of coeliac patients (Cabrera-Chávez and Calderón de la Barca 2010; Fuciños et al. 2019).

Scherf et al. (2018) demonstrated the utilization of an enzymatic treatment for gluten degradation in the creation of high-quality gluten-free products, predominantly applicable when gluten-free wheat, rye, and barley flours are incorporated in gluten-free formulations. The utilization of plant, fungal, bacterial, animal, or engineered peptidases is recommended to degrade gluten proteins and peptides into harmless fragments and obtain food products with a low celiac disease immunoreactive response. Moreover, the modification of gluten to decrease CD-immunoreactivity may be achieved by crosslinking using microbial transglutaminase. Mohan Kumar et al. (2019) demonstrated the utilization of the prolyl endoprotease of *Aspergillus niger* to induce the cleavage of the proline-rich sequences and degradation of wheat flour gluten, whereby the wheat flour gliadin content was reduced up to 90–95%. The enzymatic modification of gluten in wheat flour by the prolyl-endopeptidase of *Aspergillus niger* for use as a supplement in bread with a blend of raw and popped amaranth seeds was demonstrated by Heredia-Sandoval et al. (2016). They observed that bread supplemented with the 80% amaranth blend appeared to have 99% less immunogenic gluten than the wheat bread, thereby confirming the utilization of enzymes as an effective way to obtain gluten-reduced breads.

2.7.2 Innovative Cereal Extrusion

Extrusion is a food processing technique utilized for multiple applications – in the production of breakfast cereals, snacks, protein processing, confectioneries, and animal feed foods (Arribas et al. 2019). Traditional extruded products are mainly composed of cereals and different starches. Innovation in extrusion is based on the incorporation of alternative ingredients in the extruded product formulations to improve the nutritional and sensory quality, as well as to increase the nutritional value of the raw material. These alternative ingredients encompass whole cereal and pulse grains and fiber-rich byproducts of cereal, fruit, and vegetable processing (Oliveira et al. 2018). However, their incorporation is a challenging technological task due to their adverse effects on the expansion index, bulk density, and texture, especially hardness and crispness (Oliveira et al. 2018). Alonso dos Santos et al. (2019) demonstrated the utilization of agricultural byproducts (from rice, passion fruit, and milk processing) in the formulation of extruded gluten-free breakfast cereals. The resulting product was rich in total, soluble, and insoluble dietary fiber, which affected the product with decreased expansion and a dark color. Arribas et al. (2019) demonstrated obtaining an innovative gluten-free expanded snack product from carob fruit, pea, and rice blends. Furthermore, Masatcioglu et al. (2017) demonstrated the potential of extrusion cooking to significantly increase the enzyme-resistant starch type 3 content in high-amylose starches without the need for gelatinization, debranching and heat-moisture storage cycles. Extrusion technology has proven useful for ensuring cereal safety and the reduction of mycotoxins and antinutritional factor levels (Nikmaram et al. 2017; Ryu et al. 2019). The achieved reduction of ochratoxin A in oats and rice by twin-screw extrusion processing were in the range of 40–43% and 78–82% respectively. The addition of baking soda improved the reduction of ochratoxin A in oats, but not in rice due to the formation of a nontoxic ochratoxin A isomer (Ryu et al. 2019).

2.7.2.1 3D Printing Technology Another innovation that is closely related to extrusion technology is 3D printing technology, employed for innovative food design, and when it comes to cereals, cereal-based products of the desired shape, dimension and nutritional content (Severini et al. 2016). Unlike conventional food extrusion cooking, extrusion-based food printing comprises a digitally-controlled extrusion process to build up complex 3D food products layer by layer and obtain a product of better quality at low environmental cost (Severini et al. 2016; Sun et al. 2018). Considering the extrusion mechanisms utilized in food printing, three types of extrusion are utilized: syringe-based extrusion, air-pressure-driven extrusion, and screw-based extrusion. Moreover, according to the temperature, food printing utilizes room temperature extrusion, hot-melt extrusion, and hydrogel-forming extrusion, where room temperature extrusion is applicable to cereal processing for pasta printing using classical recipes (durum wheat semolina and water), dispensing sauce onto the surfaces of pizzas, cookies, and graphical decoration (Sun et al. 2018).

The key device in extrusion cooking is an extruder, consisting of a single/twin rotating screw located within a barrel, within which pre-ground and conditioned ingredients are converted into viscoelastic fluid under the influence of mechanical and thermal energy. Pressurized material is texturized and shaped in a die at the end of the extruder due to the pressure difference between the extruder and the atmosphere. The resulting product is well-cooked, shelf-stable, and ready for packaging. Unlike a conventional extruder, the key device in 3D printing is the extrusion-based food printer, consisting of a multi-axis stage and one or more extrusion units. The extrusion process in food printing is a digitally controlled process, able to build up complex 3D food products layer by layer. The loaded material is pushed out of the nozzle in a controlled manner – the material stream is moved according to a predefined path until the deposited layers are bonded to create a coherent solid structure (Sun et al. 2018). The printability of the dough is dependent on the rheological properties and microstructure of the dough. Dough of a higher viscoelastic modulus, higher loss factor, and higher complex viscosity and yield stress is characterized by good printability (Zhang et al. 2018).

Severini et al. (2016) demonstrated the utilization of 3D printing to obtain cereal-based snacks and hence its relevance to obtaining "personalized food products." Zhang et al. (2018) demonstrated obtaining cereal-based food structures containing probiotics by 3D printing. Severini et al. (2018) also demonstrated the utilization of 3D printing to obtain cereal-based snacks enriched with edible insects – ground larvae of yellow mealworms (*Tenebrio molitor*) as a novel source of proteins.

2.7.3 Innovative Baking

In order to meet the increasing environmental challenges and achieve a higher energy efficiency in baking, different technological solutions for baking purposes have been developing. In that sense, a combination of several energy sources (e.g. forced convection, irradiation, microwave, etc.) and their optimization is emerging and being applied to baking oven design (Papasidero et al. 2016). Ayub et al. (2018) proposed and designed a solar bakery unit and applied it to cookie baking. The solar baking system provided promising results in terms of the baking time, baking quality (good appearance), and energy utilization rate, being in the range 25–75%. The overall exergy efficiency of the system was found to be around 60%. The obtained results can be improved further by the optimization and control of baking conditions. Rondeau-Mouro et al. (2019) and Grenier et al. (2019) proposed an innovative baking method that utilizes low baking temperatures (<105 °C) and partial vacuum conditions (−20 kPa). Those baking conditions resulted in a higher oven rise (28% compared to 16%) and a lower crumb density than were reached by baking in a commercial convection oven at atmospheric pressure. Moreover, the utilization of a partial vacuum appeared to be a way to modify the gas fraction within the dough before the crust sets. Therefore, this method of baking appeared to be particularly suitable for baking gluten-free bread (Rondeau-Mouro et al. 2019). Rastogi (2019) overviewed the advantages of infrared (IR) baking over conventional baking in terms of a reduced baking time, more uniform baking, lower quality loss, versatility, simpler and more compact equipment, and energy savings. Moreover, the IR-baked products were of improved nutritional quality and more acceptable to the consumers.

2.8 Conclusion

Cereal processing has become an essential process prior to consumption; therefore this chapter has provided an understanding of cereal morphology, structure, and processing techniques with a special focus on the innovative processing of cereal grains. In recent times, traditional methods have been slowly replaced with modified and new techniques for processing cereal grains. For many countries, the sustainable processing of cereal grains with reduced losses and the higher productivity have become essential requirements. This demand creates a huge challenge for the food industry to adapt new and innovative techniques to process the cereal grains to retain bioactives and major nutrients, as well to reduce losses across the cereal processing chain, and furthermore, to produce safe, healthy cereal-based products with increased shelf lives and of retained sensorial properties.

Acknowledgment

M. Pojić would like to acknowledge the financial support of the Ministry of Education,Science and Technological Development of the Republic of Serbia (No. 451-03-68/2020-14/200222).

References

Ahmed, J. (2016). Effect of high pressure on structural and rheological properties of cereals and legume proteins. In: *Novel Food Processing: Effect on Rheological and Functional Properties* (eds. J. Ahmed, H.S. Ramaswamy, S. Kasapis, et al.), 234–264. Boca Raton: CRC Press, Taylor & Francis Group.

Aiken, R.M. and Smucker, A.J.M. (1996). Root system regulation of whole plant growth. *Annual Review of Phytopathology* 34: 325–346.

Alcázar-Alay, S.C. and Meireles, M.A.A. (2015). Physicochemical properties, modifications and applications of starches from different botanical sources. *Food Science and Technology* 35 (2): 215–236.

Alldrick, A. (2017). Chemical contamination of cereals. In: *Chemical Contaminants and Residues in Food (Second Edition)* (eds. D. Schrenk and A. Cartus), 427–449. Woodhead Publishing.

Alonso dos Santos, P., Caliari, M., Soares Soares Júnior, M. et al. (2019). Use of agricultural by-products in extruded gluten-free breakfast cereals. *Food Chemistry* 297: 124956.

Alvarenga, I.C., Ou, Z., Thiele, S. et al. (2018). Effects of milling sorghum into fractions on yield, nutrient composition, and their performance in extrusion of dog food. *Journal of Cereal Science* 82: 121–128.

Arndt, E.A. and Korolchuk, T. (2014). Whole grain flour and products including same. US Patent US 8,852,665 B2, filed 22 February 2013 and issued 7 October 2014.

Arribas, C., Cabellos, B., Cuadrado, C. et al. (2019). The effect of extrusion on the bioactive compounds and antioxidant capacity of novel gluten-free expanded products based on carob fruit, pea and rice blends. *Innovative Food Science & Emerging Technologies* 52: 100–107.

Arvanitoyannis, I.S. and Stratakos, A.C. (2010). Irradiation of cereals. In: *Irradiation of Food Commodities* (ed. I.S. Arvanitoyannis), 451–466. Academic Press.

Ayub, I., Munir, A., Ghafoor, A. et al. (2018). Solar thermal application for decentralized food baking using Scheffler reflector technology. *Journal of Solar Energy Engineering* 140 (6): 061005. https://doi.org/10.1115/1.4040206.

Bai, Y.-P., Guo, X.-N., Zhu, K.-X., and Zhou, H.-M. (2017). Shelf-life extension of semi-dried buckwheat noodles by the combination of aqueous ozone treatment and modified atmosphere packaging. *Food Chemistry* 237: 553–560.

Bashir, K., Swer, T.L., Prakash, K.S., and Aggarwal, M. (2017). Physico-chemical and functional properties of gamma irradiated whole wheat flour and starch. *LWT - Food Science and Technology* 76: 131–139.

Bhat, N.A., Wani, I.A., Hamdani, A.M. et al. (2016). Physicochemical properties of whole wheat flour as affected by gamma irradiation. *LWT - Food Science and Technology* 71: 175–183.

Bhattacharjee, P. and Singhal, R.S. (2016). Effect of irradiation on food texture and rheology. In: *Novel Food Processing: Effects on Rheological and Functional Properties* (eds. J. Ahmed, H.S. Ramaswamy, S. Kasapis, et al.), 103–125. Boca Raton: CRC Press, Taylor & Francis Group.

Bock, J.E. and Sweley, J. (2018). Incremental innovation in the milling industry: a panel discussion. *Cereal Foods World* 63 (6): 254–256.

Bommert, P. and Whipple, C. (2018). Grass inflorescence architecture and meristem determinacy. *Seminars in Cell & Developmental Biology* 79: 37–47.

Bushuk, W. (2004). Rye. In: *Encyclopedia of Grain Science*, vol. 3 (eds. C. Wrigley, C. Walker and H. Corke), 85–91. Oxford: Elsevier.

Cabrera-Chávez, F. and Calderón de la Barca, A.M. (2010). Trends in wheat technology and modification of gluten proteins for dietary treatment of coeliac disease patients. *Journal of Cereal Science* 52 (3): 337–341.

Cappa, C., Barbosa-Cánovas, G.V., Lucisano, M., and Mariotti, M. (2016). Effect of high pressure processing on the baking aptitude of corn starch and rice flour. *LWT - Food Science and Technology* 73: 20–27.

Çatal, H. and İbanoğlu, Ş. (2014). Effect of aqueous ozonation on the pasting, flow and gelatinization properties of wheat starch. *LWT - Food Science and Technology* 59 (1): 577–582.

Chittrakorn, S., Earls, D., and MacRitchie, F. (2014). Ozonation as an alternative to chlorination for soft wheat flours. *Journal of Cereal Science* 60 (1): 217–221.

Coimbra, M.A., Pinto, M.I., Martins, M.A. et al. (2012). Integrated process for extracting proteins and arabinoxylans from brewer's spent grain. WO Patent 2012/069889 A1, filed 9 December 2010 and issued 31 May 2012.

Doust, A. (2007). Architectural evolution and its implications for domestication in grasses. *Annals of Botany* 100 (5): 941–950.

Durak, A.N., Erbaş, M., and Arslan, S. (2016). Ultrasonication to inactivate the proteolytic enzymes in suni bug-damaged wheat. *Journal of Cereal Science* 71: 122–129.

Evers, T. and Millar, S. (2002). Cereal grain structure and development some implications for quality. *Journal of Cereal Science* 35: 261–284.

FAO (1992). Maize in human nutrition. http://www.fao.org/docrep/t0395e/T0395E01.htm#Structure%20of%20the%20maize%20kernel (accessed 10 March 2020).

FAO (1995). Sorghum and millets in human nutrition (FAO Food and Nutrition Series, No. 27) http://www.fao.org/3/t0818e/T0818E00.htm (accessed 20 May 2018).
FAO (1999). Fermented cereals. A global perspective. Major chemical components of cereal grains. http://www.fao.org/3/x2184e/x2184e04.htm (accessed 25 March 2020).
FAO (2020). FAOSTAT Database: Crop statistics. http://www.fao.org/faostat/en/#data/QC
Fernando, B. (2013). Rice as a source of fibre. *Journal of Rice Research* 1: e101.
Ferrari, B., Finocchiaro, F., Stanca, A.M., and Gianinetti, A. (2009). Optimization of air classification for the production of ß-glucan-enriched barley flours. *Journal of Cereal Science* 50 (2): 152–158.
Ficco, D.B.M., Borrelli, G.M., Giovanniello, V. et al. (2018). Production of anthocyanin-enriched flours of durum and soft pigmented wheats by air-classification, as a potential ingredient for functional bread. *Journal of Cereal Science* 79: 118–126.
Freitas, R.d.S.d., Faroni, L.R.D., Queiroz, M.E.L.R.d. et al. (2017). Degradation kinetics of pirimiphos-methyl residues in maize grains exposed to ozone gas. *Journal of Stored Products Research* 74: 1–5.
Fuciños, C., Estévez, N., Míguez, M. et al. (2019). Effectiveness of proteolytic enzymes to remove gluten residues and feasibility of incorporating them into cleaning products for industrial purposes. *Food Research International* 120: 167–177.
Gianfrani, C., Mamone, G., la Gatta, B. et al. (2017). Microwave-based treatments of wheat kernels do not abolish gluten epitopes implicated in celiac disease. *Food and Chemical Toxicology* 101: 105–113.
Gobbetti, M., Di Cagno, R., and De Angelis, M. (2010). Functional microorganisms for functional food quality. *Critical Reviews in Food Science and Nutrition* 50 (8): 716–727.
Gobbetti, M., De Angelis, M., Di Cagno, R. et al. (2019). Novel insights on the functional/nutritional features of the sourdough fermentation. *International Journal of Food Microbiology* 302: 103–113.
Goesaert, H., Courtin, C.M., and Delcour, J.A. (2008). Use of enzymes in the production of cereal-based functional foods and food ingredients. In: *Gluten-Free Cereal Products and Beverages* (eds. E.K. Arendt and F.D. Bello), 237–265. Academic Press.
Goldberg, G. (ed.) (2003) Plants: Diet and Health: The Report of the British Nutrition Foundation Task Force. Oxford: Blackwell.
Gómez-Caravaca, A.M., Verardo, V., Candigliota, T. et al. (2015). Use of air classification technology as green process to produce functional barley flours naturally enriched of alkylresorcinols, ß-glucans and phenolic compounds. *Food Research International* 73: 88–96.
Gozé, P., Rhazi, L., Pauss, A., and Aussenac, T. (2016). Starch characterization after ozone treatment of wheat grains. *Journal of Cereal Science* 70: 207–213.
Gozé, P., Rhazi, L., Lakhal, L. et al. (2017). Effects of ozone treatment on the molecular properties of wheat grain proteins. *Journal of Cereal Science* 75: 243–251.
Granella, S.J., Christ, D., Werncke, I. et al. (2018). Effect of drying and ozonation process on naturally contaminated wheat seeds. *Journal of Cereal Science* 80: 205–211.
Grenier, D., Lucas, T., and Le Ray, D. (2019). Enhanced aeration of part-baked bread using a novel combination of baking and partial vacuum. *Journal of Food Engineering* 248: 62–70.
Healthy Minor Cereals (2016). Deliverable D8.1: Report on the market potential of mi-nor cereal crops and consumers perceptions about them in different European regions. https://toolkit.ecn.cz/img_upload/0f728a035bc05faf5df37da80e732579/d-8.1-market-potential-of-mc_final.pdf (accessed 20 December 2019).
Hemery, Y., Holopainen, U., Lampi, A.-M. et al. (2011). Potential of dry fractionation of wheat bran for the development of food ingredients, part II: electrostatic separation of particles. *Journal of Cereal Science* 53 (1): 9–18.

Heredia-Sandoval, N.G., Calderón de la Barca, A.M., and Islas-Rubio, A.R. (2016). Gluten degradation in wheat flour with *Aspergillus niger* prolyl-endopeptidase to prepare a gluten-reduced bread supplemented with an amaranth blend. *Journal of Cereal Science* 71: 73–77.

Hetz, W., Hochholdinger, F., Schwall, M., and Feix, G. (1996). Isolation and characterisation of *rtcs* a mutant deficient in the formation of nodal roots. *The Plant Journal* 10: 845–857.

Hoseney, C. (ed.) (1994). *Principles of Cereal Science and Technology*, 3e, 1–22. St. Paul, MN: American Association of Cereal Chemists.

Humer, E., Schwarz, C., and Schedle, K. (2015). Phytate in pig and poultry nutrition. *Journal of Animal Physiology and Animal Nutrition* 99: 605–625.

Hu, X.-P., Zhang, B., Jin, Z.-Y. et al. (2017). Effect of high hydrostatic pressure and retrogradation treatments on structural and physicochemical properties of waxy wheat starch. *Food Chemistry* 232: 560–565.

Jiang, Y., Bai, X., Lang, S. et al. (2019). Optimization of ultrasonic-microwave assisted alkali extraction of arabinoxylan from the corn bran using response surface methodology. *International Journal of Biological Macromolecules* 128: 452–458.

Juliano, B.O. and Tuaño, A.P.P. (2019). Gross structure and composition of the rice grain. In: *Rice: Chemistry and Technology*, 4e (ed. J. Bao), 31–53. AACC International and Woodhead Publishing.

Kalagatur, N.K., Kamasani, J.R., Mudili, V. et al. (2018). Effect of high pressure processing on growth and mycotoxin production of *Fusarium graminearum* in maize. *Food Bioscience* 21: 53–59.

Karovičová, Z.K.-J. and Kohajdova, J. (2007). Fermentation of cereals for specific purpose. *Journal of Food and Nutrition Research* 46 (2): 51–57.

Kaur, H. and Gill, B.S. (2019). Effect of high-intensity ultrasound treatment on nutritional, rheological and structural properties of starches obtained from different cereals. *International Journal of Biological Macromolecules* 126: 367–375.

Kellogg, E.A., Camara, P.E.A.S., Ruadall, P.J. et al. (2013). Early inflorescence development in the grasses (*Poaceae*). *Frontiers in Plant Science* 4 (250): 1–16.

Kertész, A. (2009). High speed roller mill. World Intellectual Property Organization Patent WO 2009/068921 A1, filed 25 November 2008 and issued 4 June 2009.

Koehler, P. and Wieser, H. (2013). Chemistry of cereal grains. In: *Handbook of Sourdough Biotechnology* (eds. M. Gobbetti and M. Gaenzle), 11–45. New York: Springer.

Korolchuk, T. (2008). Process for producing an ultrafine-milled whole-grain wheat flour and products thereof. US patent 7419,694 B2, filed 17 December 2003 and issued 2 September 2008.

Kyozuka, J., Tokunaga, H., and Yoshida, A. (2014). Control of grass inflorescence form by the fine-tuning of meristem phase change. *Current Opinion in Plant Biology* 17C (1): 110–115.

Lamsal, B.P. and Faubion, J.M. (2009). The beneficial use of cereal and cereal components in probiotic foods. *Food Reviews International* 25 (2): 103–114.

Lazaridou, A., Vouris, D.G., Zoumpoulakis, P., and Biliaderis, C.G. (2018). Physicochemical properties of jet milled wheat flours and doughs. *Food Hydrocolloids* 80: 111–121.

Lee, J.-S., Ee, M.-L., Chung, K.-H., and Othman, Z. (2013). Formation of resistant corn starches induced by gamma-irradiation. *Carbohydrate Polymers* 97 (2): 614–617.

Lee, Y.-T., Shim, M.-J., Goh, H.-K. et al. (2019). Effect of jet milling on the physicochemical properties, pasting properties, and in vitro starch digestibility of germinated brown rice flour. *Food Chemistry* 282: 164–168.

Lestienne, I., Buisson, M., Lullien, V., and Picq, C. (2005). Losses of nutrients and anti-nutritional factors during abrasive decortication of two pearl millet cultivars (*Pennisetum glaucum*). *Food Chemistry* 100 (4): 1316–1323.

Li, H., Zhu, K., Zhou, H. et al. (2016). Comparative study of four physical approaches about allergenicity of soybean protein isolate for infant formula. *Food and Agricultural Immunology* 27 (5): 604–623.

Li, S., Luo, Z., Guan, X. et al. (2019). Effect of ultrasonic treatment on the hydration and physicochemical properties of brewing rice. *Journal of Cereal Science* 87: 78–84.

Los, A., Ziuzina, D., Boehm, D. et al. (2017). The potential of atmospheric air cold plasma for control of bacterial contaminants relevant to cereal grain production. *Innovative Food Science & Emerging Technologies* 44: 36–45.

Mahroug, H., Ribeiro, M., Rhazi, L. et al. (2019). How microwave treatment of gluten affects its toxicity for celiac patients? A study on the effect of microwaves on the structure, conformation, functionality and immunogenicity of gluten. *Food Chemistry* 297: 124986.

Masatcioglu, T.M., Sumer, Z., and Koksel, H. (2017). An innovative approach for significantly increasing enzyme resistant starch type 3 content in high amylose starches by using extrusion cooking. *Journal of Cereal Science* 74: 95–102.

McSteen, P., Laudencia-Chingcuanco, D., and Colasanti, J. (2000). A floret by any other name: control of meristem identity in maize. *Trends in Plant Science* 5 (2): 61–66.

Misra, N.N., Kaur, S., Tiwari, B.K. et al. (2015). Atmospheric pressure cold plasma (ACP) treatment of wheat flour. *Food Hydrocolloids* 44: 115–121.

Mohan Kumar, B.V., Sarabhai, S., and Prabhasankar, P. (2019). Targeted degradation of gluten proteins in wheat flour by prolyl endoprotease and its utilization in low immunogenic pasta for gluten sensitivity population. *Journal of Cereal Science* 87: 59–67.

Mukhtar, R., Shah, A., Noor, N. et al. (2017). Y-Irradiation of oat grain – effect on physicochemical, structural, thermal, and antioxidant properties of extracted starch. *International Journal of Biological Macromolecules* 104: 1313–1320.

Navarro, S.L.B., Capellini, M.C., Aracava, K.K., and Rodrigues, C.E.C. (2016). Corn germ-bran oils extracted with alcoholic solvents: Extraction yield, oil composition and evaluation of protein solubility of defatted meal. *Food and Bioproducts Processing 100 Part A*: 185–194.

Nikmaram, N., Leong, S.Y., Koubaa, M. et al. (2017). Effect of extrusion on the antinutritional factors of food products: an overview. *Food Control* 79: 62–73.

Obadi, M., Zhu, K.-X., Peng, W. et al. (2018). Shelf life characteristics of bread produced from ozonated wheat flour. *Journal of Texture Studies* 49 (5): 492–502.

OECD-FAO (2018). *OECD-FAO Agricultural Outlook 2018–2027.* Rome: OECD Publishing, Paris/Food and Agriculture Organization of the United Nations https://doi.org/10.1787/agr_outlook-2018-en (accessed 20 December 2019).

Oliveira, L.C., Alencar, N.M.M., and Steel, C.J. (2018). Improvement of sensorial and technological characteristics of extruded breakfast cereals enriched with whole grain wheat flour and jabuticaba (*Myrciaria cauliflora*) peel. *LWT - Food Science and Technology* 90: 207–214.

Padalino, L., Del Nobile, M.A., la Gatta, B. et al. (2019). Effects of microwave treatment of durum wheat kernels on quality characteristics of flour and pasta. *Food Chemistry* 283: 454–461.

Pal, P., Kaur, P., Singh, N. et al. (2016). Effect of nonthermal plasma on physico-chemical, amino acid composition, pasting and protein characteristics of short and long grain rice flour. *Food Research International* 81: 50–57.

Papageorgiou, M. and Skendi, A. (2018). Introduction to cereal processing and by-products. In: *Sustainable Recovery and Reutilization of Cereal Processing by-Products* (ed. C.M. Galanakis), 1–25. Woodhead Publishing.

Papasidero, D., Pierucci, S., and Manenti, F. (2016). Energy optimization of bread baking process undergoing quality constraints. *Energy* 116, Part 2: 1417–1422.

Pei-Ling, L., Xiao-Song, H., and Qun, S. (2010). Effect of high hydrostatic pressure on starches: a review. *Starch – Stärke* 62 (12): 615–628.

Piechowiak, T., Józefczyk, R., and Balawejder, M. (2018). Impact of ozonation process of wheat flour on the activity of selected enzymes. *Journal of Cereal Science* 84: 30–37.

Pojić, M., Spasojević, N., and Atlas, M. (2014). Chemometric approach to characterization of flour mill streams: chemical and rheological properties. *Food and Bioprocess Technology* 7: 1298–1309.

Pojić, M., Mišan, A., and Tiwari, B. (2018). Eco-innovative technologies for extraction of proteins for human consumption from renewable protein sources of plant origin. *Trends in Food Science & Technology* 75: 93–104.

Pomeranz, Y. (1982). Grain structure and end-use properties. *Journal of Food Structure* 1 (2): 107–124.

Qi, L., Li, Y., Luo, X. et al. (2016). Detoxification of zearalenone and ochratoxin A by ozone and quality evaluation of ozonised corn. *Food Additives & Contaminants: Part A* 33 (11): 1700–1710.

Rastogi, N.K. (2019). Infrared heating in baking operations. In: *Reference Module in Food Science*. Elsevier https://doi.org/10.1016/B978-0-08-100596-5.22672-3.

Ravindran, R. and Jaiswal, A.K. (2019). Wholesomeness and safety aspects of irradiated foods. *Food Chemistry* 285: 363–368.

Rondeau-Mouro, C., Godfrin, C., Cambert, M. et al. (2019). Characterization of gluten-free bread crumb baked at atmospheric and reduced pressures using TD-NMR. *Magnetic Resonance in Chemistry* 57 (9): 649–660.

Roth, M., Jekle, M., and Becker, T. (2019). Opportunities for upcycling cereal byproducts with special focus on distiller's grains. *Trends in Food Science & Technology* 91: 282–293.

Ryu, D., Kowalski, R.J., Ganjyal, G., and Lee, H.J. (2019). Reduction of ochratoxin A in oats and rice by twin-screw extrusion processing with baking soda. *Food Control* 105: 21–28.

Saldivar, S. (2003). Cereals: dietary importance. In: *Encyclopedia of Food Sciences and Nutrition* (eds. B. Caballero, L. Trugo and P. Finglas), 1027–1033. Academic Press.

Salmerón, I. (2017). Fermented cereal beverages: from probiotic, prebiotic and synbiotic towards nanoscience designed healthy drinks. *Letters in Applied Microbiology* 65 (2): 114–124.

Sandhu, H.P.S., Manthey, F.A., Simsek, S., and Ohm, J.-B. (2011). Comparison between potassium bromate and ozone as flour oxidants in breadmaking. *Cereal Chemistry* 88 (1): 103–108.

Sarangapani, C., Thirumdas, R., Devi, Y. et al. (2016). Effect of low-pressure plasma on physico–chemical and functional properties of parboiled rice flour. *LWT - Food Science and Technology* 69: 482–489.

Savi, G.D., Piacentini, K.C., Bittencourt, K.O., and Scussel, V.M. (2014). Ozone treatment efficiency on Fusarium graminearum and deoxynivalenol degradation and its effects on whole wheat grains (*Triticum aestivum* L.) quality and germination. *Journal of Stored Products Research* 59: 245–253.

Savi, G.D., Piacentini, K.C., and Scussel, V.M. (2015). Reduction in residues of deltamethrin and fenitrothion on stored wheat grains by ozone gas. *Journal of Stored Products Research* 61: 65–69.

Savi, G.D., Piacentini, K.C., Bortolotto, T., and Scussel, V.M. (2016). Degradation of bifenthrin and pirimiphos-methyl residues in stored wheat grains (*Triticum aestivum* L.) by ozonation. *Food Chemistry* 203: 246–251.

Scherf, K.A., Wieser, H., and Koehler, P. (2018). Novel approaches for enzymatic gluten degradation to create high-quality gluten-free products. *Food Research International* 110: 62–72.

Serna-Saldivar, S. and Rooney, L.W. (1995). Structure and chemistry of sorghum and millets. In: *Sorghum and Millets, Chemistry and Technology* (ed. D.A.V. Dendy), 69–124. St Paul, MN: American Association of Cereal Chemistry.

Severini, C., Derossi, A., and Azzollini, D. (2016). Variables affecting the printability of foods: preliminary tests on cereal-based products. *Innovative Food Science & Emerging Technologies* 38: 281–291.

Severini, C., Azzollini, D., Albenzio, M., and Derossi, A. (2018). On printability, quality and nutritional properties of 3D printed cereal based snacks enriched with edible insects. *Food Research International* 106: 666–676.

Shewry, P.S. and Tatham, A.S. (1999). The characteristics, structures and evolutionary relationships of prolamins. In: *Seed Protein* (eds. P.R. Shewry and R. Casey), 11–18. Springer.

Shewry, P.R., Tatham, A.S., and Kasarda, D.D. (1992). Cereal proteins in coeliac disease. In: *Coeliac Disease* (ed. M.N. Marsh), 305–348. Oxford: Blackwell Science.

Shriver, S.K. and Yang, W.W. (2011). Thermal and nonthermal methods for food allergen control. *Food Engineering Reviews* 3 (1): 26–43.

Sibakov, J., Abecassis, J., Barron, C., and Poutanen, K. (2014). Electrostatic separation combined with ultra-fine grinding to produce ß-glucan enriched ingredients from oat bran. *Innovative Food Science & Emerging Technologies* 26: 445–455.

Silventoinen, P., Sipponen, M.H., Holopainen-Mantila, U. et al. (2018). Use of air classification technology to produce protein-enriched barley ingredients. *Journal of Food Engineering* 222: 169–177.

Singh, A., Karmakar, S., Jacob, B.S. et al. (2015). Enzymatic polishing of cereal grains for improved nutrient retainment. *Journal of Food Science and Technology* 52 (6): 3147–3157.

Stoven, S., Murray, J.A., and Marietta, E. (2012). Celiac disease: advances in treatment via gluten modification. *Clinical Gastroenterology and Hepatology* 10 (8): 859–862.

Sultan, N., Wani, I.A., and Masoodi, F.A. (2018). Moisture mediated effects of Y-irradiation on physicochemical, functional, and antioxidant properties of pigmented brown rice (*Oryza sativa* L.) flour. *Journal of Cereal Science* 79: 399–407.

Sumnu, G. and Sahin, S. (2005). Recent developments in microwave heating. In: *Emerging Technologies for Food Processing* (ed. D.-W. Sun), 419–444. Elsevier.

Sun, J., Zhou, W., Yan, L. et al. (2018). Extrusion-based food printing for digitalized food design and nutrition control. *Journal of Food Engineering* 220: 1–11.

Tattiyakul, J. and Rao, M.A. (2016). Effect of high-pressure and ultrasonic processing of foods on rheological properties. In: *Novel Food Processing: Effects on Rheological and Functional Properties* (eds. J. Ahmed, H.S. Ramaswamy, S. Kasapis, et al.), 207–227. Boca Raton: CRC Press, Taylor & Francis Group.

Tiwari, B.K., Brennan, C.S., Curran, T. et al. (2010). Application of ozone in grain processing. *Journal of Cereal Science* 51 (3): 248–255.

Tolouie, H., Mohammadifar, M.A., Ghomi, H. et al. (2018). The impact of atmospheric cold plasma treatment on inactivation of lipase and lipoxygenase of wheat germs. *Innovative Food Science & Emerging Technologies* 47: 346–352.

U. Tiwari., E. Cummins (2009). Factors Influencing β-Glucan Levels and Molecular Weight in Cereal-Based Products. Cereal Chem. 86(3):290–301.

Vallons, K.J.R., Ryan, L.A.M., and Arendt, E.K. (2011). Promoting structure formation by high pressure in gluten-free flours. *LWT - Food Science and Technology* 44 (7): 1672–1680.

Verni, M., Rizzello, C.G., and Coda, R. (2019). Fermentation biotechnology applied to cereal industry by-products: nutritional and functional insights. *Frontiers in Nutrition* 6: 42.

Vollbrecht, E., Springer, P.S., Goh, L. et al. (2005). Architecture of floral branch systems in maize and related grasses. *Nature* 436: 1119–1126.

Weaver, J.E. and Zink, E. (1945). Extent and longevity of the seminal roots of certain grasses. Plant Physiology 20(3): 359–379.

Welch, R.W. (ed.) (2012). *The Oat Crop: Production and Utilization*. Springer Science & Business Media.

Wen, C., Zhang, J., Yao, H. et al. (2019). Advances in renewable plant-derived protein source: the structure, physicochemical properties affected by ultrasonication. *Ultrasonics Sonochemistry* 53: 83–98.

Wielogorska, E., Ahmed, Y., Meneely, J. et al. (2019). A holistic study to understand the detoxification of mycotoxins in maize and impact on its molecular integrity using cold atmospheric plasma treatment. *Food Chemistry* 301: 125281.

Yang, Z., Chaib, S., Gu, Q., and Hemar, Y. (2017). Impact of pressure on physicochemical properties of starch dispersions. *Food Hydrocolloids* 68: 164–177.

Yodpitak, S., Mahatheeranont, S., Boonyawan, D. et al. (2019). Cold plasma treatment to improve germination and enhance the bioactive phytochemical content of germinated brown rice. *Food Chemistry* 289: 328–339.

Zhang, L., Lou, Y., and Schutyser, M.A.I. (2018). 3D printing of cereal-based food structures containing probiotics. *Food Structure* 18: 14–22.

Zhu, F. (2018). Effect of ozone treatment on the quality of grain products. *Food Chemistry* 264: 358–366.

Zhu, F. and Li, H. (2019). Effect of high hydrostatic pressure on physicochemical properties of quinoa flour. *LWT - Food Science and Technology* 114: 108367.

3 Pseudocereals as Healthy Grains: An Overview

Muriel Henrion[1], Emilie Labat[1], and Lisa Lamothe[2]

[1] *Nestlé Research Center Orbe, Société des Produits Nestlé SA, Orbe, Switzerland*
[2] *Nestlé Research, Société des Produits Nestlé SA, Lausanne, Switzerland*

3.1 Introduction

Pseudocereals refer to plants which are grown as crops to produce a starchy grain suitable for human consumption, excluding plants already classified as cereals (species from the grass family – Gramineae), legumes, oilseeds, or nuts. Most pseudocereals have smaller seeds than the major cereal grains, but they are used in the same way. Pseudocereal seeds can be ground into flour and are often used as a replacement for cereals. However, pseudocereals are considered a minor crop on a global scale. They have been significant contributors to the human diet in certain regions in the past, while their significance has been restored in recent times. Pseudocereals may play a pivotal role in human nutrition, especially for those who have allergies to components of traditional cereals (Valcarcel-Yamani and Lannes 2012). For farmers or producers, pseudocereals can play a role in cereal rotations, reducing the buildup of grass weeds, pests, or diseases. In addition, many pseudocereals can grow in poor soils and conditions not suitable for other grain species. The three major pseudocereal crops are buckwheat (*Fagopyrum esculentum*; Polygonaceae), quinoa (*Chenopodium quinoa subsp. quinoa*; Chenopodiaceae) and amaranth (*Amaranthus caudatus*; *A. cruentus*; *A. hypochondriacus*; family: Amaranthceae). However, a large number of other species are described that have the potential to be used as pseudocereals. This chapter focuses on the origin, production, utilization, and processing of pseudocereals, and their bioactive compounds.

3.2 Pseudocereals: Origin, Production, and Utilization

Amaranth and quinoa were major crops used by the pre-Columbian cultures in Latin America. On the other hand, buckwheat originated from central Asia and was transferred to central and eastern Europe by nomadic groups. Until recently these crops were mainly cultivated on a minor scale as subsistence agriculture. Due to their good

Innovative Processing Technologies for Healthy Grains, First Edition.
Edited by Milica Pojić and Uma Tiwari.

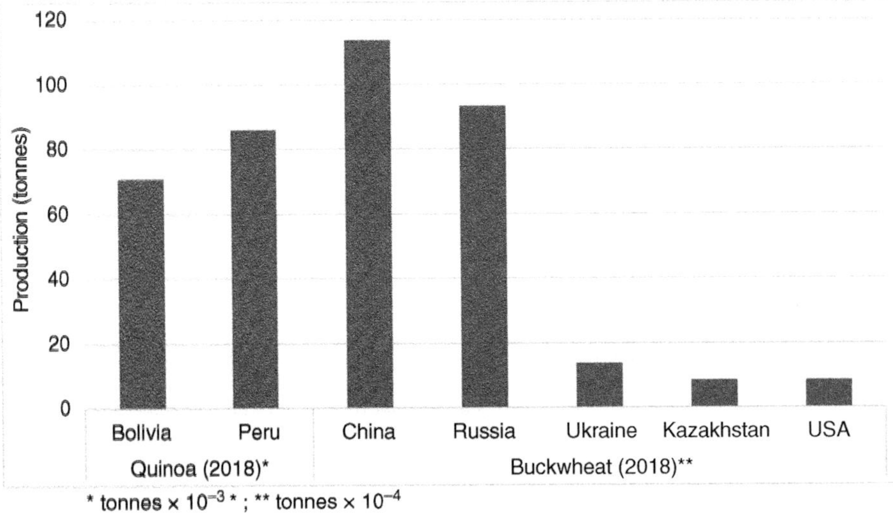

Figure 3.1 Food and Agriculture Organization (FAO) data on the total production of quinoa and buckwheat. *Source:* FAO (2020).

nutritional properties, the worldwide production of pseudocereal grains has increased, together with their commercial interest, mainly associated with the increasing demands for gluten-free products. Figure 3.1 shows the total production (tonnes) of quinoa and buckwheat in 2018.

3.2.1 Buckwheat

Buckwheat originates from Asia and has been cultivated since at least 1000 BCE. It has a strong adaptability to adverse environments and a very short growing period. Buckwheat is mainly produced in the Russian Federation, Ukraine, and China, although it is grown to a lesser extent in many other countries. In recent years, buckwheat has regained interest as an alternative crop for organic cultivation and as a raw material for "healthy" food. Common buckwheat (*F. esculentum* Moensch) is the most commonly grown species, accounting for 90% of world production. Among other species, tartary buckwheat (*Fagopyrum tataricum*) is available in the mountains regions of Asia, generally grown at a higher altitude than common buckwheat due to its frost tolerance (Arendt and Zannini 2013). Buckwheat is a dicotyledonous plant that belongs to the Caryophyllales order, similar to amaranth and quinoa, and is not taxonomically related to wheat. As stated by Arendt and Zannini (2013), buckwheat is mainly grown for the production of its seeds. The seeds are wide at the base and triangular to almost round in cross-section. They are gray-brown or brown-black in color, while their size varies according to variety. The seed comprises a thick outer hull (pericarp) and inner kernel. The hull surrounds the seed's coat, endosperm, and embryo (Figure 3.2a).

The endosperm cells have thin cell walls and consist mainly of starch. Buckwheat contains only small-sized starch granules, all similar in size, ranging from 4 to 7 μm. Buckwheat is normally ground on a stone or roller mill, either to produce whole grain flour or to obtain flour by combining streams (Ikeda 2002; Skrabanja et al. 2004). Like rice,

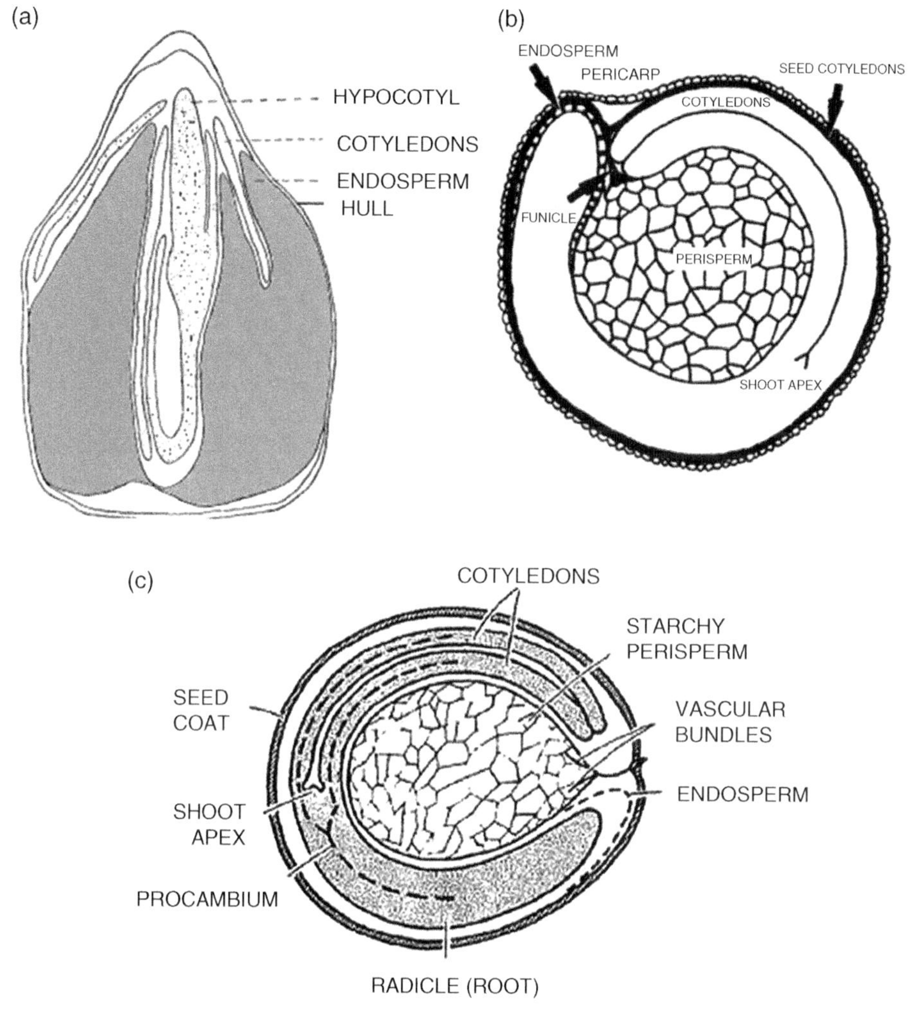

Figure 3.2 Pseudocereal seed structures: (a) buckwheat; (b) quinoa; (c) amaranth. *Source*: (a) Stevens (1912); (b) Prego et al. (1998); (c) based on Kong et al. (2009).

buckwheat flour can be prepared by dry milling or wet milling methods, which can have an impact on the functionality of the flour (Yu et al. 2018). Common buckwheat is consumed in many different applications in different countries. For example, in Europe and North America, buckwheat flour is generally mixed with wheat flour to prepare pancakes, biscuits, and noodles, whereas in Russia and Poland, the groats and the flour are used to make porridge and soup, while dumplings and noodles are commonly consumed in Japan.

3.2.2 Quinoa

The quinoa (*Chenopodium quinoa willd.*) plant belongs to the Chenopodiaceae family, which also includes spinach and beet. Quinoa is an endemic plant from South America. However, it was domesticated by people living in the Andes, particularly in Peru and Bolivia thousands of years ago, where it was considered to be a holy plant. Quinoa is rich in protein (12–23% depending on the variety), higher than that of common cereals, and has a good balance of essential amino acids with a high content of lysine, histidine, and methionine, which are generally the limiting amino acids in common cereals (Filho et al. 2017; Dakhili et al. 2019). Quinoa is highly resistant to weather, climate, and soil conditions. While its seeds and leaves both constitute the edible parts, it is the seeds that have been most investigated in terms of economic and scientific aspects. Quinoa seeds contain a central perisperm, where carbohydrate reserves are stored, which is surrounded by protein and an oil-rich embryo, endosperm, and seed coat (Figure 3.2b). The pericarp of the fruit contains saponins, which are bitter-tasting triterpenoid glycosides. The quantity of saponins is highly variable among different varieties which are distinguished as: "sweet quinoa," containing <0.11% of saponins; and "bitter quinoa," containing >0.11% of saponins. Sensory evaluation indicates that quinoa containing 110 mg/100 g saponins or less can be considered to be sweet, while this level is below the threshold for the detection of bitterness in quinoa flour (Taylor and Parker 2002). Saponins can be removed either by washing or by mechanical abrasion. This desaponification process is also called dehusking, pearling, or milling.

Quinoa desaponification leaves the nutrient-rich embryo and endosperm intact, and thus it is considered to be "whole grain." Similar to rice, quinoa seeds are consumed in soups, puffed in breakfast cereals, or by flouring them to produce baked products like cookies, bread, biscuits, pasta, crisps, tortillas, and pancakes (Bhargava et al. 2006). Quinoa sprouts are used in fresh salads (Schlick and Bubenheim 1996). In addition, quinoa seeds can be fermented for beer and/or beer-like beverage production, or for a traditional alcoholic beverage of Latin America called *chicha*. It is also used as a rich nutritional source in feeding farm animals, such as cattle, pigs, or poultry (Bhargava et al. 2006; Sezgin and Sanlier 2019).

3.2.3 Amaranth

Amaranth belongs to the genus *Amaranthus* and the major types can be divided into grain amaranth, wild vegetable amaranth, ornamental amaranth and weed amaranth. Amaranth grain was a staple food of the Aztecs, and was cultivated in Mexico and Central America (Arendt and Zannini 2013). Grain amaranth has recently been considered a promising food crop, thanks to its resistance to weather stress and high potential for biomass and grain yield. It is now cultivated in Central and South America, and some regions of Asia and Africa. The seed heads, some as long as 50 cm, resemble those of sorghum and the seeds are extremely small (0.9–1.7 mm in diameter). According to Saunders and Becker (1984), 1000 seeds weigh 0.5–1.2 g. They occur in massive numbers, sometimes more than 50 000 to a plant, and vary in color between cream, gold, pink, black, brown, yellow, and white. The seed embryo is circular, with its ends nearly touching and enclosing the perisperm (Figure 3.2c). The embryo is therefore rather large and accounts for about 25% of the grain's weight. Generally, the seed coat is smooth and thin and used as a whole seed. In contrast to quinoa and buckwheat, it is not necessary to remove the seed coat of amaranth. Grain amaranth is being utilized as a healthy food to be combined with traditional cereals grains in breakfast foods, bread, multigrain crackers, pastas, pancakes, or popped products.

3.3 Processing of Pseudocereals

3.3.1 Enzymatic Processing of Pseudocereals

The need for food processing lies in the improvement of palatability (via softening of the kernel and generation of Maillard compounds whenever a cooking step is applied), as well as the increased digestibility of the nutrients (via starch gelatinization, and/or protein denaturation). Furthermore, processing also extends the shelf life of food, and reduces antinutritional factors present in the raw form (Hotz and Gibson 2007). As with traditional cereal grains, the consumption of pseudocereals requires a pre-processing step, which is not restricted only to cooking or baking procedures, but encompasses traditional processes such as germination, malting, and/or fermentation. Pseudocereal seeds are traditionally used in the preparation of groats, as well as milled into flour to be used in a variety of dishes and food products such as noodles and breads. Pseudocereal seeds can furthermore be flaked or popped and eaten as snacks or added to muesli-type products (Lasekan and Lasekan 2012). Other traditional processing includes sourdough/sour slurry fermentation, malting for beer or fermented beverage production (e.g. amaranth seeds are fermented to produced traditional *chicha*), or even the use of the cooked greens as leafy vegetables (Mlakar et al. 2010; Siwatch and Yadav 2017).

Although pseudocereals are comparable and somewhat superior to common cereals in terms of their protein, fat, and mineral content, the presence of antinutrients, such as polyphenols, tannins, phytates, and oxalates, affects the bioavailability of the nutrient content (Mlakar et al. 2010). Phytate and lower inositol phosphates are indeed major antinutrients present in these seeds and are able to bind minerals like calcium, zinc, magnesium, and iron, making them unavailable for absorption (Mäkinen et al. 2013). Given that nutritional deficiencies in iron, zinc, iodine, and vitamin A are widespread in developing countries where plant-based foods (including legumes, but also pseudocereals) are the base of the diet, it is important to understand how processing can be used to improve nutrient bioavailability in pseudocereals.

3.3.2 Germination of Pseudocereals

Germination or malting is one of the most traditional processing methods, during which a new plant is formed from a dry seed. The process starts by water uptake in the grain or seed, leading to the acceleration of its metabolism and subsequent growth. The rise in hydrolytic enzyme activity will lead to the hydrolysis of macromolecules in the seed storage tissues, loosening the cellular tissues, and changes to the mechanical properties of the seed (Mäkinen and Arendt 2015). Consequent changes in the nutritional profile thus include modified nutrient availability and increased content of secondary metabolites with possible bioactivities. The structural breakdown that occurs increases the physical accessibility of micronutrients and the activation of hydrolytic enzymes. Their *de novo* synthesis impacts on starch and protein digestibility, as well as reducing the antinutritional factors through the activation of intrinsic phytases (Hotz and Gibson 2007; Chauhan and Singh 2013). Motta et al. (2017) evaluated the impact of boiling, steaming, and malting on the total folate (i.e. water-soluble vitamins from the B-complex group) content of amaranth, quinoa, and buckwheat. It was found that a single portion of amaranth or quinoa may contribute to up to 25% of the recommended European Food Safety Authority (EFSA) dietary reference, regardless of processing. In the case of buckwheat, malting was the preferred processing option, with malted buckwheat providing 19% of the dietary reference intake of folate, as compared with 14% when cooked. Furthermore, iron absorption may be facilitated via the reduction of certain polyphenols and tannins during germination. The increase in α-amylase activity leads to the breakdown of starch, which leads to a reduction in the viscosity of germinated/malted slurry and an improvement in the nutrient density of the meal. Nevertheless, these effects may be moderate. For example, Mäkinen et al. (2013) only observed small changes in enzymatic activity during the germination of quinoa, except for a decrease in proteolytic activity. In another study, Mäkinen and Arendt (2015) reported that malts produced from pseudocereals had low amylolytic activity, causing their poor brewing performance and high wort viscosities. For this reason, it has also been suggested that the utilization of quinoa malt in breadmaking would not be feasible due to its limited α-amylase activity (Mäkinen et al. 2013). Similarly, Chauhan and Singh (2013) reported the limited impact of the germination of amaranth on ash content, but the apparent loss of fat and increase in fiber and protein were noted due to a concomitant loss of available carbohydrates. It is worth mentioning that the impact germination may have on nutrient bioavailability may be controlled by the modulation of the temperature and time parameters. As an example, Wilhelmson et al. (2001) reported that a short germination period at 15 °C allowed retaining up to 60% of oat β-glucans, which would otherwise be largely depolymerized during the germination process.

Increased metabolic activity in the seed during germination also results in the accumulation of active oxygen species in the seed tissue, which may react with biomolecules and cause cellular damage (Mäkinen and Arendt 2015). A consequent increase in antioxidant compounds in the germinating seed represents the seed's defense mechanism against radical damage. The increased antioxidant activity during germination has been reported for amaranth, buckwheat, and quinoa (Alvarez-Jubete et al. 2010; Kim et al. 2008). Furthermore, the germination will also have an impact on flavor – for example, the increased metabolic activity leads to the generation of reducing sugars that impacts the perceived sweetness (Lasekan and Lasekan 2012). Concomitant to the generation of reducing sugars, the increase in free fatty acids and the presence of alkaline media during germination are strong prerequisites for the formation of aroma compounds when applying a heat treatment, particularly those yielding a caramel-like odor. During malting, for example, the grains are soaked and germinated before the application of a heat treatment to stop the metabolic process and to enable the development of aroma and flavor.

3.3.3 Fermentation Processing of Pseudocereals

Fermentation, including sourdough/sour slurry preparation, is another traditional food preservation process that may enhance food safety and improve nutritional quality, due to the production of nutritive factors (γ-aminobutyric acid [GABA] and bioactive peptides) and the reduction of antinutritional factors, such as phytates or saponins (Castro-Alba et al. 2019a; Rollán et al. 2019). Fermentation may improve the extraction of polyphenols, due to the presence of esterase activity in lactic acid bacteria, and subsequently increase the antioxidant capacity. Moreover, the release or synthesis of bioactive and aroma compounds that takes place during fermentation may improve the sensory profile of the product. For instance, the marked release in free amino acids occurring during fermentation will lead to the generation of volatile compounds during baking (Corsetti and Settanni 2007). Valencia et al. (1999) evaluated the impact of cooking, soaking, and fermentation on germinated and ungerminated quinoa flour. They observed that fermentation reduced phytate and its degradation products by up to 98%, with the highest impact when combined with germination. They also noted that iron solubility followed a similar trend, with a maximum impact when combining fermentation with germinated flour. Castro-Alba et al. (2019b) reported a similar finding on fermented quinoa, with up to 73% degradation of phytate. The authors also observed that fermentation was more effective on phytate reduction when applied to flour rather than grains, similar to the use of lactic acid bacteria as a starter culture versus spontaneous fermentation (Castro-Alba et al. 2019a). Interestingly, the combination of fermentation and roasting of the quinoa flours led to preferred sensory properties when consuming the fermented quinoa as porridge (Castro-Alba et al. 2019b). The fermentation of quinoa may thus allow manufacturing cereal-based products with improved nutritional quality while maintaining good sensory properties. For example, the use of 20% quinoa sourdough (obtained with autochthonous lactic acid bacteria) in white wheat bread was reported to bring improved nutritional quality and also better texture and sensory properties when compared with leavened wheat bread obtained with or without quinoa flour (Rizzello et al. 2016). The inclusion of quinoa sourdough improved protein digestibility and quality, attributed to the proteolysis by lactic acid bacteria, and likely caused the inactivation of some antinutritional factors such as trypsin inhibitor and condensed tannins. Compared with the native quinoa flour, quinoa sourdough yielded a higher specific volume of bread, a lower crumb hardness and an intense and appreciated color of the crust. When formulating pasta with 20% fermented quinoa, Lorusso et al. (2017) reported the nutritional profile of the pasta was improved without compromising technological and sensory quality. Pasta containing fermented quinoa flour was characterized by improved protein digestibility and quality, high nutritional scores, low predicted glycemic index, and high antioxidant potential. The tenacity of the pasta, as measured by hardness and fracturability parameters, was lower when using fermented quinoa flour rather than native quinoa flour, while the overall elasticity (resilience and cohesiveness) was also improved. The authors associated these changes with modification of the protein network caused by proteolysis occurring during fermentation. The fermentation of quinoa by autochthonous lactic acid bacteria furthermore enables the release of peptides with antioxidant activity through the proteolysis of native quinoa proteins, showing potential as a functional food ingredient or for pharmaceutical applications (Rizzello et al. 2017). Although most studies focus on the fermentation of quinoa, amaranth was also described as a good probiotic carrier (Matejčeková et al. 2016), while fermented buckwheat was shown to exhibit increased antioxidant capacity compared with its native counterpart (Đorđević et al. 2010).

3.3.4 Thermal Processing Methods for Pseudocereals

Along with their excellent nutrient profile and chemical composition (Alvarez-Jubete et al. 2010), pseudocereals are also a rich source of bioactive compounds, among which phenolic compounds, comprising flavonoids and phenolic acids, are major representatives (Rocchetti et al. 2019). These compounds are however known to be sensitive to heat. Thermal processing methods commonly applied to pseudocereals include cooking, roasting, and puffing or popping (Siwatch and Yadav 2017).

3.3.4.1 Cooking Pseudocereal seeds are typically consumed after boiling for 10–20 minutes in water. Cooking in excess water triggers the loss of phenolic compounds, as they are solubilized during the applied hydrothermal treatment, migrate into the cooking water, and are eventually removed when the excess water is discarded. This was precisely measured in a study by Dini et al. (2010) on bitter and sweet quinoa seeds boiled for 20 minutes in excess water. The authors found a decrease in the total phenolic content in the seeds, concomitant with an increase in phenolics in the cooking water. Nickel et al. (2016) applied three different heat treatments to prewashed quinoa grains: cooking with and without pressure, and toasting. In all three processes the water was adjusted to be evaporated/absorbed during the process to avoid any phenolic loss. The highest content of phenolic compounds was obtained after cooking under pressure, while toasting caused the greatest loss, compared with nonprocessed quinoa. These results were confirmed by Nickel et al. (2016). Interestingly, grain washing, itself commonly applied to reduce the saponin content (i.e. 15 minutes rubbing in water), also markedly increased phenolic content (Nickel et al. 2016). It was hypothesized that the washing process affected the release of conjugated phenolic compounds in the seeds (without passing into water), justifying their higher content in comparison to unprocessed grains. Moreover, Rocchetti et al. (2019) demonstrated that cooking quinoa and buckwheat seeds induced the release of specific phenolic classes, namely phenolic acids and tyrosols. When comparing fermentation and cooking, the authors revealed that flavonoid content could enable discrimination of the type of processing applied, opening interesting perspectives on the targeted health-promoting properties of buckwheat and quinoa (Rocchetti et al. 2019).

Regarding nutritional improvement via the removal of antinutrient factors, boiling is nevertheless not effective in degrading phytic acid, especially when compared with fermentation (Valencia et al. 1999). When evaluating the impact of cooking on mineral content, Mota et al. (2016a) measured losses up to 20%, with variation according to the cooking method (boiling or steaming) and the type of grain (quinoa, amaranth, or buckwheat). In another study, Motta et al. (2017) showed that boiling and steaming effected a decrease in the folate content in amaranth, but an increase in quinoa, with no significant alteration of the folate content in buckwheat. Finally, boiling and steaming effected no significant differences in the amino acid content in quinoa and buckwheat, with an exception for sulfur amino acids, while the content of certain essential amino acids in amaranth was decreased (Motta et al. 2019). These studies pinpoint how essential it is to evaluate the nutritional potential of pseudocereals in the form they are consumed.

3.3.4.2 Popping and Puffing Popping, widely used for amaranth grains, utilizes a sudden application of heat at atmospheric pressure that leads to the vaporization of the moisture inside the grain. Heat is applied until the vaporization results in enough internal pressure that the grain expands and breaks the external tissue or pericarp of the grain. Popped grains may be consumed directly or added as ingredient in snack formulations or muesli preparations. Popping is a low-cost technology, suitable for the production of traditional and/or innovative products in developing countries. During popping, a partial gelatinization of starch takes place, while the application of high temperatures also enables the formation of Maillard compounds (Lasekan and

Lasekan 2012), as well as an increase in starch and protein digestibility. On the other hand, heat-sensitive components such as vitamins or phenolic compounds may be degraded, as previously described. Nevertheless, Paucar-Menacho et al. (2018) observed that puffed amaranth seeds retained their nutritional profile to a great extent, while Muyonga et al. (2014) reported that popping or roasting did not impact on the total phenolic content either. These authors however report a reduction in protein digestibility, especially in popped amaranth, and attributed it to amino-acid degradation, the formation of disulfide bounds, and the generation of Maillard compounds (Muyonga et al. 2014).

3.3.4.3 Extrusion The utilization of extrusion cooking to incorporate pseudocereals in gluten-free snacks, baby foods, or breakfast cereals has received most attention. Diaz et al. (2013) produced corn extrudates containing up to 20% quinoa, amaranth, or kañiwa flours. They reported good lipid stability in the extrudates with added pseudocereal flours, and improved sectional expansion indexes of the extrudates, specifically with added amaranth. However, extruded quinoa and amaranth did not reach the level of expansion volume as that achieved with wheat in the study by Robin et al. (2015). It is worth mentioning that the expansion volume of starch-based extrudates is highly dependent on the extrusion parameters, which makes it difficult to compare directly between different studies. Recently Sun et al. (2019) evaluated the impact of extrusion parameters on the nutritional composition and cooking characteristics of buckwheat noodles. They reported a strong impact from the extrusion settings, and managed to minimize the phenolic losses and optimize the glycemic index of the noodles at moderate processing temperature (100–120 °C) and higher moisture cooking conditions (40%). As extrusion is a high-shear technology that will greatly impact on the supramolecular organization of starch and protein and thus the digestibility of both, the glycemic indexes obtained for extruded foods should be taken with caution. The formulation and supply of gluten-free products suitable for patients with celiac disease should go hand in hand with an adequate glycemic index (Guerra-Matias and Arêas 2005).

3.3.5 Pseudocereals in Gluten-Free Processing

The development of gluten-free cereal products has always been a challenge for the food industry, due to the absence of the viscoelastic properties conferred by gluten that influence the texture and sensory properties of bread, pasta, and biscuits. In this context, there is lack in availability of gluten-free products in the market, while the products that are already available are of low quality and poor nutritional value (Alvarez-Jubete et al. 2009). In bread-making, gluten is mainly replaced by a mixture of low nutritional value including starch or starch-rich flours (i.e. rice flour) and hydrocolloids, as described by Machado Alencar et al. (2015). Pseudocereals have received much attention from the food industry as suitable raw material for gluten-free products and alternative sources of proteins, dietary fibers, antioxidants, minerals, etc. For example, amaranth, buckwheat, or quinoa are nutritionally relevant replacements for the part of starch additives in product formulations such as gluten-free breads, pasta, breakfast cereals, and cookies. Several successful attempts of using pseudocereals for gluten-free baking have been reported (Machado Alencar et al. 2015; Turkut et al. 2016; Jan et al. 2018; Molinari et al. 2018). Bread formulated with amaranth and quinoa was of an improved nutritional profile due to its rich protein, lipid, and mineral content. The utilization of amaranth, quinoa, and buckwheat grains enabled the production of breads with high vitamin E recovery and higher vitamin E content in comparison with the gluten-free control (Alvarez-Jubete et al. 2010). Moreover, Machado Alencar et al. (2015) showed that bread containing pseudocereals and sweeteners exhibited specific volume, firmness, and water activity similar to those of the control bread formulated mainly on rice flour and starch. Gluten functionality is also of key importance in the pasta industry. Most gluten-free pasta on the market is based on starch. Lorenzo et al. (2018) described how to replace gluten functionality in gluten-free pasta using quinoa flour, zein, and other biopolymers. The authors showed that replacing corn with quinoa flour increases the protein content in the dough, which greatly influences both its viscoelastic characteristics and the drying process kinetics. Kerpes et al. (2017) presented some challenges while developing gluten-free beer with pseudocereals, especially due to the low enzymatic load of those raw materials, but also proposed technological solutions. The utilization of pseudocereals in gluten-free manufacturing has huge potential, especially for improvement in the "naturality" of gluten-free products. Pseudocereals in general and amaranth in particular are good sources of high-quality proteins, including an excellent amino acid profile. This is of great importance as a cheap alternative source of proteins for human consumption, with a low environmental impact, and with the potential to satisfy the increasing vegetarianism and veganism trends that the food industry is facing (Pojić et al. 2018).

3.4 Emerging Significance of Pseudocereals

Interest in the major pseudocereals (e.g. buckwheat, amaranth, and quinoa) has increased because they are important protein sources in many diets, especially in gluten-free diets. Buckwheat is particularly popular among patients with celiac disease, because it does not cause allergic reactions in the intestinal mucosa (Guerra-Matias and Arêas 2005). Quinoa has been shown to be well-tolerated by celiac patients at doses as high as 50 g/day (Zevallos et al. 2014). Beyond their well-known protein quality, the results from recent pre-clinical studies suggest that the phytochemicals from pseudocereals exhibit beneficial health effects such as the prevention and reduction of oxidative stress and anti-inflammatory, antihypertensive, and cardiovascular disease prevention (Liu 2004; Liu 2007; Golzarand et al. 2015). Certain health benefits attributed to whole grains are associated with their phytochemical content, of which between 75 and 85% is found in the bound form (Okarter and Liu 2010). The phytochemical profiles of pseudocereals are very similar to those of cereal grains, and emerging science suggests that phenolics, both in their conjugated and bound forms, also have beneficial impacts on immune responses and gut health (Tang and Tsao 2017). Thus, the major pseudocereal grains discussed here constitute a set of promising raw materials for many uses in the food industry, including nutraceuticals.

3.4.1 Nutritional Value of Pseudocereals

3.4.1.1 Buckwheat The macronutrient composition and quality of buckwheat is very similar to that of cereals. Two buckwheat species grown worldwide – common buckwheat and tartary buckwheat – are generally considered to have great potential to confer health benefits due to their high content of flavonoids and other components such as rutin (Table 3.1). Buckwheat contains around 12% protein content with an amino acid profile that is characterized with higher lysine, arginine, and aspartic acid content, and a lower glutamic acid and proline content compared with common cereals (Zhang et al. 2012).

Buckwheat is also very widely used in gluten-free product formulations. However, it has been reported that individuals with celiac disease that also suffer from food-related allergies claim to have adverse gastrointestinal effects and allergic reactions (urticaria, asthma, angioedema, and allergic rhinitis) after buckwheat consumption (Wieslander and Norbäck 2001; Zhu 2016). Some buckwheat proteins, those in the lower molecular weight range between 16 and 24 kDa, have been shown to elicit allergic reactions, which is why it is considered a potent food allergen (Tanaka et al. 2002; Park et al. 2000). Buckwheat allergy is an IgE-mediated acute reaction that can follow the ingestion of buckwheat or occupational exposure (Wieslander and Norbäck 2001). Those protein allergens are resistant to heating and pepsin digestion. Another allergen identified in buckwheat seeds is one of higher molecular weight, 56 kDa, resembling the legume-like 11S storage protein. However, its allergenicity could be reduced by inducing protein glycation via thermal processing and the Maillard reaction between the protein and polysaccharides (Zhu 2016).

Buckwheat also contains considerable levels of trypsin inhibitors, which can impair protein digestion, as well as phytate in the aleurone layer and the embryo, which impairs micronutrient absorption (Steadman et al. 2001; Skrabanja et al. 2004). However, these antinutritional factors can be significantly reduced with various processing methods, such as boiling, microwave heating, and/or high-hydrostatic-pressure processing (Deng et al. 2015). The lipid content in buckwheat is generally low, between 2 and 4%, and, similarly to that of amaranth and quinoa, has oleic as the main fatty acid (Golijan et al. 2019). Although the total amount is low, buckwheat lipids are nutritionally superior to those of common cereal grains because they have a higher proportion of unsaturated fatty acids (Steadman et al. 2001). Buckwheat differs from amaranth and quinoa in its starch properties. The starch content in buckwheat grains is similar to that of wheat (≈ 70%), its amylose content averages around 20%, and, in

Table 3.1 Content of flavonoids and phenolic compounds in pseudocereal grains.

Pseudocereal grains	Rutin	Quercetin	Total flavonoids	Total phenolics	Source
Buckwheat	6060–18670	14.8	6.65–22.74	15874–71359	Kim et al. (2008); Qin et al. (2010); Vollmannova et al. (2013); Ghimeray et al. (2017)
Quinoa	170–368	n/a	2238	459–1839	Paśko et al. (2008); Vollmannova et al. (2013)
Amaranth	80–508	68 ± 3	676	147–2870	Czerwiński et al. (2004); Paśko et al. (2008); Kalinova and Dadakova (2009); Vollmannova et al. (2013)

Values expressed in g/kg dry matter.

its native state, buckwheat starch can contain between 18 and 33% resistant starch (Qin et al. 2010). A pre-clinical study done on mice fed with a high-fat diet supplemented with resistant starch from buckwheat showed a significant reduction in the blood markers of cardiovascular disease and inflammation and had a prebiotic effect on microbiota composition (Zhou et al. 2019). However, the resistant starch content in buckwheat has been shown to be significantly reduced by thermal processing of high intensity, such as extrusion cooking (Sun et al. 2019). On the other hand, a study by Wang and Bai (2017) showed a slight increase in the resistant starch content of buckwheat by inducing starch retrogradation using ultrasonic treatment in combination with cooling. Other thermal processing methods combined with cooling cycles have also shown a tendency to increase resistant starch content; however, this effect is not specific to pseudocereal starches.

The dietary fiber content in buckwheat ranges from 4 to 7% depending on whether it contains the hull or not. Furthermore, between 20 and 30% of its total dietary fiber content is soluble, being a higher content than that of monocot cereals (Wefers and Bunzel 2015; Kiewlicz et al. 2020). Moreover, an analysis of nonstarch polysaccharides in buckwheat indicated that a large amount of pectic polysaccharides was the main constituent in both insoluble and soluble fiber fractions (Wefers and Bunzel 2015). A study of Préstamo et al. (2003) showed a significant increase in *Lactobacillus* species and bifidobacteria along with a concomitant decrease in pathogenic bacteria in Wistar rats fed with a buckwheat-based diet for 30 days. These pre-clinical results indicated that buckwheat fibers have the potential to act as prebiotics. This particular composition of nonstarch polysaccharides is common among the major pseudocereals discussed here, and is one of the main differences compared with the nonstarch polysaccharides in cereals grains, which are mainly composed of heteroxylans.

3.4.1.2 Quinoa Much like amaranth, quinoa is also known for its protein content, which can range from 8 to 22% (Repo-Carrasco et al. 2003). The quality of proteins is high, characterized by a significant amount of essential amino acids, especially lysine, tryptophan, and cysteine, higher than that of common cereals and similar to that of legumes (Mota et al. 2016b). Both *in vitro* and clinical research have shown that quinoa grains are well-tolerated by celiac disease patients (Mickowska et al. 2012; Peñas et al. 2014; Zevallos et al. 2014), which makes it a good alternative for gluten-free product development. A high content of vitamins E, B, and C, as well as minerals such as calcium, iron, manganese, magnesium, copper, and potassium, have also been reported for quinoa seeds (Konishi et al. 2000). *In vivo* trials have shown that the digestibility of quinoa proteins is high, averaging around 92%, and can be further increased after the removal of the saponin compounds that coat the grains (James 2009). However, most of the quinoa that is commercially available today is washed, or has its pericarp removed by abrasion, in order to remove the saponins, which render the quinoa grains bitter. It is worth mentioning that a pre-clinical study demonstrated that a quinoa pericarp extract, rich in saponins, had a beneficial effect on reducing diet-induced hypercholesterolemia and liver cholesterol levels in Wistar rats (Konishi et al. 2000). These results indicate that the byproduct of saponin removal from quinoa can be utilized as a functional ingredient targeting cardiovascular diseases. The starch content of quinoa grains is very similar to that of common cereals, but with a low amylose content (less than 10%), short-chain amylopectin, and small starch granules (˂2.5 μm) – all properties that are associated with fast gelatinization during thermal processing, as well as high or rapid digestibility when consumed

(Wolter et al. 2013; Srichuwong et al. 2017). Quinoa grains also contain a significant amount of dietary fiber, ranging from 7 to 14%, out of which around 20% is soluble dietary fiber, while the remaining insoluble fraction is not highly lignified (Lamothe et al. 2015). A pre-clinical study by Liu et al. (2018) showed that quinoa consumption alleviated the dysbiosis of gut microbiota, as well as the clinical symptoms arising from dextran sulfate sodium (DSS)-induced inflammatory bowel syndrome. These results indicated the potential of quinoa fiber to be used as an ingredient to improve gut health via dietary approaches. Beyond macronutrients, one portion of quinoa, or amaranth, can contribute ≥25% of the dietary reference value for folates (Motta et al. 2017), which makes these pseudocereals suitable for the diets of women in their reproductive years. Furthermore, quinoa grains are also rich in antioxidant compounds such as polyphenols, tocopherols and carotenes (Nsimba et al. 2008; Tang et al. 2015). Although the lipid content in quinoa grains is not so high, averaging around 5.3%, its overall quality is also considered better than that of common cereals due to a high index of polyunsaturation, 3.7–4.9, similar to that of soybean oil (Vidueiros et al. 2015; Wood et al. 1993). The unsaturated fraction of quinoa lipids is mainly composed of linoleic acid (52%), oleic acid (23%), and α-linolenic acid (11%).

3.4.1.3 Amaranth Amaranth has the highest content of protein among all the major pseudocereals. Depending on the variety, its protein content can range from 13 to 21%, and the main protein fractions are constituted by albumin, 11S-globulin, globulin-P, and glutelin. Furthermore, amaranth protein has a high content of essential amino acids, especially lysine, methionine, cysteine, and histidine (Mota et al. 2016b), which tend to be lacking in other sources of vegetable protein. Furthermore, the digestibility of amaranth proteins has been measured at around 85%, which can be improved by thermal processing in specific ways that do not have a detrimental impact on the profile of essential amino acids (Muyonga et al. 2014). Despite their high quality, amaranth protein isolates or concentrates are not very common functional ingredients in the food industry, due to the poor solubility of amaranth proteins in aqueous solvents (Scilingo et al. 2002). The lipid content in amaranth grains can range from 3 to 7%; *A. caudatus* has the higher quantities, compared with other varieties such as *A. hypocondriacus*. Around 80% of the lipids in amaranth are unsaturated, with oleic and linoleic acid as the main fatty acids in this fraction. Due to the particular composition of amaranth oil, the enzymatic interesterification to create structured lipids that resemble lipids from breastmilk was explored (Pina-Rodriguez and Akoh 2009). The results from this work showed the potential of the underutilized amaranth oil as an ingredient to improve the lipid profiles of infant formulas produced with vegetable oils. Amaranth grains have also been shown to be high in tocopherols, which are known for their antioxidant capacity, and squalene, which has been shown to decrease serum cholesterol and is an ingredient commonly used in skincare (Zhang et al. 2019). Beyond protein and lipids, the carbohydrate profile of amaranth grains comprises 50–70% starch (Srichuwong et al. 2017). The amaranth starch granules are of very small size (1–3 μm) and are high in amylopectin, which results in rapid gelatinization during cooking (Fuentes et al. 2019). This particular granular morphology results in starches that have a low resistant starch content and, when cooked, have high and rapid digestibility. Studies have shown that amaranth-based products have high glycemic index values and can elicit higher insulin responses than white bread, and thus are not suitable for diabetics (Guerra-Matias and Arêas 2005). On the other hand, amaranth contains a considerable amount of dietary fiber (8–11%), and up to 30% of its fiber content is soluble.

Nonstarch polysaccharides in amaranth grain are mainly composed of highly branched xyloglucans and a minor portion of pectins (Lamothe et al. 2015). This amount of dietary fiber, in combination with its composition, makes amaranth a good companion to cereal grains as ingredients that can help increase dietary fiber consumption.

3.5 Functional Ingredients of Pseudocereals

3.5.1 Phenolic Compounds

Whole grain and dietary fiber consumption are associated with reduced risks of developing cardiovascular diseases and an reduced incidence of type 2 diabetes. It has previously been suggested that the mechanism by which whole grains or dietary fiber help in the prevention of these diseases is associated with the phenolic compounds localized in the outer parts of the grain. The major pseudocereals reviewed here contain similar quantities and types of phenolic compounds to cereal grains, and their effects on human health have been studied. A study by de la Rosa et al. (2009) reported that amaranth (*A. hypocondriacus*) grain contains three types of polyphenols (rutin, isoquercetin, and nicotiflorin) and three types of flavonoids (4-hydroxybenzoic acid, syringic acid, and vanillic acid). Polyphenols such as rutin and nicotiflorin have been associated with the prevention of cardiovascular and Alzheimer's diseases, as well as a decrease in memory impairment (Cervantes-Laurean et al. 2006; Huang et al. 2007). *In vitro* studies have shown that amaranth grains and flours have a lower antioxidant capacity than oats, but higher than soy, being positively correlated to the total phenolic content measured in the grains (Czerwiński et al. 2004). Buckwheat is also known for its high levels of polyphenols and is particularly rich in rutin, which has been shown to reduce blood pressure and the absorption of cholesterol (Zhu 2016). As opposed to amaranth, the *in vitro* antioxidant capacity of buckwheat was reported to be higher than that of oats and barley (Zduńczyk et al. 2006). Quinoa grains are a good source of betalains – more specifically, betaxanthins, for which high-antioxidant and free-radical-scavenging properties have been reported (Escribano et al. 2017). Furthermore, the results from pre-clinical studies (Table 3.2) have shown that amaranth grains or their phenolic-rich fractions have a positive impact on lipid profiles in animals fed with high-fat or high-cholesterol diets.

Wu et al. (2019) investigated the antiobesity effects of amaranth, compared with lycopene and red sorghum, in high-fat-diet-induced obese mice. The amaranth resulted in reduced body weight and improved glucose tolerance, and effectively ameliorated lipid metabolism. In addition, amaranth sprouts and cookies made from amaranth flour have been shown to reduce the activity of angiotensin-converting enzyme, as a marker for hypertension and thrombosis (Aphalo et al. 2015; Ontiveros et al. 2020).

Besides their impact on lipid profiles, amaranth extracts have also been studied for their impact on glucose metabolism. A pre-clinical study conducted by Zambrana et al. (2018) showed that a single dose of an alcohol extract from amaranth improved glucose tolerance in rats (Table 3.2). Long-term administration of the extract also had a significant impact on insulin sensitivity and reduced serum HbAc1 levels. A potential explanation for this effect on glucose metabolism may be related to the inhibition of α-amylase, the main digestive enzyme for starch in the digestive tract. In agreement, methanolic extracts from two different varieties of amaranth grain were also shown to reduce the activity of α-amylase by 28 and 50% (Conforti et al. 2005). A reduction in the activity of carbohydrate digestive enzymes would result in

Table 3.2 *In vitro* and pre-clinical studies of health effects of amaranth and buckwheat extracts or products.

Pseudocereal grain	Sample type	Major finding	Source
Amaranth	Amaranth alcohol extracts	Improvement of glucose tolerance, insulin sensitivity and reduced levels of HbAc1 in rats	Zambrana et al. (2018)
	Amaranth	Reduction in body weight, improved glucose tolerance and lipid metabolism in high-fat-diet-induced obese mice	Wu et al. (2019)
	Amaranth sprouts and amaranth cookies	Reduced *in vitro* activity of angiotensin-converting enzyme	Aphalo et al. (2015)
Buckwheat	Buckwheat flour crackers	No impact on post-prandial glycemia or insulinemia, modulation of satiety hormones in type-2 diabetic patients	Stringer et al. (2013)
	Buckwheat bran extracts	Inhibition of sucrose activity, reduced blood glucose levels after consumption of a sucrose solution in mice	Hosaka et al. (2011)
	100 g Tartary buckwheat/day	Reduced fasting insulin, reduced total cholesterol and LDL-cholesterol in type-2 diabetic patients	Qiu et al. (2016)

concomitant lower glycemic and insulin responses. Stringer et al. (2013) conducted a study investigating the impact of buckwheat crackers on the post-prandial responses of satiety hormones in type-2 diabetic subjects. The results indicated that although buckwheat crackers did not show a significant reduction in post-prandial glycemia or insulinemia, the modulation of satiety hormones was significant. Another study on type-2 diabetic patients showed that a diet supplemented with 110 g/day of tartary buckwheat for four weeks resulted in significant decreases in fasting insulin, total cholesterol, and low-density lipoprotein cholesterol (Qiu et al. 2016). In a study by Hosaka et al. (2011), buckwheat bran extracts (BBE) inhibited sucrase activity *in vitro* more effectively than extracts prepared from buckwheat whole flour. Also, the administration of BBE 30 minutes prior to the consumption of a sucrose solution resulted in significant reductions of blood glucose levels in mice. However, the administration of pure rutin, one of the abundant polyphenols identified in BBE, did not lower blood glucose level. These results indicated that components of BBE other than rutin may be responsible for the inhibitory activity against sucrase *in vivo* and suggest that BBE may have a therapeutic use in the management of type 2 diabetes.

Overall, studies investigating the physiological impact of pseudocereals have mainly focused on amaranth and buckwheat. However, profiles of phenolic compounds from quinoa are similar to amaranth, so similar effects may be expected. In addition, the majority of these investigations have used *in vitro* or pre-clinical models and further research is needed to substantiate the postulated health benefits.

3.5.2 Bioactive Peptides

As discussed in the previous section, pseudocereals are rich in protein of high nutritional value and a lot of research has been conducted on the bioactive properties of their intrinsic peptides as well. For example, lunasin is a peptide found in soy protein with a unique 43-amino-acid sequence that has been tested for anticarcinogenic properties in a mammalian cell culture model and in a skin cancer mouse model (de Lumen 2005). Silva-Sánchez et al. (2008) reported the content of a lunasin-like peptide present in amaranth seeds at a concentration of 11.1 μg lunasin equivalent/g of amaranth seed protein – identified in glutelin albumin, prolamin, and globulin amaranth protein fractions, that induced apoptosis in immortalized cells from cervical cancer (HeLa cell line). Lunasin has also been detected in quinoa proteins, at a concentration of 1–5 mg lunasin/kg of quinoa seed (Ren et al. 2017). The same study demonstrated that quinoa lunasin inhibited the production of tumor necrosis factor-α (TNF-α), as well as interleukin-6 (IL-6), on lipopolysaccharide-stimulated RAW264.7 macrophages. In addition, other bioactive peptides with antihypertensive properties have been identified in amaranth protein. This effect has been shown in an *in vivo* study by Fritz et al. (2011) where amaranth protein hydrolyzates exhibited an inhibitory effect of angiotensin-I-converting enzyme and resulted in a reduction of blood pressure in spontaneously hypertensive rats. More recent studies have suggested the antithrombotic effects of peptides from amaranth protein hydrolyzates, due to the inhibitory effect they exhibited on fibrin coagulation (Sabbione et al. 2016). Quinoa protein peptides have not been studied as extensively as peptides from amaranth proteins. However, *in vitro* studies suggested that quinoa peptides may have significant effects on inhibiting carbohydrate digestive enzymes and antioxidant properties. Buckwheat proteins also showed a strong supplemental effect with other proteins of improving the dietary amino acid balance, with the special biological effects of lowering cholesterol and hypertension (Zhang et al. 2012). Other properties also attributed to buckwheat peptides include antibacterial, trypsin-inhibiting, hypocholesterolemic, and antidiabetic effects (Zhou et al. 2015). Similar to amaranth peptides, an inhibitory effect on angiotensin-I-converting enzyme has also been identified in buckwheat protein extracts (Ma et al. 2006). It is quite evident that there is a general lack of clinical evidence for the postulated health benefits arising from bioactive peptides in the major pseudocereals reviewed here. Therefore, along with further characterization of these compounds, their effects should be explored in human studies in order to obtain substantiation for their benefits. Still, the body of research that is available shows great promise and is proof of the emerging significance of these grains.

3.6 Conclusion and Future Perspectives

Pseudocereals are nutritionally relevant foods that account for a large part of the diet in developing countries. Despite their high protein and fiber content, the presence of antinutritional factors and reduced mineral intake in the traditional regions of their consumption have encouraged the extended use of traditional processes such as germination, malting, and fermentation to maximize nutrient availability. Currently, the application of gluten-free alternatives is being extended into bread-making and the efficient use of sterilizing and cooking techniques such as popping/puffing and extrusion. Both technologies are considered low-cost, which is economically relevant to promoting pseudocereals in developing local markets. The increased consumption of gluten-free products has transformed the cereal product markets in Europe and in the United States. A growing shift to vegetarian diets has imposed the need for the utilization of breakthrough technologies to maximize the nutritional uniqueness of pseudocereals and create innovative products. For instance, molecular biology and protein engineering techniques have already been used to improve the functional properties of amaranth proteins. Such techniques, including molecular tools and the production of recombinant proteins may represent interesting future prospects to address the nutritional and nutraceutical potential of these unique crops.

References

Alvarez-Jubete, L., Hose, M., Hansen, Å. et al. (2009). Impact of baking on vitamin E content of pseudocereals amaranth, quinoa, and buckwheat. *Cereal Chemistry* 86: 511–515.

Alvarez-Jubete, L., Wijngaard, H., Arendt, E.K., and Gallagher, E. (2010). Polyphenol composition and in vitro antioxidant activity of amaranth, quinoa buckwheat and wheat as affected by sprouting and baking. *Food Chemistry* 119 (2): 770–778.

Aphalo, P., Martínez, E.N., and Añón, M.C. (2015). Amaranth sprouts: a potential health promoting and nutritive natural food. *International Journal of Food Properties* 18 (12): 2688–2698.

Arendt, E.K. and Zannini, E. (eds.) (2013). *Cereal Grains for the Food and Beverage Industries*. Woodhead Publishing.

Bhargava, A., Shukla, S., and Ohri, D. (2006). *Chenopodium quinoa* – an Indian perspective. *Industrial Crops and Products* 23 (1): 73–87.

Castro-Alba, V., Lazarte, C.E., Perez-Rea, D. et al. (2019a). Fermentation of pseudocereals quinoa, canihua, and amaranth to improve mineral accessibility through degradation of phytate. *Journal of the Science of Food and Agriculture* 99: 5239–5248.

Castro-Alba, V., Lazarte, C.E., Perez-Rea, D. et al. (2019b). Effect of fermentation and dry roasting on the nutritional quality and sensory attributes of quinoa. *Food Science & Nutrition* 7: 3902–3911.

Cervantes-Laurean, D., Schramm, D.D., Jacobson, E.L. et al. (2006). Inhibition of advanced glycation end product formation on collagen by rutin and its metabolites. *The Journal of Nutritional Biochemistry* 17 (8): 531–540.

Chauhan, A. and Singh, S. (2013). Influence of germination on physico–chemical properties of amaranth (Amaranthus Spp.). flour. *International Journal of Agricultural Science and Food Technology* 4: 215–220.

Conforti, F., Statti, G., Loizzo, M.R. et al. (2005). *In vitro* antioxidant effect and inhibition of α-amylase of two varieties of *Amaranthus caudatus* seeds. *Biological and Pharmaceutical Bulletin* 28 (6): 1098–1102.

Corsetti, A. and Settanni, L. (2007). Lactobacilli in sourdough fermentation. *Food Research International* 40: 539–558.

Czerwiński, J., Bartnikowska, E., Leontowicz, H. et al. (2004). Oat (*Avena sativa* L.) and amaranth (*Amaranthus hypochondriacus*). Meals positively affect plasma lipid profile in rats fed cholesterol-containing diets. *The Journal of Nutritional Biochemistry* 15 (10): 622–629.

Dakhili, S., Abdolalizadeh, L., Hosseini, S.M. et al. (2019). Quinoa protein: composition, structure and functional properties. *Food Chemistry* 299: 125161.

Deng, Y., Padilla-Zakour, O., Zhao, Y., and Tao, S. (2015). Influences of high hydrostatic pressure, microwave heating, and boiling on chemical compositions, antinutritional factors, fatty acids, in vitro protein digestibility, and microstructure of buckwheat. *Food and Bioprocess Technology* 8 (11): 2235–2245.

Diaz, J.M.R., Kirjoranta, S., Tenitz, S. et al. (2013). Use of amaranth, quinoa and kañiwa in extruded corn-based snacks. *Journal of Cereal Science* 58 (1): 59–67.

Dini, I., Tenore, G.C., and Dini, A. (2010). Antioxidant compound contents and antioxidant activity before and after cooking in sweet and bitter *Chenopodium quinoa* seeds. *LWT - Food Science and Technology* 43: 447–451.

Đorđević, T.M., Šiler-Marinković, S.S., and Dimitrijević-Branković, S.I. (2010). Effect of fermentation on antioxidant properties of some cereals and pseudo cereals. *Food Chemistry* 119 (3): 957–963.

Escribano, J., Cabanes, J., Jiménez-Atiénzar, M. et al. (2017). Characterization of betalains, saponins and antioxidant power in differently colored quinoa (*Chenopodium quinoa*) varieties. *Food Chemistry* 234: 285–294.

FAO (2020). FAO STAT database: Crop production. http://www.fao.org/faostat/en/#data/QC (accessed 17 March 2020).

Filho, A.M.M., Pirozi, M.R., Borges, J.T.D.S. et al. (2017). Quinoa: nutritional, functional, and antinutritional aspects. *Critical Reviews in Food Science and Nutrition* 57 (8): 1618–1630.

Fritz, M., Vecchi, B., Rinaldi, G., and Añón, M.C. (2011). Amaranth seed protein hydrolysates have in vivo and in vitro antihypertensive activity. *Food Chemistry* 126 (3): 878–884.

Fuentes, C., Perez-Rea, D., Bergenståhl, B. et al. (2019). Physicochemical and structural properties of starch from five Andean crops grown in Bolivia. *International Journal of Biological Macromolecules* 125: 829–838.

Ghimeray, A.K., Sharma, P., Phoutaxay, P. et al. (2017). Far infrared irradiation alters total polyphenol, total flavonoid, antioxidant property and quercetin production in tartary buckwheat sprout powder. *Journal of Cereal Science* 59 (2): 167–172.

Golijan, J., Milinčić, D.D., Petronijević, R. et al. (2019). The fatty acid and triacylglycerol profiles of conventionally and organically produced grains of maize, spelt and buckwheat. *Journal of Cereal Science* 90: 102845.

Golzarand, M., Bahadoran, Z., Mirmiran, P. et al. (2015). Dietary phytochemical index is inversely associated with the occurrence of hypertension in adults: a 3-year follow-up (the Tehran lipid and glucose study). *European Journal of Clinical Nutrition* 69 (3): 392–408.

Guerra-Matias, A.C. and Arêas, J.A.G. (2005). Glycemic and insulinemic responses in women consuming extruded amaranth (*Amaranthus cruentus* L). *Nutrition Research* 25 (9): 815–822.

Hosaka, T., Nii, Y., Tomotake, H. et al. (2011). Extracts of common buckwheat bran prevent sucrose digestion. *Journal of Nutritional Science and Vitaminology* 57 (6): 441–445.

Hotz, C. and Gibson, R.S. (2007). Traditional food-processing and preparation practices to enhance the bioavailability of micronutrients in plant-based diets. *The Journal of Nutrition* 137 (4): 1097–1100.

Huang, J.-L., Fu, S.-T., Jiang, Y.-Y. et al. (2007). Protective effects of Nicotiflorin on reducing memory dysfunction, energy metabolism failure and oxidative stress in multi-infarct dementia model rats. *Pharmacology Biochemistry and Behavior* 86 (4): 741–748.

Ikeda, K. (2002). Buckwheat composition, chemistry, and processing. *Advances in Food and Nutrition Research* 44: 395–434.

James, L.E.A. (2009). Quinoa (*Chenopodium quinoa* Willd.): composition, chemistry, nutritional, and functional properties. *Advances in Food and Nutrition Research* 58: 1–31.

Jan, K.N., Panesar, P.S., and Singh, S. (2018). Optimization of antioxidant activity, textural and sensory characteristics of gluten-free cookies made from whole Indian quinoa flour. *LWT - Food Science and Technology* 93: 573–582.

Kalinova, J. and Dadakova, E. (2009). Rutin and total quercetin content in amaranth (*Amaranth* spp.). *Plant Foods for Human Nutrition* 64: 68.

Kerpes, R., Fischer, S., and Becker, T. (2017). The production of gluten-free beer: degradation of hordeins during malting and brewing and the application of modern process technology focusing on endogenous malt peptidases. *Trends in Food Science & Technology* 67: 129–138.

Kiewlicz, J. and Rybicka, I. (2020). Minerals and their bioavailability in relation to dietary fiber, phytates and tannins from gluten and gluten-free flakes. *Food Chemistry* 305: 125452.

Kim, S.-J., Zaidul, I.S.M., Suzuki, T. et al. (2008). Comparison of phenolic compositions between common and tartary buckwheat (*Fagopyrum*) sprouts. *Food Chemistry* 110: 814–820.

Kong, X., Bao, J., and Corke, H. (2009). Physical properties of *Amaranthus* starch. *Food Chemistry* 113: 371–376.

Konishi, Y., Arai, N., Umeda, J. et al. (2000). Cholesterol lowering effect of the methanol insoluble materials from the quinoa seed pericarp. In: *Hydrocolloids* (ed. K. Nishinari), 417–422. Elsevier Science.

Lamothe, L.M., Srichuwong, S., Reuhs, B.L., and Hamaker, B.R. (2015). Quinoa (*Chenopodium quinoa* W.). and amaranth (*Amaranthus caudatus* L.). provide dietary fibres high in pectic substances and xyloglucans. *Food Chemistry* 167: 490–496.

Lasekan, O. and Lasekan, A. (2012). Effect of processing and flavour fine-tuning techniques on the volatile flavour constituents of pseudocereals and some minor cereals. *Journal of Food, Agriculture and Environment* 10: 73–79.

Liu, R.H. (2004). Potential synergy of phytochemicals in cancer prevention: mechanism of action. *The Journal of Nutrition* 134 (12): 3479S–3485S.

Liu, R.H. (2007). Whole grain phytochemicals and health. *Journal of Cereal Science* 46 (3): 207–219.

Liu, W., Zhang, Y., Qiu, B. et al. (2018). Quinoa whole grain diet compromises the changes of gut microbiota and colonic colitis induced by dextran sulfate sodium in C57BL/6 mice. *Scientific Reports* 8 (1): 14916.

Lorenzo, G., Sosa, M., and Califano, A. (2018). Alternative proteins and pseudocereals in the development of gluten-free pasta. In: *Alternative and Replacement Foods* (eds. A.M. Holban and A.M. Grumezescu), 433–458. Academic Press.

Lorusso, A., Verni, M., Montemurro, M. et al. (2017). Use of fermented quinoa flour for pasta making and evaluation of the technological and nutritional features. *LWT - Food Science and Technology* 78: 215–221.

de Lumen, B.O. (2005). Lunasin: a cancer-preventive soy peptide. *Nutrition Reviews* 63 (1): 16–21.

Ma, M.-S., Bae, I.Y., Lee, H.G., and Yang, C.-B. (2006). Purification and identification of angiotensin I-converting enzyme inhibitory peptide from buckwheat (*Fagopyrum esculentum* Moench). *Food Chemistry* 96 (1): 36–42.

Machado Alencar, N.M., Steel, C.J., Alvim, I.D. et al. (2015). Addition of quinoa and amaranth flour in gluten-free breads: temporal profile and instrumental analysis. *LWT - Food Science and Technology* 62 (2): 1011–1018.

Mäkinen, O.E. and Arendt, E.K. (2015). Nonbrewing applications of malted cereals, pseudocereals, and legumes: a review. *Journal of the American Society of Brewing Chemists* 73 (3): 223–227.

Mäkinen, O.E., Zannini, E., and Arendt, E.K. (2013). Germination of oat and quinoa and evaluation of the malts as gluten free baking ingredients. *Plant Foods for Human Nutrition* 68 (1): 90–95.

Matejčeková, Z., Liptáková, D., and Valík, L. (2016). Evaluation of the potential of amaranth flour for lactic acid fermentation. *Journal of Pharmacy and Nutrition Sciences* 6: 1–6.

Mickowska, B., Socha, P., Urminska, D., and Cieślik, E. (2012). Immunodetection, electrophoresis and amino acid composition of alcohol soluble proteins extracted from grains of selected varieties of pseudocereals, legumes, oat, maize and rice. *Cereal Research Communications* 41 (1): 160–169.

Mlakar, S.G., Turinek, M., Jakop, M. et al. (2010). Grain amaranth as an alternative and perspective crop in temperate climate. *Journal for Geography* 5 (1): 135–145.

Molinari, R., Costantini, L., Timperio, A.M. et al. (2018). Tartary buckwheat malt as ingredient of gluten-free cookies. *Journal of Cereal Science* 80: 37–43.

Mota, C., Nascimento, A.C., Santos, M. et al. (2016a). The effect of cooking methods on the mineral content of quinoa (*Chenopodium quinoa*), amaranth (*Amaranthus* sp.) and buckwheat (*Fagopyrum esculentum*). *Journal of Food Composition and Analysis* 49: 57–64.

Mota, C., Santos, M., Mauro, R. et al. (2016b). Protein content and amino acids profile of pseudocereals. *Food Chemistry* 193: 55–61.

Motta, C., Delgado, I., Matos, A.S. et al. (2017). Folates in quinoa (*Chenopodium quinoa*), amaranth (*Amaranthus* sp.) and buckwheat (*Fagopyrum esculentum*): influence of cooking and malting. *Journal of Food Composition and Analysis* 64: 181–187.

Motta, C., Castanheira, I., Bryan Gonzales, G. et al. (2019). Impact of cooking methods and malting on amino acids content in amaranth, buckwheat and quinoa. *Journal of Food Composition and Analysis* 76: 58–65.

Muyonga, J.H., Andabati, B., and Ssepuuya, G. (2014). Effect of heat processing on selected grain amaranth physicochemical properties. *Food Science & Nutrition* 2 (1): 9–16.

Nickel, J., Pio Spanier, L., Torma Botelho, F. et al. (2016). Effect of different types of processing on the total phenolic compound content, antioxidant capacity, and saponin content of *Chenopodium quinoa* Willd grains. *Food Chemistry* 209: 139–143.

Nsimba, R.Y., Kikuzaki, H., and Konishi, Y. (2008). Antioxidant activity of various extracts and fractions of *Chenopodium quinoa* and *Amaranthus* spp. seeds. *Food Chemistry* 106 (2): 760–766.

Okarter, N. and Liu, R.H. (2010). Health benefits of whole grain phytochemicals. *Critical Reviews in Food Science and Nutrition* 50 (3): 193–208.

Ontiveros, N., López-Teros, V., de Jesús Vergara-Jiménez, M. et al. (2020). Amaranth-hydrolyzate enriched cookies reduce the systolic blood pressure in spontaneously hypertensive rats. *Journal of Functional Foods* 64: 103613.

Park, J.W., Kang, D.B., Kim, C.W. et al. (2000). Identification and characterization of the major allergens of buckwheat. *Allergy* 55 (11): 1035–1041.

Paśko, P., Sajewicz, M., Gorinstein, S., and Zachwieja, Z. (2008). Analysis of selected phenolic acids and flavonoids in *Amaranthus cruentus* and *Chenopodium quinoa* seeds and sprouts by HPLC. *Acta Chromatographica* 20 (4): 661–672.

Paucar-Menacho, L.M., Dueñas, M., Peñas, E. et al. (2018). Effect of dry heat puffing on nutritional composition, fatty acid, amino acid and phenolic profiles of pseudocereals grains. *Polish Journal of Food and Nutrition Sciences* 68 (4): 289–297.

Peñas, E., Uberti, F., di Lorenzo, C. et al. (2014). Biochemical and immunochemical evidences supporting the inclusion of quinoa (*Chenopodium quinoa* Willd.) as a gluten-free ingredient. *Plant Foods for Human Nutrition* 69 (4): 297–303.

Pina-Rodriguez, A.M. and Akoh, C.C. (2009). Synthesis and characterization of a structured lipid from amaranth oil as a partial fat substitute in milk-based infant formula. *Journal of Agricultural and Food Chemistry* 57 (15): 6748–6756.

Pojić, M., Mišan, A., and Tiwari, B. (2018). Eco-innovative technologies for extraction of proteins for human consumption from renewable protein sources of plant origin. *Trends in Food Science & Technology* 75: 93–104.

Prego, I., Maldonado, S., and Otegui, M. (1998). Seed structure and localization of reserves in Chenopodium quinoa. *Annals of Botany* 82 (4): 481–488.

Préstamo, G., Pedrazuela, A., Peñas, E. et al. (2003). Role of buckwheat diet on rats as prebiotic and healthy food. *Nutrition Research* 23 (6): 803–814.

Qin, P., Wang, Q., Shan, F. et al. (2010). Nutritional composition and flavonoids content of flour from different buckwheat cultivars. *International Journal of Food Science & Technology* 45: 951–958.

Qiu, J., Liu, Y., Yue, Y. et al. (2016). Dietary tartary buckwheat intake attenuates insulin resistance and improves lipid profiles in patients with type 2 diabetes: a randomized controlled trial. *Nutrition Research* 36 (12): 1392–1401.

Ren, G., Zhu, Y., Shi, Z., and Li, J. (2017). Detection of lunasin in quinoa (*Chenopodium quinoa* Willd.) and the *in vitro* evaluation of its antioxidant and anti-inflammatory activities. *Journal of the Science of Food and Agriculture* 97 (12): 4110–4116.

Repo-Carrasco, R., Espinoza, C., and Jacobsen, S.-E. (2003). Nutritional value and use of the Andean crops quinoa (*Chenopodium quinoa*) and kañiwa (*Chenopodium pallidicaule*). *Food Reviews International* 19 (1–2): 179–189.

Rizzello, C.G., Lorusso, A., Montemurro, M., and Gobbetti, M. (2016). Use of sourdough made with quinoa (*Chenopodium quinoa*) flour and autochthonous selected lactic acid bacteria for enhancing the nutritional, textural and sensory features of white bread. *Food Microbiology* 56: 1–13.

Rizzello, C.G., Lorusso, A., Russo, V. et al. (2017). Improving the antioxidant properties of quinoa flour through fermentation with selected autochthonous lactic acid bacteria. *International Journal of Food Microbiology* 241: 252–261.

Robin, F., Théoduloz, C., and Srichuwong, S. (2015). Properties of extruded whole grain cereals and pseudocereals flours. *International Journal of Food Science & Technology* 50 (10): 2152–2159.

Rocchetti, G., Miragoli, F., Zacconi, C. et al. (2019). Impact of cooking and fermentation by lactic acid bacteria on phenolic profile of quinoa and buckwheat seeds. *Food Research International* 119: 886–894.

Rollán, G.C., Gerez, C.L., and LeBlanc, J.G. (2019). Lactic fermentation as a strategy to improve the nutritional and functional values of pseudocereals. *Frontiers in Nutrition* 6: 98.

de la Rosa, A.P.B., Fomsgaard, I.S., Laursen, B. et al. (2009). Amaranth (*Amaranthus hypochondriacus*) as an alternative crop for sustainable food production: phenolic acids and flavonoids with potential impact on its nutraceutical quality. *Journal of Cereal Science* 49 (1): 117–121.

Sabbione, A.C., Nardo, A.E., Añón, M.C., and Scilingo, A. (2016). Amaranth peptides with antithrombotic activity released by simulated gastrointestinal digestion. *Journal of Functional Foods* 20: 204–214.

Saunders, R.M. and Becker, R. (1984). *Amaranthus*: a potential food and feed resource. In: *Advances in Cereal Science and Technology* (ed. Y. Pomeranz), 357–397. St. Paul, MN: American Association of Cereal Chemists.

Schlick, G. and Bubenheim, D.L. (1996). Quinoa: candidate crop for NASA's controlled ecological life support systems. In: *Progress in New Crops* (ed. e.J. Janick), 632–640. Arlington, VA: ASHS Press.

Scilingo, A.A., Ortiz, S.E.M., Martínez, E.N., and Añón, M.C. (2002). Amaranth protein isolates modified by hydrolytic and thermal treatments. Relationship between structure and solubility. *Food Research International* 35 (9): 855–862.

Sezgin, A.C. and Sanlier, N. (2019). A new generation plant for the conventional cuisine: quinoa (*Chenopodium quinoa* Willd.). *Trends in Food Science & Technology* 86: 51–58.

Silva-Sánchez, C., De La Rosa, A.P.B., León-Galván, M.F. et al. (2008). Bioactive peptides in amaranth (*Amaranthus hypochondriacus*) seed. *Journal of Agricultural and Food Chemistry* 56 (4): 1233–1240.

Siwatch, M. and Yadav, R.B. (2017). Pseudocereals: nutritional quality, processing and potential health benefits. *Current Nutrition & Food Science* 13 (4): 296–302.

Skrabanja, V., Kreft, I., Golob, T. et al. (2004). Nutrient content in buckwheat milling fractions. *Cereal Chemistry* 81 (2): 172–176.

Srichuwong, S., Curti, D., Austin, S. et al. (2017). Physicochemical properties and starch digestibility of whole grain sorghums, millet, quinoa and amaranth flours, as affected by starch and non-starch constituents. *Food Chemistry* 233: 1–10.

Steadman, K.J., Burgoon, M.S., Lewis, B.A. et al. (2001). Minerals, phytic acid, tannin and rutin in buckwheat seed milling fractions. *Journal of the Science of Food and Agriculture* 81 (11): 1094–1100.

Stevens, N.E. (1912). The morphology of the seed of buckwheat. *Botanical Gazette* 53 (1): 59–66.

Stringer, D.M., Taylor, C.G., Appah, P. et al. (2013). Consumption of buckwheat modulates the post-prandial response of selected gastrointestinal satiety hormones in individuals with type 2 diabetes mellitus. *Metabolism* 62 (7): 1021–1031.

Sun, X., Yu, C., Fu, M. et al. (2019). Extruded whole buckwheat noodles: effects of processing variables on the degree of starch gelatinization, changes of nutritional components, cooking characteristics and in vitro starch digestibility. *Food & Function* 10 (10): 6362–6373.

Tanaka, K., Matsumoto, K., Akasawa, A. et al. (2002). Pepsin-resistant 16-kD buckwheat protein is associated with immediate hypersensitivity reaction in patients with buckwheat allergy. *International Archives of Allergy and Immunology* 129 (1): 49–56.

Tang, Y. and Tsao, R. (2017). Phytochemicals in quinoa and amaranth grains and their antioxidant, anti-inflammatory, and potential health beneficial effects: a review. *Molecular Nutrition & Food Research* 61 (7): 1600767.

Tang, Y., Li, X., Chen, P.X. et al. (2015). Characterisation of fatty acid, carotenoid, tocopherol/tocotrienol compositions and antioxidant activities in seeds of three *Chenopodium quinoa* Willd genotypes. *Food Chemistry* 174: 502–508.

Taylor, J.R.N. and Parker, M.L. (2002). Quinoa. In: *Pseudocereals and less Common Cereals: Grain Properties and Utilization Potential* (eds. P.S. Belton and J.R.N. Taylor), 93–122. Berlin, Heidelberg: Springer.

Turkut, G.M., Cakmak, H., Kumcuoglu, S., and Tavman, S. (2016). Effect of quinoa flour on gluten-free bread batter rheology and bread quality. *Journal of Cereal Science* 69: 174–181.

Valcarcel-Yamani, B. and da Lanne, S.C. (2012). Applications of quinoa (Chenopodium Quinoa Willd.) and amaranth (Amaranthus spp.) and their influence in the nutritional value of cereal based foods. *Food and Public Health* 2 (6): 265–275.

Valencia, S., Svanberg, U., Sandberg, A.-S., and Ruales, J. (1999). Processing of quinoa (*Chenopodium quinoa*, Willd): effects on *in vitro* iron availability and phytate hydrolysis. *International Journal of Food Sciences and Nutrition* 50 (3): 203–211.

Vidueiros, S.M., Curti, R.N., Dyner, L.M. et al. (2015). Diversity and interrelationships in nutritional traits in cultivated quinoa (*Chenopodium quinoa* Willd.) from Northwest Argentina. *Journal of Cereal Science* 62: 87–93.

Vollmannova, A., Margitanová, E., Tóth, T. et al. (2013). Cultivar influence on total polyphenol and rutin contents and total antioxidant capacity in buckwheat, amaranth, and quinoa seeds. *Czech Journal of Food Sciences* 31 (6): 589–595.

Wang, L. and Bai, X. (2017). The producing technology of resistant starch from buckwheat using ultrasonic treatment. *Agrotechnology* 6 (161): 2.

Wefers, D. and Bunzel, M. (2015). Characterization of dietary fiber polysaccharides from dehulled common buckwheat (*Fagopyrum esculentum*) seeds. *Cereal Chemistry* 92 (6): 598–603.

Wieslander, G. and Norbäck, D. (2001). Buckwheat allergy. *Allergy* 56 (8): 703–704.

Wilhelmson, A., Oksman-Caldentey, K., Laitila, A. et al. (2001). Development of a germination process for producing high β-glucan, whole grain food ingredients from oat. *Cereal Chemistry* 78 (6): 715–720.

Wolter, A., Hager, A.-S., Zannini, E., and Arendt, E.K. (2013). In vitro starch digestibility and predicted glycaemic indexes of buckwheat, oat, quinoa, sorghum, teff and commercial gluten-free bread. *Journal of Cereal Science* 58 (3): 431–436.

Wood, S.G., Lawson, L.D., Fairbanks, D.J. et al. (1993). Seed lipid content and fatty acid composition of three quinoa cultivars. *Journal of Food Composition and Analysis* 6 (1): 41–44.

Wu, T., Gao, Y., Hao, J. et al. (2019). Lycopene, amaranth, and sorghum red pigments counteract obesity and modulate the gut microbiota in high-fat diet fed C57BL/6 mice. *Journal of Functional Foods* 60: 103437.

Yu, D., Chen, J., Ma, J. et al. (2018). Effects of different milling methods on physicochemical properties of common buckwheat flour. *LWT - Food Science and Technology* 92: 220–226.

Zambrana, S., Lundqvist, L., Veliz, V. et al. (2018). *Amaranthus caudatus* stimulates insulin secretion in Goto-Kakizaki rats, a model of diabetes mellitus type 2. *Nutrients* 10 (1): 94.

Zduńczyk, Z., Flis, M., Zieliński, H. et al. (2006). In vitro antioxidant activities of barley, husked oat, naked oat, triticale, and buckwheat wastes and their influence on the growth and biomarkers of antioxidant status in rats. *Journal of Agricultural and Food Chemistry* 54 (12): 4168–4175.

Zevallos, V.F., Herencia, L.I., Chang, F. et al. (2014). Gastrointestinal effects of eating quinoa (*Chenopodium quinoa* Willd.) in celiac patients. *The American Journal of Gastroenterology* 109 (2): 270–278.

Zhang, Z.-L., Zhou, M.-L., Tang, Y. et al. (2012). Bioactive compounds in functional buckwheat food. *Food Research International* 49 (1): 389–395.

Zhang, Z., Kang, Y., and Che, L. (2019). Composition and thermal characteristics of seed oil obtained from Chinese amaranth. *LWT - Food Science and Technology* 111: 39–45.

Zhou, X., Wen, L., Li, Z. et al. (2015). Advance on the benefits of bioactive peptides from buckwheat. *Phytochemistry Reviews* 14 (3): 381–388.

Zhou, Y., Zhao, S., Jiang, Y. et al. (2019). Regulatory function of buckwheat-resistant starch supplementation on lipid profile and gut microbiota in mice fed with a high-fat diet. *Journal of Food Science* 84 (9): 2674–2681.

Zhu, F. (2016). Chemical composition and health effects of Tartary buckwheat. *Food Chemistry* 203: 231–245.

4

Advances in Conventional Cereal and Pseudocereal Processing

Vijaykrishnaraj Muthugopal Sasthri, Nivedha Krishnakumar, and Pichan Prabhasankar

Flour Milling, Baking and Confectionery Technology Department, CSIR-Central Food Technological Research Institute, Mysore, Karnataka, India

4.1 Introduction

The processing of cereal and pseudocereal grains has evolved to satisfy the needs of growing numbers of increasingly demanding consumers and market requirements, thus ensuring a variety of cereal products that are safe and healthy. Processing is defined as a set of methods and techniques used to transform raw materials into final products. Since prehistoric times, the crude processing of food materials including fermentation and sun-drying has been used to preserve foodstuffs and make them suitable for consumption. Nowadays, there is more emphasis toward clean-label, high-quality, and value-added foods that are convenient to use (Awuah et al. 2007). Moreover, the demand for foods with health-promoting properties has increased concerns about the effects of processing on the functional components of food products. Many health-conscious consumers prefer minimally processed food. However, the consumption of raw and minimally processed food is often associated with safety issues such as the presence of pathogens and the residues of chemical contaminants (pesticides, mycotoxins, etc.). Therefore, the emergence of novel non-thermal processing and its application in cereal and pseudocereal science and technology to produce safe and nutritious food which increases shelf life is increasingly gaining in importance. Nevertheless, all cereals are subjected to certain types of processing in order to make them edible and palatable for consumers (Oghbaei and Prakash 2016).

Innovative Processing Technologies for Healthy Grains, First Edition.
Edited by Milica Pojić and Uma Tiwari.

4.2 Conventional Grain Processing

Cereals and pseudocereals undergo *primary* and *secondary processing*. The *primary processing* includes milling, often combined with fractionation, pearling, and malting.

- Primary processing of grains is necessary for the removal of the outer layers (often utilized for food or feed purposes), and thereby facilitates subsequent processing stages.
- Secondary processing, which includes baking, extrusion and/or puffing, flaking, frying, steaming, roasting, etc., makes cereal grains more palatable, digestible, and suitable for consumption. For example, pasta, bread and baking products, breakfast cereals, and cereal-based confectionary products are products of secondary processing.

During processing several changes may take place that affect the nutritional and functional properties of cereal or pseudocereal products, which influence the availability and digestibility of macro- and micronutrients, and shelf life (Owens et al. 1997). Although whole grains (cereal or pseudocereal) are one of the prime sources of bioactive compounds and dietary fibers, they are still not preferred by the majority of the population, due to their appearance, sensory properties, or lack of familiarity with cooking methods, etc. (Kantor et al. 2001). For example, whole cereal grain bread or pseudocereal-based bread are less familiar among the population compared with white bread made from refined wheat flour. Moreover, refined cereals are in high demand due to their texture, long shelf life, appearance, and special techno-functional properties required for baking. The conventional processing techniques for grains usually occur in various stages, and are largely dependent upon the structure of the grain, as well as the physical and mechanical properties of the grain caryopsis. Furthermore, conventional grain processing also includes mechanical and thermal processing operations to gain a product of desirable quality.

4.2.1 Mechanical Processing

Mechanical cereal processing includes traditional processes such as dry milling (applied to wheat, rye, and corn), pearling (applied to rice, oat, and barley), and wet milling (applied to corn and wheat), in which milling is often combined with fractionation (i.e. sieving) (Papageorgiou and Skendi 2018).

4.2.1.1 Milling Milling is the principal processing procedure of the cereal industry, which can be performed in a wet or dry milling process. Dry milling is used for flour production in a process that separates the outer fibrous layers of the kernel and germ from the starchy endosperm. Flour milling is commonly performed for wheat, rye, and corn, and requires grinding the grains, followed by sifting and purifying (Gómez et al. 2009; Papageorgiou and Skendi 2018). With successive grinding stages, a range of coarse, medium, and fine fractions and flour are obtained. In order to maintain uniformity in the milling fractions, the removal of bran and germ particles is performed using an airflow and sieving process (Kent and Evers 1994). Pearling is also a type of dry milling process, in which the gradual removal of the seed coat (testa and pericarp), aleurone, and subaleurone layers, and the germ from the rice, oat, or barley kernel is performed by an abrasive technique to obtain the polished grain (Papageorgiou and Skendi 2018). Pseudocerals are mostly milled using the dry milling method. Quinoa grains are small-sized grains, having a high content of lysine and methionine, and are thus considered a nutritious and healthy food choice. Caperuto et al. (2000) observed the highest recovery of breakage and reduction in quinoa flour, with an average particle size of 187.7 μm. They reported that with higher recovery of flour, the protein level reduced in the whole meal from 125 g/kg to 35.5 g/kg in the flour and observed low level of lysine, but an increase in methionine.

Another milling type is known as wet milling, which is used for gluten and starch production. Unlike dry milling, wet milling is a process of grinding the soaked grains and subsequently separating the grain compounds, mainly starch, proteins, and fibers. It involves both physical and chemical changes, while dry milling is a mere size reduction operation. In wet milling, the protein matrix holding the starch granules together is destroyed, releasing the starch granules from the protein network (Kent and Evers 1994).

The production process of rice flour can be carried out in a dry, wet, or semi-dry milling process, whereby the major differences between the processes lie in the physicochemical characteristics of the resulting flour (Ngamnikom and Songsermpong 2011). Wet milling is highly preferred because the output is fine fractions of flour with minimum damaged starch and less amylopectin fragmentation compared with a dry milling process. Wet ground rice flour is the most suitable raw material for the production of traditional rice-based products (Suksomboon and Naivikul 2006; Ngamnikom and Songsermpong 2011). However, wet grinding utilizes excessive machinery and manpower for soaking, grinding, filtering, drying, and sieving, high water and energy consumption, and wastewater treatment. Both dry and wet grinding processes may result in low flour quality and utilize high energy consumption, which means that they are generally considered to be nonsustainable processes. Therefore, the innovative freeze grinding process was proposed as an advanced and viable alternative to the traditional wet and dry grinding processes (Ngamnikom and Songsermpong 2011). By soaking the rice samples in liquid nitrogen before dry grinding using a hammer mill, roller mill, and pin mill, it was reported that freeze grinding may result in reduced average particle size and cause less starch damage while producing a higher flour yield compared with the dry milling process. Moreover, by the utilization of freeze grinding, the specific energy consumption was significantly reduced, compared with the wet grinding process (Ngamnikom and Songsermpong 2011).

General problems associated with conventional mechanical cereal grain processing are:

- Processing reduces the nutritional value of the milling products, but at the same time it is important to reduce anti-nutrient compounds (Chowdhury and Punia 1997).
- The extent and impact of the milling process highly influences the composition and proportions of the nutrients in the flour and bran (Greffeuille et al. 2006).
- In conventional processing, the degree of milling (DOM) is influenced by the grain's size, shape, and hardness, and affects the sensory and textural properties of the milling product (Tran et al. 2018).

However, innovation in the improvement of traditional milling equipment has contributed to the enhancement of the final product quality, higher raw material recovery, and energy-efficient processing.

4.2.1.2 Kneading Kneading is a mechanical operation in making dough, in which two or more ingredients are mixed together (e.g. mixing flour with water) to form the final product. When these ingredients are combined and kneaded, the flour particles are hydrated and disrupted, the gluten proteins swell, and a viscoelastic mass is formed – the gluten network. During further kneading, the gluten is sheared and stretched, inducing its deformation into filaments encompassing the whole dough mass (Baudouin et al. 2020).

4.2.2 Thermal Processing

Most grains consumed around the world are subjected to some type of thermal processing. This may include baking (200–220 °C) bread or baking for pasta production, in which semolina is extruded and dried. Pasta is subjected to a lower temperature regime (40–70 °C) than other grain products (Kent and Evers 1994). Thermal processing imparts temperature changes on food and affects various food components, including proteins. The general consequences to the respective proteins of cereals and pseudocereals when they are subjected to thermal processing are the Maillard reactions, denaturation, and aggregation.

The Maillard reaction is a heat-induced browning reaction between the carbonyl groups of the reducing sugars and the amino groups of the free or protein-bound amino acids present in the cereals and pseudocereals (Helou et al. 2016). However, the occurrence of the Maillard reaction between the free asparagine and carbonyl groups of the reducing sugars during baking can cause the formation of carcinogenic substances such as acrylamide and hydroxymethylfurfural. Therefore, different technological solutions have been suggested to eliminate the formation of these compounds, such as adjustments to the attributes of the raw materials and modifications to the formulation (avoiding ammonium-based baking agents, low doses of NaCl, adding cysteine) and the processing parameters (prolonging the fermentation time, utilizing deck ovens rather than convection ovens, reducing the baking temperature, and prolonging heat treatment at lower temperatures) (Xu et al. 2019).

4.2.2.1 Puffing and Popping Puffing and popping are known as the the simplest and quickest traditional dry heat methods for preparing snacks and ready-to-eat breakfast cereal products. In the popping process, cereal grains are exposed to high temperatures for short period of time, whereby heated vapor produced inside the kernel causes starch gelatinization and a sudden expansion of the endosperm, breaking the outer bran layer and allowing its partial escape. In the puffing process, a controlled expansion of the kernel occurs, while the vapor pressure escapes through the micropores of the kernel due to high pressure or a thermal gradient (Mishra et al. 2014). Although a wide range of cereals (e.g. rice, wheat, corn, sorghum, barley, etc.) is used for popping or puffing, the efficacy and the quality of the final product are dependent on the variety and characteristics of the kernel, such as its moisture content, the composition of the grain, its physical characteristics, and the type of endosperm, as well as on the method of popping or puffing (Mishra et al. 2014). Popping and puffing can be accomplished by using dry heat (Hoke et al. 2005) such as sand roasting (the temperature of the sand is about 250 °C), salt roasting, gun puffing, or hot oil frying (200–220 °C), as well as by using a heating medium such as hot air, or microwave radiation (Jaybhaye et al. 2014). Recently, Lee et al. (2019) demonstrated that the utilization of an explosive puffing process can help with the reduction of ochratoxin A in rice and oats. They observed a reduction in the level of ochratoxin A (15–28% for rice; 38–52% for oat kernels) following the application of explosive puffing at 0.5, 0.7, and 0.9 MPa on rice and oat kernels.

4.2.2.2 Roasting and Drying Roasting is carried out when cereal or pseudocereal grains are placed in a hot airflow and heat is transferred to the surface, mainly by convection. The water vapor formed is carried away from the drying surface in the airstream. The temperature for roasting usually ranges from 140 to 160 °C, applied for 20–30 minutes (Zielinski et al. 2009).

For drying, many types of equipment such as cabinet tray dryers, tunnel dryers, conveyor dryers, bin dryers, fluidized bed dryers, etc. are commonly used. During the heating process, the digestibility of nonendosperm protein may be reduced, and resistant starch can form. Resistant starch is similar to dietary fiber, resisting digestion and absorption in the small intestine, and acting as a substrate for colonic microflora. Schlörmann et al. (2020) indicated that roasting up to 160 °C resulted in oat products with improved sensory properties and a favorable nutritional composition, without extensive acrylamide formation. The fat, protein, starch, and β-glucan content was not affected by roasting, whereas roasting negligibly affected dietary fibers.

4.2.2.3 Extrusion Cooking Extrusion cooking is a multi-step and multi-functional process that enables the production of a large number of food products such as pasta, cereal flakes, bread sticks, flatbreads, snacks, textured vegetable protein, etc. (Altan et al. 2008; Singh et al. 2007). Figure 4.1 shows the schemes of two types of food extruder. The extrusion process represents a combination of thermal and mechanical

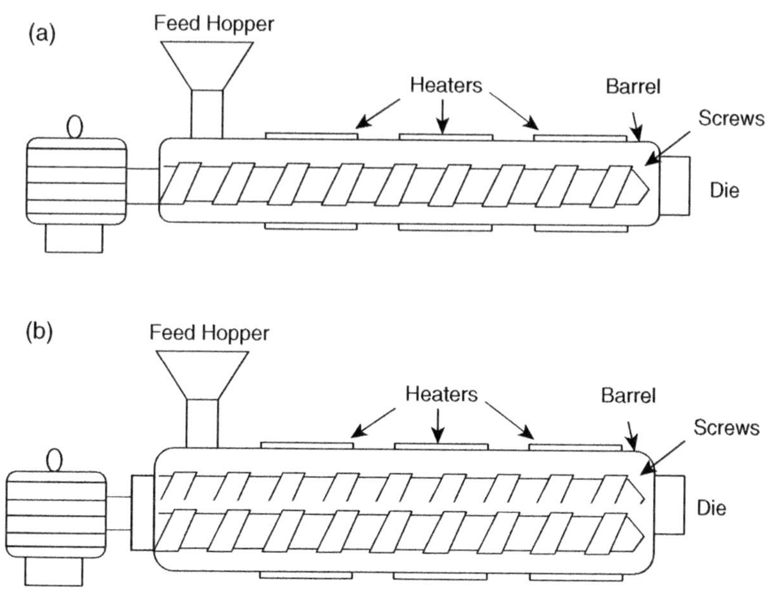

Figure 4.1 Schemes of (a) single-screw food extruder; (b) twin-screw food extruder. *Source:* taken from Dalbhagat et al. 2019.

processing that involves the application of high heat (100–170 °C), high pressure (10–20 bar), and shear forces to uncooked material composed mostly of cereal, during which several different processes occur: material transport, mixing, shearing, particle size reduction, melting, texturizing, caramelizing, and shaping (Kaur et al. 2015; Nikmaram et al. 2017). The extrusion process results in plasticizing and cooking starchy materials in an enclosed barrel with a single screw or two screws, followed by a number of biomolecular changes, including the gelatinization of starch, cross-linking the proteins, and the generation of flavors. Moreover, extrusion cooking influences a decrease in insoluble dietary fiber, and an increase in soluble dietary fiber, due to the disruption of covalent and noncovalent bonds in the carbohydrate and protein moieties, resulting in smaller and more soluble molecular fragments (Rashid et al. 2015). Furthermore, the interaction between the properties of the material being processed and the processing parameters, such as the feed moisture, the screw speed, the temperature, and the screw configuration, turns extrusion into a multiple-input/output process (Ganjyal et al. 2003; Ditudompo et al. 2013). During extrusion, the rheological properties of the melt at the die exit and during expansion play an important role (Moraru and Kokini 2003). These changes are associated with the physico-chemical changes occurring in the starch and other biopolymers.

The extrusion of cereal-based products has advantages over other common processing methods due to its low cost, high speed, high productivity, versatility, unique product shapes, and energy efficacy (Faraj et al. 2004). Moreover, the potential of extrusion cooking to improve the safety of cereals by reducing mycotoxin levels and antinutritional factors has been demonstrated (Nikmaram et al. 2017; Khaneghah et al. 2018; Ryu et al. 2019). The decrease in the mycotoxin level is dependent on the type of extruder, the type of screw, the die configuration, the initial mycotoxin concentration, the barrel temperature, the screw speed, the moisture content of the material, and the use of additives (e.g. glucose, ammonia, or sodium bisulphite). Castells et al. (2005) reported reductions of 100, 95, and 83% for fumonisins, aflatoxins, and zearalenone respectively, while lower reduction rates were observed for deoxynivalenol, ochratoxin A, and moniliformin, being 55, 40, and 30% respectively. Ryu et al. (2019) reported a reduction in ochratoxin A in oat and rice products obtained from artificially contaminated raw material by twin-screw extrusion processing with baking soda. They observed reductions in ochratoxin A after extrusion cooking in oat flakes and rice flour of 40–43 and 78–82% respectively, while the reductions achieved in ochratoxin A during extrusion with added baking soda were up to 65 and 72% in oats and rice respectively. Kaur et al. (2015) reported the reductions in antinutritional factors in wheat, rice, barley, and oat bran, namely phytate, polyphenols, oxalates, and trypsin inhibitors, of 54.51, 73.38, 36.84, and 72.39%, respectively. Innovation in the production of extruded products including either cereals or pseudocereals, or byproducts as ingredients in the formulation of extruded breakfast cereals, has attracted the consumer's interest for its health benefits (Caldwell et al. 2016). Alonso dos Santos et al. (2019) demonstrated the extrusion of rice, passion fruit, and milk by-products, providing solution for waste reutilization and increasing the selection of extruded gluten-free breakfast cereal products.

4.2.2.4 Flaking Flaked cereals include two major groups of flakes: flakes made directly from whole-grain kernels or parts of kernels, and flakes made from finely ground, previously extruded cereal material. Usually, corn flakes are not manufactured from whole corn kernels, due to problems with fat oxidation. Dry milling removes the germ and the bran from the corn kernels and leaves large endosperm particles known as flaking grits (3.4–7.8 mm in size). The production of cornflakes comprises of several steps: pressure-cooking the flaking grits with other ingredients (sugar, malt, salt, and water), drying the cooked grits at 66 °C to lower their moisture to 20%, and the equilibration of the cooked and dried grits for 6–24 hours. The subsequent steps comprise flaking the mass through a pair of counter-rotating rollers, and toasting the soft flakes to develop a crisp texture, brown color, and characteristic flavor (at 270–330 °C for 90 seconds). The moisture content of the finished flaked product is 1–3%. Additional steps may include spraying the flakes with selected nutrients to increase the nutritive value of the final product and/or aromas and sweeteners to enhance the flavor. Unlike corn, other types of cereal flakes (wheat, rice, barley, etc.) are produced from whole cereal kernels containing germ, bran, and endosperm, by the same principle as described for cornflakes (Caldwell et al. 2016; Serna-Saldivar 2016). Innovative solutions in the production of flaked products include the utilization of alternative, nontraditional raw materials, such as wild rice, as demonstrated by Sumczynski et al. (2018).

4.2.2.5 Bread Baking The industrialization of bread baking is considered to be a formative step in the creation of the modern world (Cauvain 2012). Bread is the major staple worldwide, made from a variety of grains, including wheat, maize, oats, rice, rye, barley, millet, quinoa, sorghum, and other cereals and pseudocereals. Bread baking is a complex process that includes physical, chemical, and biochemical changes in the product, such as volume expansion, the evaporation of water, the formation of a porous structure, the denaturation of protein, the gelatinization of starch, crust formation, and a browning reaction. During baking, the formation of crust color due to the Maillard reaction and the caramelization of sugars, the production of flavor and aroma compounds, the formation of toxic products (e.g. acrylamide) and a decrease in the nutritional value of the proteins occur (Purlis 2010). A typical browning of the bread crust can be observed when the crust temperature reaches a critical browning temperature, in the range 110–120 °C (Mundt and Wedzicha 2007; Purlis and Salvadori 2009), whereas no crust browning is observed at the early stage of baking at 205–235 °C (Zhang et al. 2016). Moreover, crust browning occurs when the moisture content of the bread is lower than 25% (Zhang et al. 2016). In order to mitigate acrylamide formation, and reduce its harmful effect on human health, different approaches have been suggested. Mildner-Szkudlarz et al. (2019) and Fu et al. (2018) reported a significant decrease of acrylamide formation in bread when it was formulated with polyphenols. Bread formulated with polyphenols ((+)-catechin, quercetin, gallic, ferulic, and caffeic acids) had decreases in acrylamide formation from 16.2 to 95.2%, as reported by Mildner-Szkudlarz et al. (2019). Ferulic acid exhibited the highest level of acrylamide inhibition, whereas quercetin exhibited a promoting effect on the formation of acrylamide (+9.8%) when its concentration was increased. The inhibitory effect of polyphenols on acrylamide formation was explained by the presence of 4-vinylguaiacol, a degradation derivative with strong antioxidant activity in heterogeneous systems. However, the addition of polyphenols adversely affected the bread's volatile profile, decreased yeast fermentation, enhanced lipid oxidation products, and inhibited enzymes in the bread. When the bread was formulated with (−)-epigallocatechin gallate from green tea, a decrease in acrylamide formation of 37% was achieved, without significantly affecting the textural properties of the bread's crumb (Fu et al. 2018). However, an increased lightness and yellowness and decreased redness of the bread crust, and a decreased granule size and porosity of the bread's crumb were noticed. Nachi et al. (2018) and Bartkiene et al. (2018) demonstrated the potential of inoculated lactic acid bacteria (LAB) strains in sourdoughs to reduce acrylamide formation in bread.

4.3 Bioprocessing of Cereals and Pseudocereals

Recent advances in biotechnology, and growing concern for environmental issues and the sustainability of natural resources, are rapidly transforming different industries to become more environmentally friendly and bio-based. In general, biotechnology includes the application of living organisms and their components at an industrial scale to make or modify products (Yang 2007). In that sense, cereal bioprocessing implies enzyme-assisted processing, fermentation technologies, and biorefinery.

4.3.1 Enzyme-assisted Cereal and Pseudocereal Processing

Enzymes have long been recognized as improving the processing behavior or the properties of cereal foods, such as the taste, texture, and shelf-life, with limited detriment to nutritional value (Singh et al. 2015). Grains contain *endogenous* enzymes, naturally present within the grain kernel, mainly located in the aleurone layer, bran, and germ, as well as *exogenous* enzymes, produced by the microbiota on grain surface. These enzymes contribute to the quality and processing characteristics of the grain's raw material, especially in the presence of moisture and with time. The technological qualities of cereals and cereal-based products are dependent on the activity of endogenous and exogenous enzymes, which is affected by the cultivars, the climatic conditions during growing and harvesting, the storage conditions, and processing by milling. Endogenous enzymes (mainly amylolytic enzymes, proteases, cellulases, hemicellulases, lipases, esterases, phytases, oxidases, xylanases, etc.) are considered to impair the processing properties of flour and other grain products, with the exception of malting. During malting, the activation of the grain's enzyme system is promoted – namely α- and β-amylases, which are generally needed for starch hydrolysis during wort production (Poutanen 1997, 2020). The different technological approaches that are generally taken involve the minimization of the effects of endogenous enzymes, such as the inactivation of lipases, lipoxygenases, and peroxidases of oats to prevent the easy hydrolysis of oat lipids and the formation of rancidity (Poutanen 2020). On the other hand, microbial enzymes provide another source of enzymes for cereal grain processing, due to their ability to change the structural and functional properties of cereal foods, especially in the baking industry, and other areas of cereal processing (Poutanen 1997, 2020) (Table 4.1). In that sense, Mohan Kumar et al. (2019) reported that *Aspergillus niger* prolyl endoprotease (AN-PEP) can specifically cleave the proline glutamine sequence in gluten proteins and reduce the immunogenicity of the wheat flour. Flour modified in this way was subsequently used for the formulation of hypoimmunogenic pasta, characterized with a reduction in gluten content of up to 99.95%. Kumar and Prabhasankar (2017) reported the enzymatic modification of *Triticum durum* semolina, *Triticum dicoccum* semolina, and *Triticum aestivum* flour to improve their nutritional qualities. The utilization of an α-amylase inhibitor and branching enzyme effected reductions in the *in vitro* starch digestibility and the

Table 4.1 Examples of reported enzyme-induced product quality improvements in baking.

Property	Target improvement	Enzymes used
Volume	Larger volume, high-fiber baking, improving the baking quality of flour	α-Amylases, hemicellulases, cellulases, lipases, (proteases)
Stability	Anti-staling effects, extended shelf life, improved freshness	α-Amylases, hemicellulases,
Texture	Softer crumb, fine and regular pore structure, stability of frozen doughs, better crispness, less hygroscopicity	Hemicellulases, α-amylases, proteases, (lipases)
Color	Browning effect, improved crust color, bleaching effect	α-Amylases, hemicellulases, lipoxygenases
Flavor	Production of fermentation substrates and aroma precursors	α-Amylases, proteases, lipoxygenases, lipases, glucose oxidases
Overall quality	Compensating for recipe changes, potassium bromate replacement, sodium metabisulphite replacement, emulsifier replacement, replacement of vital gluten, reduced-fat baking	α-Amylases, hemicellulases, proteases. Lipoxygenases, glucose oxidases
Nutritional properties	Increased total and soluble dietary fiber content	Hemicellulases

(*Source*: taken from Poutanen 1997).

estimated glycemic index of the noodles prepared from the modified raw material, which could be beneficial for people with diabetes.

4.3.1.1 Enzymes in Baking The utilization of exogenous enzymes in baking improves the handling properties of the dough, affects the increase in bread volume, improves the crumb structure, and increases the shelf life (Poutanen 1997). Microbial enzymes such as thermostable α-amylase, pullanases, isoamylases, etc. have long been used in industrial baking. These enzymes are produced by several microbes, including both bacteria and fungi under proper environmental conditions. Among the food-grade biocatalysts, fungal enzymes have acquired a special place in comparison to bacterial enzymes. Apart from the application of enzymes in conventional industrial baking with refined flour as the standard reference, the innovative use of enzymes for gluten-free bread, nonwheat bread, and bread from unrefined wheat flours has been demonstrated (Bender and Schönlechner 2020; Parenti et al. 2020). The application is mainly based on cross-linking activity and forming a network structure in the protein via ε-(γ-glutamyl)-lysine bonds, which affect the stabilization of batter and the improvement of the batter's handling and rheological properties (Miwa 2020). Moreover, they affect crumb softness, bread volume, crust browning, and the maintenance of freshness. Commonly used enzymes in the production of gluten-free products are amylases, cyclodextrin glycosyltransferases, and/or the protein-complexing transglutaminase, as well as glucose oxidase, laccase, and proteases. The effect of enzymes as applied in gluten-free baking is dependent on the type of flour, the presence and the type of arabinoxylans, proteins and hydrocolloids that modify the availability of water in the batter (for example), and the batter pH and/or temperature (Bender and Schönlechner 2020).

4.3.1.2 Enzymes in Other Areas of Cereal Processing Unprocessed cereal grains require different pretreatment steps in which enzymes could be applied, in addition to conventional methods. By affecting the macro-, micro-, and molecular structure, the content and quality of nutrients, phytochemicals, and toxic compounds, enzyme pretreatment improves processability, safety, stability, or technical and nutritional functionality (Poutanen 2020). Enzymes can be used for mycotoxin reduction in a process in which the enzymes biotransform mycotoxin into nontoxic metabolites. Aflatoxins can be degraded by aflatoxin oxidase, peroxidase, laccase, F420H2-dependent reductase, manganese peroxidase, aflatoxin degradation enzyme, and myxobacteria aflatoxin degradation enzyme; zearalenone by laccase, lactono hydrolase, and 2cys-peroxiredoxin; Fumonisin B1 by carboxylesterase and aminotransferase, carboxylesterase B and aminotransferase, and fumonisin esterase. However, Loi et al. (2017) showed that enzymatic mycotoxin degradation depends on the source of the enzyme, its concentration, and the applied conditions (Loi et al. 2017). Susanna and Prabhasankar (2015) developed hypoimmunogenic pasta of an acceptable quality using a combination of xylanase, protease, and transglutaminase, which can be used as an alternative to gluten-free products. Singh et al. (2015) demonstrated how enzymes can be used for polishing cereal grains. In this process, the depolymerisation of the carbohydrates of bran occurs due to the employment of cell-wall-degrading enzymes, affecting phenolic mobilization and dietary fiber solubilization. Enzyme biopolishing is performed by carbohydrate-cleaving enzymes, namely cellulases (e.g. endoglucanase, exoglucanase, and betaglucosidases), xylanases, β-glucanases, and esterases (Singh et al. 2015). Innovative strategies of applying enzymes not only in cereal processing, but in other branches of the industry, are designed to achieve zero waste (Singh et al. 2015; Bilal and Iqbal 2019). For this reason, the valorization of waste streams in a wide range of different products is enabled, such as biofuels, bioactive compounds, biodegradable plastics, prebiotics, sweeteners, sugars, surfactants, etc. (Bilal and Iqbal 2019). Such results signify the applicability and effectiveness of enzymatic methods, but the most efficient conditions for the application of enzymes on grains are still yet to be defined.

4.3.2 Fermentation in Cereal Processing

Fermentation is one of the most commonly used traditional technologies, and suitable for improving the nutritional value and sensory attributes of cereal-based products in simple and economically viable ways (Ogunremi et al. 2017). A variety of beverages, dough, and gruels from barley, maize, millet, oat, rice, sorghum, and wheat are the results of fermentation processes with multifunctional strains (e.g. LAB and yeasts), all of which are approved by the Food and Drug Administration (FDA) and the European Food Safety Authority (EFSA), and designated as Generally Recognized as Safe (GRAS) and Presumed to be Safe (having Qualified Presumption of Safety, QPS), respectively (Ogunremi et al. 2017). Fermented products are in various forms such as fermented beverages, gruels, porridges, soups, etc. (Beuchat 1997), typically in Africa and Asia, but also in Latin America and the Pacific Islands. In Western countries, food fermentations are often integrated in marketing strategies to construct nutritional claims, in response to the increased attention paid by consumers to a healthy way of life, and to address specific sensory characteristics (Humblot and Guyot 2008). The utilization of nonwheat grain flours and nonconventional flours, mainly from pseudocereals, offers innovative product solutions characterized by peculiar flavor and better nutritional value (Coda et al. 2010, 2014; Moroni et al. 2012).

However, the traditional fermentation processes are not well-standardized, having negative effects on the properties of the final products. Therefore, the chosen starter cultures have a critical impact on the final quality of the fermented cereal-based products, and they must metabolize a wide spectrum of carbohydrates in cereals, be tolerant to typical stress conditions during cereal fermentation, and be able to secrete inhibitory metabolites rapidly (Ogunremi et al. 2017).

From a nutritional point of view, the fermentation process offers the opportunity to increase nutrient and energy density and to decrease antinutritional factors, such as tannins and phytic acid (Coda et al. 2010; Moroni et al. 2012). Phytic acid, in particular, can be present in high concentrations in some nonwheat cereals and pseudocereals, especially when milled as whole grains. Sourdough fermentation increases phytase activity indirectly, creating more favorable pH conditions for the activity of endogenous phytases, or directly, through the enzymatic activity of LAB which degrade phytates and increase mineral absorption (Rizzello et al. 2010). The use of nonwheat cereal and pseudocereal flours often corresponds to a lower hydrolysis index (HI) compared with wheat products, as the consequence of their higher concentrations of dietary fiber and resistant starch (Coda et al. 2010). Moreover, when sourdough fermentation is applied, a further decrease in the rate of starch hydrolysis and HI due to the biological acidification is observed in several cereal, leguminous, and pseudocereal flours (Coda et al. 2011). Therefore, the improvement of fermentation technology for flours that are alternatives to wheat can be considered an important opportunity for innovation in the field and for satisfying consumer demands for more natural and healthy food.

4.3.2.1 Sourdough Technology Even though the market for novel bakery products with alternative cereals or pseudocereals is increasing, the use of such flours is restricted due to their poor baking quality, as well as the final sensory quality of the baked products (Gallagher et al. 2004). It has been shown that the fermentation of alternative flours by LAB can improve both sensory and baking qualities, and plays an important role in providing wholesome food with attractive flavor and texture in developing countries.

Sourdough technology is an established biotechnological process for improving and diversifying the sensory quality of bread, with high applicability to the fermentation of cereals other than wheat, and pseudocereals. Studies of the sourdough microbiota of cereals and pseudocereals have revealed a large biodiversity in their microflora that has increased interest in starter cultures, contributing to the enhancement of nutritional and sensory qualities (Corsetti and Settanni 2007). Sourdough technology appeared to be useful in fighting against noncommunicable diseases – obesity, cardiovascular disease, and diabetes, by enabling sugar and salt reduction in bakery products (Sahin et al. 2019; Belz et al. 2019). In a sourdough system, the production of polyols from LAB and yeasts to enhance sweetness occurs – the activity of *Lactobacillus sanfranciscensis*, *Leuconostoc mesenteroides*, and *Leuconostoc citreum* yields mannitol, and *Leuconostoc oenos* yields erythritol, while *Candida milleri* produces xylitol in the presence of xylose, thus contributing to the sweetness of the bakery products. Moreover, exopolysaccharides produced by LAB and/or yeasts improve the texture and structure of bakery products, thus acting as a bulking ingredient, which is otherwise the role of sugar (Sahin et al. 2019). During sourdough fermentation, an accumulation of taste-active compounds occurs, such as ornithine, which enhances the "roasted" flavor of the bread crust, as well as glutamate, which enhances the umami taste, which masks an insufficiently salty taste (Zhao et al. 2015). Apart from that, the presence of antifungal compounds – organic acids and antifungal peptides in the sourdough system – contributes to an increase in the shelf life of bread with a low salt content (Belz et al. 2019). In the future, it can be foreseen that sourdough processing could be used to improve traditional technology, as well as to design foods with particular characteristics.

4.3.2.2 Fermentation for Cereal-based Beverages Fermented cereal-based beverages are already very well-known in different parts of the world, due to the fact that they represent a type of traditional product, such as kvass, boza, bushera, pozol, togwa, mahewu, etc. Fermented cereal beverages are based on a suspension of grains that is subjected to fermentation with fermentative microorganisms, commonly with LAB, by which their sensory and nutritional properties and their shelf life are improved. In this process, a decrease in the level of carbohydrates and nondigestible poly- and oligosaccharides occurs. Fermentation improves the availability of certain amino acids, B vitamins, and minerals, especially iron, zinc, and calcium, due to the enzymatic degradation of antinutritional factors such as phytates. Innovation in this sector is associated with the development of nontraditional cereal-based probiotic beverages, whose popularity is increasing nowadays due to either medical reasons (the incidence of lactose intolerance and allergies to cow's milk) or lifestyle choice. Fermented cereal-based beverages act as probiotics, prebiotics, and synbiotics, with the potential to improve gastrointestinal health (Basinskiene and Cizeikiene 2020).

4.3.3 Biorefinery Processing

The inevitability of fossil oil depletion around the globe has led the scientific community to look toward sustainable growth using renewable raw materials, which decreases dependence on petroleum. In this context, the concept of a *biorefinery* offers a performance competitive with traditional petrochemical refineries (Lynd et al. 1999). The *biorefinery* refers to the production of biofuels (e.g. bioethanol and biodiesel), bioenergy, and valuable biochemicals (e.g. organic acids) from renewable biomass sources with little or no waste, in thermochemical (e.g. pyrolysis), biochemical (e.g. fermentation, esterification), mechanical (e.g. size reduction), or microbial processes. This concept is similar to the concept of petrochemical processing, except for the fact that the biorefinery utilizes renewable biomass feedstock instead of oil. However, the major drawback of this concept is the variability of the quality of the feedstock utilized for the biorefinery due to geographic and seasonal variations, causing inconsistency in the yield of the end products. Moreover, the biorefinery process, like any other processing, generates several side products and waste products, except in the case of ethanol or lactic acid production (Cherubini and Ulgiati 2010; ElMekawy et al. 2013).

4.3.3.1 Cereal-crop-based Biorefinery The cereal-crop-based biorefinery concept has attracted huge interest in recent years, not only in research, but also in industry. Apart from the main products of a biorefinery, several side products and waste products are also generated. Existing cereal biorefineries use dry cereals as raw materials (e.g. wheat, rice, and maize) for the production of first-generation biofuels, while second-generation biofuels are produced from lignocellulosic residues (ElMekawy et al. 2013). Moreover, cereal byproducts contain large amounts of ingredients for the production of a wide range of valorized products, with cellulose, hemicelluloses, and lignin as principal components (ElMekawy et al. 2013; González-García et al. 2018).

Cellulose can be converted to bioethanol, and hemicelluloses to xylooligosaccharides, whilst lignin can be used to obtain resin precursors, heavy metal sequestrants, antimicrobial agents, aromatic compounds, syngas products, etc. (González-García et al. 2018).

González-García et al. (2018) demonstrated the utilization of waste products from breweries (brewers' spent grains and barley straw) as potential feedstock for the joint production of bioethanol and xylooligosaccharides, and identified that the large requirement for steam and the production of the enzymes required for bioethanol production affected the environmental profile of this process. Koutinas et al. (2014) demonstrated a comprehensive wheat-based biorefinery concept that uses feed stocks from different wheat processing streams for bioconversion and the production of value-added products (bioethanol, biodegradable plastics, succinic acid) to maximize the economic effects of production (Figure 4.2). Dorado et al. (2009) demonstrated the utilization of wheat milling byproducts for the production of succinic acid in a fermentative process by *Actinobacillus succinogenes*. Wheat bran fraction was subjected to two solid-state fermentations with *Aspergillus awamori* and *Aspergillus oryzae* to obtain a hydrolysate, which was subsequently used as a medium for *A. succinogenes* fermentations, resulting in the production of 50.6 g/L succinic acid. The importance of this process for its potential integration into a wheat-milling process was emphasized (Dorado et al. 2009).

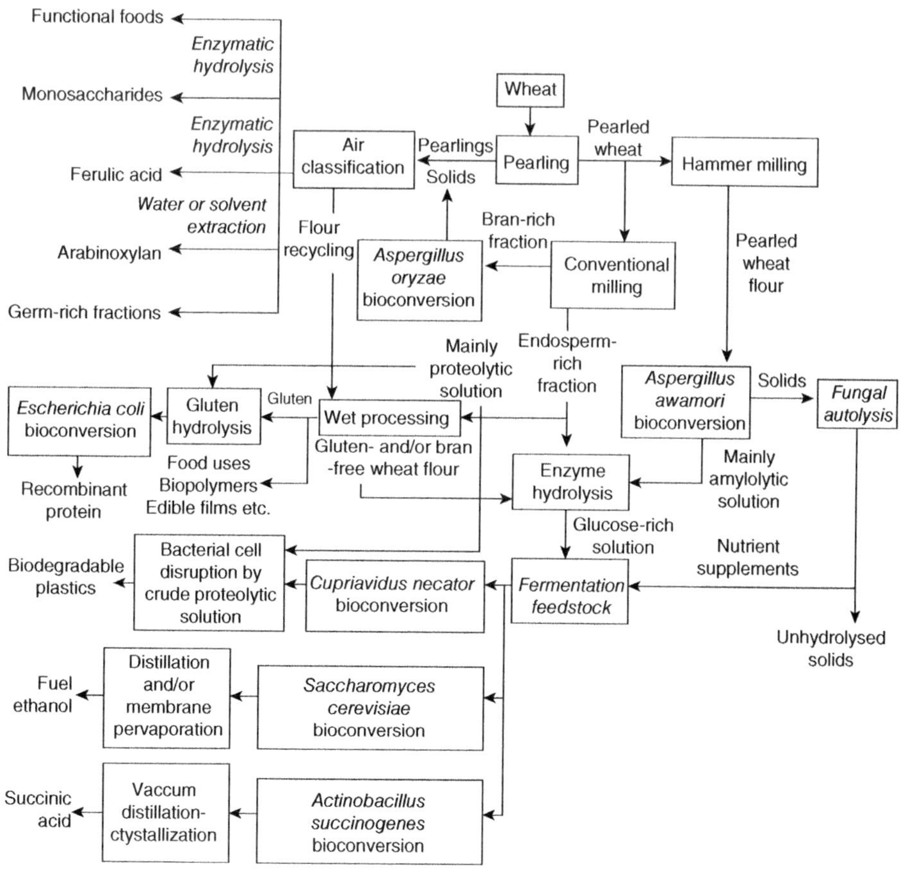

Figure 4.2 Scheme of a wheat-based biorefinery. *Source*: taken from Koutinas et al. 2014.

The hemicellulosic hydrolysate from biorefineries was utilized by *Nesterenkonia sp.* strain for the production of α-amylase, volatile fatty acids, acetone, butanol, ethanol, and single cell protein. It was also found that supplementing the hydrolysates with sweet sorghum grain and adding biocompatible surfactants is one possible approach for improving α-amylase activity. Moreover, the crude enzyme could efficiently be utilized in the production of ethanol from sorghum grains instead of the commercial α-amylase without any reduction in yield in the hydrolysis or fermentation stages (Lolasi et al. 2018). Ferri et al. (2020) demonstrated wheat bran valorization through a biorefinery concept for the production of ferulic acid by enzymatic hydrolysis without the use of strong acid, alkali, or toxic compounds. The ferulic acid yield was in the range of 0.82–1.05 g/kg bran, obtained by rehydrating the bran by autoclaving or by steam explosion using a bran : water ratio of 1:20, followed by enzymatic pretreatment with endopeptidase and endoamylase to remove protein and sugars, and a final enzymatic hydrolysis with xylanase and feruloyl esterase to hydrolyze hemicellulose and esterified phenolic acids from the bran cell walls.

4.4 The Impact of Processing on the Nutritional Composition of Cereals and Pseudocereals

Any processing treatment alters the nutritional quality of the cereals, depending on its type and the severity of conditions. This is reflected either in a reduction of nutrients, phytochemicals, and antinutrients, and/or in an improvement in the digestibility or availability of nutrients. In order to obtain the maximum nutritional and health benefits from cereal consumption, it is of the utmost importance to understand these changes and to select appropriate processing techniques accordingly (Oghbaei and Prakash 2016).

4.4.1 The Impact of Thermal Processing

Apart from the changes in structural conformation and functionality due to protein denaturation and partial starch gelatinization induced by prolonged heat treatment, the bioavailability of both macro- and micronutrients is affected by thermal processing (Oghbaei and Prakash 2016; Paulik et al. 2019). Generally, thermal processing improves the digestibility and bioavailability of macro- and micronutrients. Thus, the digestibility and the bioavailability of iron can increase upon thermal processing, due to the loss of antinutrients (Raghuvanshi et al. 2011). However, thermal processing, namely pressure cooking and microwave treatment, reduced the zinc bioaccessibility from cereal grains. Pressure cooking decreased the bioaccessibility of zinc in finger millet and rice by 63 and 57% respectively. The decrease in the bioaccessibility of zinc due to thermal treatment could be attributed to the interactions of zinc with proteins, and/or other grain components that hinder its absorption (Hemalatha et al. 2007).

Cooking kodo or finger millet by roasting or boiling resulted in a reduction in antioxidant activity, associated with oxidation and degradation reactions. The effects of steaming and roasting on the nutraceutical and antioxidant properties of little millet showed as an increase in the total activity of the phenolic, flavonoid, tannin, and antioxidant content. The high antioxidant properties of roasted millet may be associated with the formation of higher Maillard products during high-temperature, short-time processing (Pradeep and Guha 2011).

4.4.2 The Impact of Malting and Germination

The germination and malting of cereal grains is considered a "green food engineering" technology, generally associated with an accumulation of bioactive compounds (e.g. vitamins, g-aminobutyric acid [GABA] and polyphenols) and a decrease in antinutrient content (Gan et al. 2017). Moreover, due to the activation of enzyme complexes, degradation occurs in the main macronutrients, such as carbohydrates, protein, and fatty acids, accompanied with an increase in simple sugars, free amino acids, and organic acids (Gan et al. 2017). In a study, the antioxidant capacity of the fraction containing free phenolic acids was increased (twofold) after 96 hours of malting in finger millet, whereas the antioxidant capacity of the fraction containing bound phenolic acids was decreased. It was explained either by the action of induced esterases on phenolic acid-polysaccharide and/or phenolic acid-protein complexes, resulting in the liberation of phenolic acids, or by *de novo* synthesis of phenolic acids (Rao and Muralikrishna 2002).

4.4.3 The Impact of Mechanical Processing

The impact of milling processes on the quality of the final products depends on the type of milling process applied: (i) processes in which the whole grain is converted into flour without the separation of the anatomical parts of the kernels, or (ii) processes in which the bran, germ, and outer layers are gradually separated – commonly applied at the industrial level (Oghbaei and Prakash 2016). In this process, the application of successive grinding and sieving operations, together with the uneven distribution of cereal constituents by kernel cross-section, results in a different distribution of chemical, biochemical, nutritional, and rheological properties between milling products (Pojić et al. 2014; Oghbaei and Prakash 2016). Since nutrients are generally present in higher concentrations in the outer part of the grain, refined flours, which mainly comprise parts of the starchy endosperm, are lower in micronutrients, proteins, and fibers than whole-grain flours and tail-end milling streams. The difference in composition is also reflected in the content of antinutritive factors that are associated with fiber (e.g. phytic acid, tannins, polyphenols, trypsin inhibitors, etc.), as well as minerals, especially calcium, iron, and magnesium, and vitamins, especially thiamine, biotin, vitamin B6, folic acid, riboflavin, niacin, and pantothenic acid (Oghbaei and Prakash 2016). The higher the degree of milling (DOM), the greater the loss of nutrients. Moreover, by removing the pericarp layer of millet grains, which is rich in polyphenols and antioxidant compounds, a decrease occurred in the radical scavenging and antioxidant activities (Chandrasekara et al. 2012). The fractionation of kodo millet into husk and endosperm also decreased the 2,2-diphenyl-1-picrylhydrazyl (DPPH)-quenching activity (Hegde and Chandra 2005). Paulik et al. (2019) examined the modification of wheat biopolymers during milling, due to thermal and mechanical forces occurring during general grinding. They observed that the functional changes in starch were solely based on mechanical stress during grinding, whereas thermal stress could cause an increase in the onset of gelatinization in wheat flour. These findings indicated the changes in the structure or conformation of proteins that led to the rise in gelatinization onset and flour hydration. That the changes in the composition of milling products occur due to milling process explains the promotion of whole-grain consumption. Therefore, the mechanical processing of cereal grains must be performed under optimized conditions to protect their quality and potential health benefits.

4.5 Conclusion and Perspectives of Emerging Technologies in Cereal Processing

The cereal processing sector has seen evidence of various innovations that are related to: (i) the selection of innovative raw materials and ingredients, (ii) the utilization of innovative processing technologies, (iii) launching innovative cereal-based products, and (iv) increasing sustainability by the valorization of cereal processing byproducts. Nowadays, the cereal processing sector is looking toward nontraditional raw materials, such as whole-grain flour from buckwheat, quinoa, amaranth, chia, teff, etc., to formulate bread, breakfast cereals, instant porridges, snack products, energy bars, ready-to-bake mixes, and beverages. These products utilize ingredients obtained through conventional processes, although the recent focus is toward alternative technological solutions to produce safe, nutritious, and additive-free products with preserved techno-functional properties. Therefore, in recent years nonthermal processes have gained in importance to replace or to complement traditional processes, to increase the shelf life of the products, increase safety, and preserve nutritional value. The advantages of non-thermal processing technologies are associated with the potential to retain flavor, reduce the loss of nutrients and their functional properties, and reduce the antinutritional factors while maintaining the structural properties of the final products (Spilimbergo and Bertucco 2003). The significance of novel technologies as alternative processing methods in the cereal sector is due to their potential to assist in food safety assurance. For example, ionizing radiation and ionized air technologies are applied in grain processing for disinfection and decontamination; while ultrasound, high-pressure processing, pulsed electric field (PEF), and microwave technologies exhibit the potential for modifying the physico-chemical properties of major cereal components (e.g. starch and protein), thereby also affecting the allergenicity of cereal proteins and the inactivation of enzymes (Keklik et al. 2012; Zeng et al. 2016; Li et al. 2019). However, the majority of these technologies are in the early stages of development and commercial application, especially in the cereal processing sector. Although their scalability has been emphasized, their cost-effectiveness and economic viability are still unknown. Furthermore, the lack of industrial-scale equipment – with the exception of PEF and high-hydrostatic-pressure (HHP) technologies – and relevant legislation are major challenges for the food industry (Pojić et al. 2018). However, use of advanced technologies for grain processing provides a high potential to alter the allergenicity of protein, and that would subsequently increase health benefits for consumers.

References

Alonso dos Santos, P., Caliari, M., Júnior, M.S.S. et al. (2019). Use of agricultural by-products in extruded gluten-free breakfast cereals. *Food Chemistry* 297: 124956.

Altan, A., McCarthy, K.L., and Maskan, M. (2008). Extrusion cooking of barley flour and process parameter optimization by using response surface methodology. *Journal of the Science of Food and Agriculture* 88: 1648–1659.

Awuah, G.B., Ramaswamy, H.S., and Economides, A. (2007). Thermal processing and quality: principles and overview. *Chemical Engineering and Processing: Process Intensification* 46 (6): 584–602.

Bartkiene, E., Bartkevics, V., Lele, V. et al. (2018). A concept of mould spoilage prevention and acrylamide reduction in wheat bread: application of lactobacilli in combination with a cranberry coating. *Food Control* 91: 284–293.

Basinskiene, L. and Cizeikiene, D. (2020). Cereal-based nonalcoholic beverages. In: *Trends in Non-Alcoholic Beverages* (ed. C.M. Galanakis), 63–99. Academic Press.

Baudouin, F., Nogueira, T.L., van der Mijnsbrugge, A. et al. (2020). Mechanochemical activation of gluten network development during dough mixing. *Journal of Food Engineering* 283: 110035.

Belz, M.C.E., Axel, C., Arendt, E.K. et al. (2019). Improvement of taste and shelf life of yeasted low-salt bread containing functional sourdoughs using Lactobacillus amylovorus DSM 19280 and Weisella cibaria MG1. *International Journal of Food Microbiology* 302: 69–79.

Bender, D. and Schönlechner, R. (2020). Innovative approaches towards improved gluten-free bread properties. *Journal of Cereal Science* 91: 102904.

Beuchat, L.R. (1997). Traditional fermented foods. In: *Food Microbiology-Fundamentals and Frontiers* (eds. M.P. Doyle, L.R. Beuchat and T.J. Montville), 629–648. Washington, DC: ASM Press.

Bilal, M. and Iqbal, H.M.N. (2019). Sustainable bioconversion of food waste into high-value products by immobilized enzymes to meet bio-economy challenges and opportunities – a review. *Food Research International* 123: 226–240.

Caldwell, E.F., McKeehen, J.D., and Kadan, R.S. (2016). Cereals: breakfast cereals. In: *Encyclopedia of Food Grains*, 2e (eds. C. Wrigley, H. Corke, K. Seetharaman and J. Faubion), 262–267. Academic Press.

Caperuto, L.C., Amaya-Farfan, J., and Camargo, C.R.O. (2000). Performance of quinoa (*Chenopodium quinoa* Willd) flour in the manufacture of gluten-free spaghetti. *Journal of the Science of Food and Agriculture* 81: 95–101.

Castells, M., Marín, S., Sanchis, V., and Ramos, A.J. (2005). Fate of mycotoxins in cereals during extrusion cooking: a review. *Food Additives & Contaminants* 22 (2): 150–157.

Cauvain, S. (2012). Breadmaking: an overview. In: *Breadmaking* (ed. S.P. Cauvain), 9–31. Woodhead Publishing.

Chandrasekara, A., Naczk, M., and Shahidi, F. (2012). Effect of processing on the antioxidant activity of millet grains. *Food Chemistry* 133 (1): 1–9.

Cherubini, F. and Ulgiati, S. (2010). Crop residues as raw materials for biorefinery systems – a LCA case study. *Applied Energy* 87 (1): 47–57.

Chowdhury, S. and Punia, D. (1997). Nutrient and antinutrient composition of pearl millet grains as affected by milling and baking. *Food/Nahrung* 41: 105–107.

Coda, R., Nionelli, L., Rizzello, C.G. et al. (2010). Spelt and emmer flours: characterization of the lactic acid bacteria microbiota and selection of mixed starters for bread making. *Journal of Applied Microbiology* 108 (3): 925–935.

Coda, R., Rizzello, C.G., Trani, A., and Gobbetti, M. (2011). Manufacture and characterization of functional emmer beverages fermented by selected lactic acid bacteria. *Food Microbiology* 28 (3): 526–536.

Coda, R., Di Cagno, R., Gobbetti, M., and Rizzello, C.G. (2014). Sourdough lactic acid bacteria: exploration of non-wheat cereal-based fermentation. *Food Microbiology* 37: 51–58.

Corsetti, A. and Settanni, L. (2007). Lactobacilli in sourdough fermentation. *Food Research International* 40 (5): 539–558.

Dalbhagat, C.G., Mahato, D.K., and Mishra, H.N. (2019). Effect of extrusion processing on physicochemical, functional and nutritional characteristics of rice and rice-based products: a review. *Trends in Food Science & Technology* 85: 226–240.

Ditudompo, S., Takhar, P.S., Ganjyal, G.M., and Hanna, M.A. (2013). The effect of temperature and moisture on the mechanical properties of extruded cornstarch. *Journal of Texture Studies* 44 (3): 225–237.

Dorado, M.P., Lin, S.K.C., Koutinas, A. et al. (2009). Cereal-based biorefinery development: utilisation of wheat milling by-products for the production of succinic acid. *Journal of Biotechnology* 143 (1): 51–59.

ElMekawy, A., Diels, L., De Wever, H., and Pant, D. (2013). Valorization of cereal based biorefinery byproducts: reality and expectations. *Environmental Science & Technology* 47 (16): 9014–9027.

Faraj, A., Vasanthan, T., and Hoover, R. (2004). The effect of extrusion cooking on resistant starch formation in waxy and regular barley flours. *Food Research International* 37: 517–525.

Ferri, M., Happel, A., Zanaroli, G. et al. (2020). Advances in combined enzymatic extraction of ferulic acid from wheat bran. *New Biotechnology* 56: 38–45.

Fu, Z., Yoo, M.J.Y., Zhou, W. et al. (2018). Effect of (−)-epigallocatechin gallate (EGCG) extracted from green tea in reducing the formation of acrylamide during the bread baking process. *Food Chemistry* 242: 162–168.

Gallagher, E., Gormley, T.R., and Arendt, E.K. (2004). Recent advances in the formulation of gluten-free cereal-based products. *Trends in Food Science & Technology* 15 (3–4): 143–152.

Gan, R.-Y., Lui, W.-Y., Wu, K. et al. (2017). Bioactive compounds and bioactivities of germinated edible seeds and sprouts: an updated review. *Trends in Food Science & Technology* 59: 1–14.

Ganjyal, G.M., Hanna, M.A., and Jones, D.D. (2003). Modeling selected properties of extruded waxy maize cross-linked starches with neural networks. *Journal of Food Science* 68 (4): 1384–1388.

Gómez, M., Pardo, J., Oliete, B., and Caballero, P.A. (2009). Effect of the milling process on quality characteristics of rye flour. *Journal of the Science of Food and Agriculture* 89: 470–476.

González-García, S., Morales, P.C., and Gullón, B. (2018). Estimating the environmental impacts of a brewery waste-based biorefinery: bio-ethanol and xylooligosaccharides joint production case study. *Industrial Crops and Products* 123: 331–340.

Greffeuille, V., Abecassis, J., Lapierre, C., and Lullien-Pellerin, V. (2006). Bran size distribution at milling and mechanical and biochemical characterization of common wheat grain outer layers: a relationship assessment. *Cereal Chemistry* 83: 641–646.

Hegde, P.S. and Chandra, T.S. (2005). ESR spectroscopic study reveals higher free radical quenching potential in kodo millet (*Paspalum scrobiculatum*) compared to other millets. *Food Chemistry* 92 (1): 177–182.

Helou, C., Jacolot, P., Niquet-Léridon, C. et al. (2016). Maillard reaction products in bread: a novel semi-quantitative method for evaluating melanoidins in bread. *Food Chemistry* 190: 904–911.

Hemalatha, S., Platel, K., and Srinivasan, K. (2007). Influence of heat processing on the bioaccessibility of zinc and iron from cereals and pulses consumed in India. *Journal of Trace Elements in Medicine and Biology* 21 (1): 1–7.

Hoke, K., Houšová, J., and Houška, M. (2005). Optimum conditions of rice puffing. *Czech Journal of Food Sciences* 23 (1): 1–11.

Humblot, C. and Guyot, J.P. (2008). Other fermentations. In: *Molecular Techniques in the Microbial Ecology of Fermented Foods* (eds. L. Cocolin and D. Ercolini), 208–224. New York: Springer.

Jaybhaye, R.V., Pardeshi, I.L., Vengaiah, P.C., and Srivastav, P.P. (2014). Processing and technology for millet based food products: a review. *Journal of Ready to Eat Food* 1 (2): 32–48.

Kantor, L.S., Variyam, J.N., Allshouse, J.E. et al. (2001). Choose a variety of grains daily, especially whole grains: a challenge for consumers. *The Journal of Nutrition* 131 (2): 473S–486S.

Kaur, S., Sharma, S., Singh, B. et al. (2015). Effect of extrusion variables (temperature, moisture) on the antinutrient components of cereal brans. *Journal of Food Science and Technology* 52: 1670–1676.

Keklik, N.M., Krishnamurthy, K., and Demirci, A. (2012). Microbial decontamination of food by ultraviolet (UV) and pulsed UV light. In: *Microbial Decontamination in the Food Industry* (eds. A. Demirci and M.O. Ngadi), 344–369. Woodhead Publishing.

Kent, N.L. and Evers, A.D. (eds.) (1994). *Technology of Cereals*, 4e. Elsevier Science Ltd.

Khaneghah, A.M., Fakhri, Y., and Sant'Ana, A.S. (2018). Impact of unit operations during processing of cereal-based products on the levels of deoxynivalenol, total aflatoxin, ochratoxin A, and zearalenone: a systematic review and meta-analysis. *Food Chemistry* 268: 611–624.

Koutinas, A.A., Du, C., Lin, C.S.K., and Webb, C. (2014). Developments in cereal-based biorefineries. In: *Advances in Biorefineries* (ed. K. Waldron), 303–334. Woodhead Publishing.

Kumar, S.B. and Prabhasankar, P. (2017). Enzyme treated flours in noodle processing: a study on an innovative technology. *Journal of Food Measurement and Characterization* 11 (3): 1174–1187.

Lee, H.J., Kim, S., Suh, H.J., and Ryu, D. (2019). Effects of explosive puffing process on the reduction of ochratoxin a in rice and oats. *Food Control* 95: 334–338.

Li, Q., Wu, Q.Y., Jiang, W. et al. (2019). Effect of pulsed electric field on structural properties and digestibility of starches with different crystalline type in solid state. *Carbohydrate Polymers* 207: 362–370.

Loi, M., Fanelli, F., Liuzzi, V.C. et al. (2017). Mycotoxin biotransformation by native and commercial enzymes: present and future perspectives. *Toxins* 9 (4): 111.

Lolasi, F., Amiri, H., Asadollahi, M.A., and Karimi, K. (2018). Using sweet sorghum bagasse for production of amylases required for its grain hydrolysis via a biorefinery platform. *Industrial Crops and Products* 125: 473–481.

Lynd, L.R., Wyman, C.E., and Gerngross, T.U. (1999). Biocommodity engineering. *Biotechnology Progress* 15 (5): 777–793.

Mildner-Szkudlarz, S., Różańska, M., Piechowska, P. et al. (2019). Effects of polyphenols on volatile profile and acrylamide formation in a model wheat bread system. *Food Chemistry* 297: 125008.

Mishra, G., Joshi, D.C., and Panda, B.K. (2014). Popping and puffing of cereal grains: a review. *Journal of Grain Processing and Storage* 1 (2): 34–46.

Miwa, N. (2020). Innovation in the food industry using microbial transglutaminase: keys to success and future prospects. *Analytical Biochemistry* 597: 113638.

Mohan Kumar, B.V., Sarabhai, S., and Prabhasankar, P. (2019). Targeted degradation of gluten proteins in wheat flour by prolyl endoprotease and its utilization in low immunogenic pasta for gluten sensitivity population. *Journal of Cereal Science* 87: 59–67.

Moraru, C.I. and Kokini, J.L. (2003). Nucleation and expansion during extrusion and microwave heating of cereal foods. *Comprehensive Reviews in Food Science and Food Safety* 2 (4): 147–165.

Moroni, A.V., Zannini, E., Sensidoni, G., and Arendt, E.K. (2012). Exploitation of buckwheat sourdough for the production of wheat bread. *European Food Research and Technology* 235 (4): 659–668.

Mundt, S. and Wedzicha, B.L. (2007). A kinetic model for browning in the baking of biscuits: effects of water activity and temperature. *LWT - Food Science and Technology* 40 (6): 1078–1082.

Nachi, I., Fhoula, I., Smida, I. et al. (2018). Assessment of lactic acid bacteria application for the reduction of acrylamide formation in bread. *LWT - Food Science and Technology* 92: 435–441.

Ngamnikom, P. and Songsermpong, S. (2011). The effects of freeze, dry, and wet grinding processes on rice flour properties and their energy consumption. *Journal of Food Engineering* 104: 632–638.

Nikmaram, N., Leong, S.Y., Koubaa, M. et al. (2017). Effect of extrusion on the anti-nutritional factors of food products: an overview. *Food Control* 79: 62–73.

Oghbaei, M. and Prakash, J. (2016). Effect of primary processing of cereals and legumes on its nutritional quality: a comprehensive review. *Cogent Food & Agriculture* 2: 1136015.

Ogunremi, O.R., Banwo, K., and Sanni, A.I. (2017). Starter-culture to improve the quality of cereal-based fermented foods: trends in selection and application. *Current Opinion in Food Science* 13: 38–43.

Owens, F.N., Secrist, D.S., Hill, W.J., and Gill, D.R. (1997). The effect of grain source and grain processing on performance of feedlot cattle: a review. *Journal of Animal Science* 75 (3): 868–879.

Papageorgiou, M. and Skendi, A. (2018). Introduction to cereal processing and by-products. In: *Sustainable Recovery and Reutilization of Cereal Processing By-Products* (ed. C.M. Galanakis), 1–25. Woodhead Publishing.

Parenti, O., Guerrini, L., and Zanoni, B. (2020). Techniques and technologies for the bread-making process with unrefined wheat flours. *Trends in Food Science & Technology* 99: 152–166.

Paulik, S., Jekle, M., and Becker, T. (2019). Mechanically and thermally induced degradation and modification of cereal biopolymers during grinding. *Polymers* 11: 448.

Pojić, M., Spasojević, N., and Atlas, M. (2014). Chemometric approach to characterization of flour mill streams: chemical and rheological properties. *Food and Bioprocess Technology* 7: 1298–1309.

Pojić, M., Mišan, A., and Tiwari, B. (2018). Eco-innovative technologies for extraction of proteins for human consumption from renewable protein sources of plant origin. *Trends in Food Science & Technology* 75: 93–104.

Poutanen, K. (1997). Enzymes: an important tool in the improvement of the quality of cereal foods. *Trends in Food Science and Technology* 8 (9): 300–306.

Poutanen, K.S. (2020). Cereal raw material pretreatment. In: *Breakfast Cereals and How They Are Made*, 3e (eds. A.A. Perdon, S.L. Schonauer and K.S. Poutanen), 97–107. AACC International Press.

Pradeep, S.R. and Guha, M. (2011). Effect of processing methods on the nutraceutical and antioxidant properties of little millet (*Panicum sumatrense*) extracts. *Food Chemistry* 126 (4): 1643–1647.

Purlis, E. (2010). Browning development in bakery products – A review. *Journal of Food Engineering* 99 (3): 239–249.

Purlis, E. and Salvadori, V.O. (2009). Bread baking as a moving boundary problem. Part 1: mathematical modelling. *Journal of Food Engineering* 91 (3): 428–433.

Raghuvanshi, R.S., Singh, S., Bisht, K., and Singh, D.P. (2011). Processing of mungbean products and its nutritional and organoleptic evaluation. *International Journal of Food Science and Technology* 46 (7): 1378–1387.

Rao, S.M.V.S.S.T. and Muralikrishna, G. (2002). Evaluation of the antioxidant properties of free and bound phenolic acids from native and malted finger millet (Ragi, *Eleusine coracana* Indaf-15). *Journal of Agricultural and Food Chemistry* 50 (4): 889–892.

Rashid, S., Rakha, A., Anjum, F.M. et al. (2015). Effects of extrusion cooking on the dietary fibre content and water solubility index of wheat bran extrudates. *International Journal of Food Science and Technology* 50: 1533–1537.

Rizzello, C.G., Nionelli, L., Coda, R. et al. (2010). Effect of sourdough fermentation on stabilisation, and chemical and nutritional characteristics of wheat germ. *Food Chemistry* 119 (3): 1079–1089.

Ryu, D., Kowalski, R.J., Ganjyal, G., and Lee, H.J. (2019). Reduction of ochratoxin A in oats and rice by twin-screw extrusion processing with baking soda. *Food Control* 105: 21–28.

Sahin, A.W., Zannini, E., Coffey, A., and Arendt, E.K. (2019). Sugar reduction in bakery products: current strategies and sourdough technology as a potential novel approach. *Food Research International* 126: 108583.

Schlörmann, W., Zetzmann, S., Wiege, B. et al. (2020). Impact of different roasting conditions on sensory properties and health-related compounds of oat products. *Food Chemistry* 307: 125548.

Serna-Saldivar, S.O. (2016). Maize: foods from maize. In: *Encyclopedia of Food Grains*, 2e (eds. C. Wrigley, H. Corke, K. Seetharaman and J. Faubion), 97–109. Academic Press.

Singh, B., Sekhon, K.S., and Singh, N. (2007). Effects of moisture, temperature and level of pea grits on extrusion behaviour and product characteristics of rice. *Food Chemistry* 100: 198–202.

Singh, A., Karmakar, S., Jacob, B.S. et al. (eds.) (2015). Enzymatic polishing of cereal grains for improved nutrient retainment. *Journal of Food Science and Technology* 52 (6): 3147–3157.

Spilimbergo, S. and Bertucco, A. (2003). Non-thermal bacterial inactivation with dense CO_2. *Biotechnology and Bioengineering* 84 (6): 627–638.

Suksomboon, A. and Naivikul, O. (2006). Effect of dry-and wet-milling processes on chemical, physicochemical properties and starch molecular structures of rice starches. *Kasetsart Journal (Natural Science)* 40: 125–134.

Sumczynski, D., Koubová, E., Šenkárová, L., and Orsavová, J. (2018). Rice flakes produced from commercial wild rice: chemical compositions, vitamin B compounds, mineral and trace element contents and their dietary intake evaluation. *Food Chemistry* 264: 386–392.

Susanna, S. and Prabhasankar, P. (2015). Effect of different enzymes on immunogenicity of pasta. *Food and Agricultural Immunology* 26 (2): 231–247.

Tran, K.N., Witt, T., Gidley, M.J., and Fitzgerald, M. (2018). Accounting for the effect of degree of milling on rice protein extraction in an industrial setting. *Food Chemistry* 253: 221–226.

Xu, J., Wang, W., and Li, Y. (2019). Dough properties, bread quality, and associated interactions with added phenolic compounds: a review. *Journal of Functional Foods* 52: 629–639.

Yang, S.T. (2007). Bioprocessing–from biotechnology to biorefinery. In: *Bioprocessing for Value-Added Products from Renewable Resources* (ed. S.-T. Yang), 1–24. Elsevier.

Zeng, F., Gao, Q.Y., Han, Z. et al. (2016). Structural properties and digestibility of pulsed electric field treated waxy rice starch. *Food Chemistry* 194: 1313–1319.

Zhang, L., Putranto, A., Zhou, W. et al. (2016). Miniature bread baking as a timesaving research approach and mathematical modeling of browning kinetics. *Food and Bioproducts Processing* 100, Part A: 401–411.

Zhao, C.J., Kinner, M., Wismer, W., and Ganzle, M.G. (2015). Effect of glutamate accumulation during sourdough fermentation with *Lactobacillus reuteri* on the taste of bread and sodium-reduced bread. *Cereal Chemistry* 92 (2): 224–230.

Zielinski, H., Michalska, A., Amigo-Benavent, M. et al. (2009). Changes in protein quality and antioxidant properties of buckwheat seeds and groats induced by roasting. *Journal of Agricultural and Food Chemistry* 57 (11): 4771–4776.

5
Healthy Grain Products

Aleksandra Mišan, Anamarija Mandić, Tamara Dapčević Hadnađev, and Bojana Filipčev

Institute of Food Technology, University of Novi Sad, Novi Sad, Serbia

5.1 Introduction to Different Types of Healthy Grain Products and Their Specific Features

Cereals and pseudocereals have been recognized as very useful raw materials for the design of healthy foods due to multiple beneficial health effects. In this respect, healthy grains can be used either directly or as a source of specific ingredients for product development. They contain an array of functional ingredients with pro-health effects, such as dietary fibers (insoluble, soluble), prebiotic carbohydrates, resistant starch, proteins, minerals, vitamins, antioxidants, etc. (Ötles and Cagindi 2006). Cereal starches can be used as encapsulation materials for sensitive substances (Brennan and Cleary 2005; Charalampopoulos et al. 2002). Bakery goods have been also targeted as promising vehicles for the delivery of functional ingredients to consumers owing to their status as global dietary staples (Siró et al. 2008). The fortification of refined wheat flour with B vitamins, minerals, and lysine was an early form of improvement in the functionality of cereal-based food (Rosell 2003). This trend was further continued by the nutritional enhancement of wheat bread with proteins, complex carbohydrates, and fiber (Rosell 2003). Likewise, making bread with different substitutions of pseudocereal flour for wheat flour increased bioactive and antioxidant activity in the final products (Alvarez-Jubete et al. 2009).

The emerging need for food with the potential to improve the health and well-being of consumers has led to different concepts in the design of healthy food: the concept of functional food and the integrative (holistic) approach. The scientific literature on grain-based food products with increased health potential is plentiful. The majority of presented designs fit the functional food concept (adding one or more ingredients that provide a beneficial physiological effect), but recent attempts have been more in line with the holistic approach (undergoing less detrimental processing and preserv-

Innovative Processing Technologies for Healthy Grains, First Edition.
Edited by Milica Pojić and Uma Tiwari.

ing the natural structure and composition of the food). As seen from Table 5.1, the concept of functional foods has resulted in the development of a wide range of products supplemented with one or more beneficial ingredients – various macro-, micro-, and phytonutrients, or cereal-based or noncereal ingredients.

5.1.1 Healthy Grain Products with Enhanced Dietary Fiber Content

The incorporation of various forms of dietary fiber (soluble: oligosaccharides, pectins, β-glucans, psyllium fiber, and galactomannan gums, and insoluble: cellulose, hemicellulose, and lignin) in cereal products is one of the most studied topics in cereal science and technology. Foschia et al. (2013) provided an overview of the changes to quality in common cereal products, like bread, pasta, cakes, and extruded snacks, upon supplementation with ordinary forms of dietary fiber used in the industry (e.g. inulin, fructo-oligosaccharides, β-glucans, arabinoxylans, and resistant starch). The fiber additions caused alterations in:

- *Dough properties:* Increased water absorption (WA) and development time; decreased mixing stability, dough development during proofing, and dough extensibility; and formation of a stiffer dough.
- *Bread quality:* Decreased volume and height, harder crumb, loss of crust crispiness, increased crumb density, and changed color and taste.
- *Pasta quality:* Changes in WA and swelling index, mostly increased cooking loss, changed optimum cooking time and temperature, increased gumminess and chewiness, decreased firmness, and impaired appearance (color, surface) and taste.
- *Extruded snack quality:* Most often detrimental changes in expansion volume, density, firmness, and crispiness, but increased slowly-digestible starch content.
- *Cake quality:* Most frequently changes in batter viscosity and elasticity, decreased cake volume, increased crumb firmness and gumminess, and decreased cohesiveness.

However, the listed observations are not univocal, because the actual quality changes seemed to depend on the type (i.e. molecular structure and physico-chemical properties) and amount of fiber added, the interactions of the dietary fibers with starch, the effect on the stability of the gas cells, etc.

5.1.2 Healthy Grain Products with Enhanced Bioactive Compounds

The enhancement of the antioxidant potential of cereal products has been another frequently addressed topic in the scientific literature. Dziki et al. (2014) overviewed the various attempts to increase the antioxidant activity of wheat bread by enriching it with other cereals, cereal byproducts, pseudocereals, spices, herbs, waste plant material, and other sources of natural phenolic antioxidants such as molasses, fruit and vegetable extracts or powders, etc. (Figure 5.1). In addition to being gluten-free, pseudocereals are a source of good quality protein, dietary fiber and lipids, and are rich in unsaturated fatty acids. Moreover, they contain a significant amount of other bioactive components such as saponins, phytosterols, squalene, fagopyritols, and polyphenols.

Table 5.1 Healthy grain and bakery products according to the functional foods concept.

ENRICHED PRODUCTS	FORTIFIED PRODUCTS	ENHANCED PRODUCTS	ALTERED PRODUCTS
WHEAT-FLOUR N/A	**Micronutrient biofortification**[1,2] **Vitamin B group**[3,4] **Micronutrients**[5]	N/A	N/A
BREAD **(Non)cereal fibers**[6,7] **Prebiotic compounds**[8] **High-protein ingredients of plant origin**[9] **High-protein ingredients of animal origin**[10] **Alternative flours (legumes, tubers, pseudocereals)**[11,12,13,14] **ω-3 fatty acids**[15,16] **By-products**[17,18] **Medicinal plants**[19]	**Minerals** (Ca, Mg, Fe, Zn)[20,21] **Cereal-based fiber**[22,23] **Proteins** **Bioactives**[24] **Medicinal plants**[25,26]	**Fermented wheat bran and germ**[27,28,29] **GABA enhanced bread**[30,31]	**Bread with reduced salt content**[32,33,34] **Bread with reduced fat content**[35] **Sugar-free biscuits, cookies**[36,37]
BISCUITS, COOKIES **Medicinal plants**[26] **By-products**[38,39] **Alternative flours**[40,41]	**Lysine**[42]	N/A	**Reduced fat content**[43,35,44]
PASTA **(Non)cereal fibers**[45,46] **High-protein ingredients of plant origin**[47] **High-protein ingredients of animal origin**[48] **Alternative flours**[49]	**By-products**[50]	N/A	N/A

(*Source*: [1]Goloran et al. 2019; [2]Kaur et al. 2020; [3]Garrod et al. 2019; [4]Tiong et al. 2015; [5]Akhtar et al. 2011; [6]Fendri et al. 2016; [7]Fu et al. 2015; [8]Rubel et al. 2015; [9]Hoehnel et al. 2019; [10]da Rosa Machado and Thys 2019; [11]Millar et al. 2019; [12]Miñarro et al. 2012; [13]Monnet et al. 2019; [14]Olojede et al. 2020; [15]Cox et al. 2011; [16]Gökmen et al. 2011; [17]Pojić et al. 2015; [18]Purić et al. 2020; [19]Đurović et al. 2020; [20]Bryszewska et al. 2019; [21]Rebellato et al. 2017; [22]Benítez et al. 2018; [23]Pontonio et al. 2020; [24]Lin et al. 2019; [25]Bolarinwa et al. 2019; [26]Pestorić et al. 2017; [27]Katina et al. 2012; [28]Pontonio et al. 2017; [29]Rizzello et al. 2010; [30]Coda et al. 2010; [31]Diana et al. 2014; [32]Cauvain 2019; [33]Pasqualone et al. 2019; [34]Sinesio et al. 2019; [35]Fernandes and de las Mercedes Salas-Mellado 2017; [36]Gallagher et al. 2003; [37]Laguna et al. 2013; [38]Alongi et al. 2019; [39]Šarić et al. 2016; [40]Filipčev et al. 2011; [41]Zhao et al. 2019; [42]Virág et al., 2013; [43]Dapčević Hadnađev et al. 2015; [44]Moriano et al. 2019; [45]Hassan et al. 2020; [46]Peressini et al. 2020; [47]Zarzycki et al. 2020; [48]Biró et al. 2019; [49]Littardi et al. 2020; [50]Cedola et al. 2020)

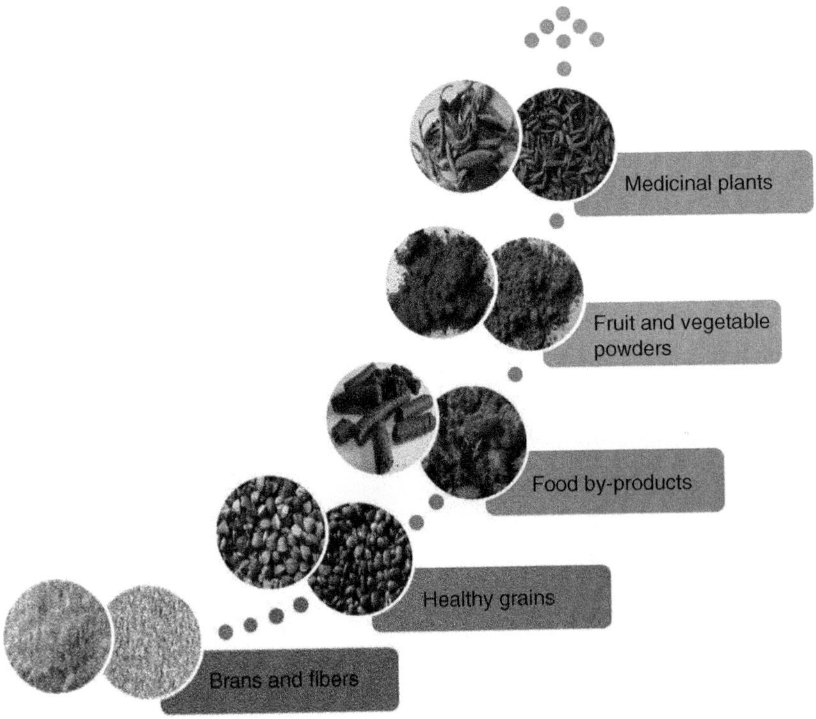

Figure 5.1 Versatility of ingredients for healthy grain products. *Source*: Courtesy of Aleksandra Mišan.

Therefore, they are suitable for consumers with special needs – elderly, children, high-performance athletes, diabetics, celiacs, and people intolerant to gluten or lactose (Valcárcel-Yamani and Lannes 2012).

The amount of antioxidant functional ingredients added to bread is limited due to a compromise between nutritional and sensory properties. The acceptable doses mostly range between 10 and 30% (on flour basis, i.e. the amount of additive as a percentage of the amount of flour) for (pseudo)cereal ingredients and around 5% for ingredients of noncereal origin (Dziki et al. 2014). A detailed overview of the addition of medicinal plants as sources of antioxidants to bread and biscuits was made by Pestorić et al. (2017). The optimal amounts of medicinal plants in bakery formulations reach up to 7% (on flour basis). Apart from the color, herbal supplements affected the dough behavior, bread aroma, taste, and crumb firmness. In biscuits, herbal ingredients tended to provide a softer, more fragile crumb structure, but improved the oxidative stability, and proved effective in delaying the onset of rancidity during storage (Pestorić et al. 2017).

In a number of studies, sourdough technology has been indicated as a very promising tool to enhance the functional potential of cereal foods. It was reported that sourdough technology may confer many valuable functional properties to the end product: a lowered glycemic response, the improved bioavailability of phytochemicals and the dietary fiber complex, an improved uptake of minerals, and enhancement with the beneficial products of microbial fermentation (peptides, γ-amino butyric acid, and exopolysaccharides) (Gobetti et al. 2014). Bread produced with fermented bran had a higher volume and better crumb softness. In addition, the pretreated bran had a higher content of soluble fiber and bioactive compounds such as folates and free phenolic acids (Katina et al. 2012).

From a holistic perspective, healthy products are those based on minimally processed whole-grain ingredients with a retained natural food structure (Fardet 2014). Recent interdisciplinary studies addressed the issue of improving the health-promoting potential of cereals, which has led to the development of new products with a higher nutrient density (Delcour et al. 2012; Poutanen et al. 2014). New technological approaches for the less severe fractionation of cereal grains and minimal loss of the outer anatomical parts of grains have been established (Delcour et al. 2012), and have resulted in "low-pericarp-and-low-crease-material-wholegrain flour." This whole-grain flour contains as much peripheral tissue as possible of the wheat grain, while only the outermost layers, which are detrimental to technological quality and safety, are eliminated. New protocols have been defined for producing novel bioactive ingredients from cereal grains using dry processing operations – the high-bioactive aleurone layer is isolated from bran, micronized bran, and cryogenic ultra-finely ground bran (Delcour et al. 2012). Moreover, cereal-based ingredients with enhanced prebiotic potential were produced using enzymes to separate and modify the structures of the cereal fractions (production of arabinoxylan oligosaccharides by solubilization, and partial hydrolysis of wheat bran arabinoxylan) or using ball milling of wheat and rye bran (Delcour et al. 2012). The design of these novel ingredients accounts for the influence of structure on the bioaccessibility of targeted bioactive compounds and nutrients. Structural issues are also important in the case of making bread with barley and oats, because the molecular weight (MW) and concentration of β-glucan are the key factors that determine its actual physiological effect. To minimize the reduction in MW of barley β-glucan, different modifications of the bread-making process have been suggested: shortening the proofing time, omitting the fermentation step with barley flour, using coarse barley flour, etc. (Rieder et al. 2012).

5.2 Nutritional Profile and Health Benefits of Healthy Grain Products

Whole-grain cereals and foods have been focal points of scientific and commercial research during the past 10 years, as a significant number of epidemiological studies have shown their protective role against the risks of many chronic diseases (Jacobs et al. 1998). McRae (2017) published a review of meta-analyses conducted in relation to the health benefits of whole grain consumption, and concluded that there is evidence that the intake of whole grains can be beneficial in the prevention of type 2 diabetes, cardiovascular disease, and colorectal, pancreatic, and gastric cancers. The same author suggests that the consumption of 2–3 servings per day (~45 g) of whole grains may be a justifiable public health goal.

Referring to the Whole Grains Council (2018) definition, "whole grains or foods made from them contain all the essential parts and naturally-occurring nutrients of the entire grain seed in their original proportions." Numerous studies have confirmed that dietary fibers, vitamins, minerals, and bioactive compounds, such as phenolic compounds, phytosterols, tocols, and carotenoids are concentrated much more in outer layers of grains, which are discarded during the production of refined grains (Sedej et al. 2011). The phenolic compounds of whole-grains, consisting of lignans, alkylresorcinols, and phenolic acids, have been considered the major compounds that provide the protective health effects, as it has been proven that they are metabolized and absorbed in humans (Andreasen et al. 2001; Jacobs et al. 2002; Ross et al. 2003a,b).

5.2.1 Nutritional Profile of Bran

Along with germ, bran is an integral part of whole grains, and is often produced as a byproduct of milling in the production of refined grains. Bran is the outermost fraction of the cereal grain, which consists of multiple layers and presents 10–15% of the weight of the kernel in wheat (Anson et al. 2012). The rice grain contains 5% bran, 3–8% embryo; 12–18.5% of the combined total of the brand and the embryo is oil (Saikia and Deka 2011). Referring to Lloyd et al. (2000), rice bran oil contains high amounts of unsaponifiable materials (3–5%) and a particular antioxidant complex (tocopherols, tocotrienols, and γ-oryzanol). These molecules have shown cardio-metabolic protection benefits, such as antidiabetic and antihypertensive effects, lipid-lowering effects due to cholesterol synthesis downregulation, and increased fecal excretion. Moreover, rice bran phytochemicals have been described as mitigating oxidative stress by increasing antioxidant enzymes and by reducing the production of oxygen radicals, and possessing anti-inflammatory activity (Perez-Ternero et al. 2017).

5.2.1.1 Ferulic Acid Ferulic acid is the most abundant hydroxycinnamic acid in the plant world, and its dehydrodimers are important structural components in plant cell walls (Mišan et al. 2016). The ferulic acid content of wheat varies in the range 0.8–2 g/kg (dry weight basis), while corn bran presents one of the best sources of this potent antioxidant (Dapčević-Hadnađev et al. 2018). Ferulic acid can exist in free, soluble, conjugated, and bound forms. Bound ferulic acid is the predominant form (>93% of total) in corn, wheat, oats, and rice, with a ratio of free to soluble to conjugated and bound ferulic acid in corn and wheat of 0.1 : 1.0 : 100 (Gani et al. 2012). According to Adom and Liu (2002), the total ferulic acid content declines in the following order: corn > wheat > oats > rice. In rice, ferulic acid forms glycosides and esters with sterols, and that mixture is called γ-oryzanol. Ferulic acid esters of 4,4′-dimethylsterols (cycloartenol and 24-methylenecycloartanol) and of 4-desmethylsterols (campesterol, β-sitosterol, and campestanol) have been identified as major components of γ-oryzanol (Xu and Godber 1999). γ-Oryzanol has been shown to lower blood cholesterol, to reduce levels of cholesterol in the liver, and to provide cardio-metabolic protection (Perez-Ternero et al. 2017). Apart from that, γ-oryzanol has been proven to ameliorate the acute stress induced by behavioral anxiety testing in mice (Akter et al. 2018).

5.2.1.2 Lignin Following the outermost pericarp, the testa layer presents a hydrophobic tissue that is rich in lignin and alkylresorcinols. Lignin, an amorphous, highly branched polyphenolic macromolecule of complex structure with high MW, consisting primarily of phenyl propanoid units (*p*-coumaryl alcohol, coniferyl alcohol, and/or sinapyl alcohol) is a major component of whole-grain cereals, and may account for 3–7% of the bran (Bondia-Pons et al. 2009). A significant number of published results provide evidence of lignin's protective role against the development of different diseases: obesity, diabetes, thrombosis, viral infections, and cancer, which could be explained by their antioxidant capacity (Vinardell and Mitjans 2017).

5.2.1.3 Alkyresorcinols Alkyresorcinols are phenolic lipids that are present exclusively in bran and specifically in the outer layer of the seed coat, and that is why they have been suggested as specific dietary biomarkers for whole-grain consumption. Compared to simple phenolic acids, which are polar compounds, alkylresorcinols are amphiphilic, meaning they have both polar and nonpolar properties. This bioactive compound consists of a phenolic ring with two hydroxyl groups and a hydrophobic and mostly saturated alkyl chain that contains an odd number of carbon atoms, from 15 to 25 (Ross et al. 2003a). The highest amounts were reported in rye and common and durum wheat, and the lowest in barley (Andersson et al. 2008; Landberg et al. 2006). In rye, the quantitatively most abundant homologs are C17:0, C19:0, and C21:0, whereas C19:0 and C21:0 are the most abundant homologs in wheat (Ross et al. 2003b). The ratio between C17:0 and C21:0 has been suggested to distinguish between different cereals because the ratio is close to 1.0 for rye, 0.1 for common wheat, and 0.001 for durum wheat (Landberg et al. 2006).

5.2.1.4 Lignans Lignans are compounds composed of two coupled phenylpropanoid units linked by the central carbons of their side chains, and are present in a wide variety of plant foods, including whole grains. For example, rye lignans are present at a concentration of 2 mg/100 g of rye grain (Bondia-Pons et al. 2009), and have been shown to be converted by the intestinal microflora to the mammalian lignans enterodiol and enterolactone. They all have a polyphenolic structure and have antioxidant effects. Due to having a similar structure to estradiol, lignans are classified as phytoestrogens (Adlercreutz 2007).

5.2.2 Nutritional Profile of the Aleurone Layer

The innermost layer in wheat bran, the aleurone layer (50% of the bran weight), contains bioactive compounds: phytates (910–1930 mg/100 g of wheat grain), niacin inclusions, and proteins (Anson et al. 2012). Phytic acid (myo-inositol hexaphosphate) has been considered as an antinutrient, as it binds minerals to form complexes which are insoluble at physiological pH and, consequently, exhibit low bioavailability (Mandić et al. 2011). However, regarding its relatively high binding affinity toward minerals, especially toward iron, phytic acid has been recognized as a potent antioxidant capable of interrupting the reactions of the Haber–Weiss cycle, thus preventing the formation of hydroxyl radicals, which in turn can prevent lipid peroxidation (Sakač et al. 2010).

5.2.2.1 β-Glucans β-glucans are polysaccharides specific to the cell walls of the aleurone layer and endosperm in barley and oat kernels. In barley they are more concentrated in the endosperm, while in oats they are concentrated in the aleurone layer (Bhatty 1993). They are linear polymers of glucose molecules connected by 70% β-(1–4) and 30% β-(1–3) linkages. The most significant amounts of β-glucans are found in barley (3–11%) and oats (3–7%), with lesser amounts reported in rye (1–2%) and wheat (<1%), and only trace amounts in corn, sorghum, rice, and other cereals of importance as food (Wood 1992). Based on scientific evidence that β-glucans present the main component responsible for the cholesterol-lowering effects of oat bran (Bell et al. 1999), in 1997 the Food and Drug Administration (FDA) in the US allowed a health claim stating that diets low in saturated fat and cholesterol that include soluble fiber from whole oats "may" or "might" reduce the risk of heart disease (Gani et al. 2012).

5.2.2.2 Avenanthramides Located mainly in the aleurone layer, avenanthramides are polyphenols found exclusively in oats. They are amides of 5-hydroxyanthranilic acid that have been shown to possess distinct antioxidant and anti-atherosclerotic activities. They are reported to possess an array of bioactivities including: anti-inflammation, antiproliferation, antioxidation, antipruritic, and vasodilator activities. Their bioavailability in humans and potential to modulate different signaling pathways associated with cancer, diabetes, inflammation, and cardiovascular diseases has been reported as well (Vishwas et al. 2018).

5.2.2.3 Arabinoxylans Arabinoxylans are the main noncellulose and nonstarch polysaccharides in cereals (Figure 5.2).

Arabinoxylans belong to pentosans, as they consist of a backbone of β-(1,4)-linked xylose residues, which are replaced by arabinose residues on the C(O)-2 and/or C(O)-3 positions (Dornez et al. 2009). Phenolic acids, such as ferulic acid, can be ester-linked on the C(O)-5 position of arabinose. Oxidative crosslinking between different ferulic acid residues forms inter/intra-chain diferulic acid bridges (Mendis and Simsek 2014). The hydrolysis of arabinoxylans results in their hydrolysis products – arabinoxylan oligosaccharides. Arabinoxylans and arabinoxylan oligosaccharides belong to a class of dietary fibers associated with many health benefits, including immunomodulatory activity, cholesterol-lowering activity, attenuate type II diabetes, the enhanced absorption of certain minerals, the fecal bulking effect, the prebiotics effect, etc. (Mendis and Simsek 2014).

5.2.3 Anthocyanin and Carotenoid-Pigmented Grains

Apart from the above-mentioned bioactive compounds, anthocyanin-pigmented grains such as purple or blue wheat, purple or blue corn, and red or black rice contain comparable levels of anthocyanins to fruits and vegetables, and hold promise for the

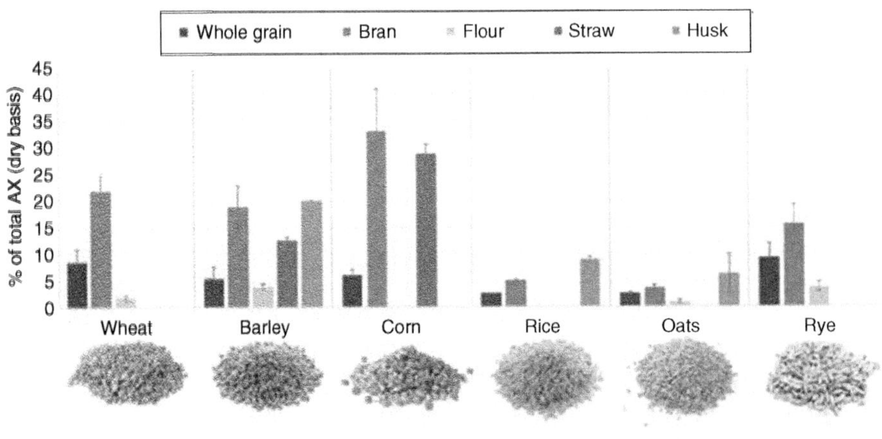

Figure 5.2 The average content of arabinoxylans in different grains and their anatomic parts. *Source*: taken from Bastos et al. 2018.

development of functional ingredients (Abdel-Aala et al. 2018). Anthocyanins are accumulated in the aleurone or pericarp layer and are blue, purple, or combination of these colors, while flavonoids, such as the yellow C-glycosides of flavones, flavonols, flavanonols, proanthocyanidins, and the reddish-colored phlobaphenes, are mainly located in the outer layer of the grains (Lachman et al. 2017).

Carotenoids, such as lutein, β-carotene, β-apocarotenal, β-cryptoxanthin, zeaxanthin, antheraxanthin, taraxanthin (lutein-5,6-epoxide), triticoxanthin, and flavoxanthin, are deeply colored yellow, orange, or red pigments that may be present in wheat and rice grains (Lachman et al. 2017; Melini et al. 2019). Referring to Lachman et al. (2017), lutein is the most abundant carotenoid in wheat, followed by zeaxanthin, antheraxanthin, α-carotene, and β-carotene, while β-cryptoxanthin is a minor component or not detected at all.

The accumulation of the above-mentioned pigments in grains can represent an important target in breeding programmes that aim to increase the concentrations of bioactive components in the final products (Lachman et al. 2017). The milling fractions of these grains, especially those rich in anthocyanins, could serve as a natural colorant and/or antioxidant in the food and nonfood industries, and hence their antioxidant capacity would be important in applications (Abdel-Aala et al. 2018).

Scientific evidence suggests that many chronic diseases around the world are preventable through lifelong practices of adhering to healthy dietary patterns, engaging in physical activity, and maintaining an acceptable weight (Neuhouser 2019). Healthy dietary patterns were defined in *The 2015 Dietary Guidelines Advisory Committee Scientific Report* as diets high in fruits, vegetables, whole grains, low and nonfat dairy, and lean protein (Millen et al. 2016). A reformulation of the cereal-based products that are available on the market can be expected not only regarding a higher exploitation of "whole-grain" flours, but also because the World Health Organization (WHO 2018) has advised that the focus of product reformulation should be on reducing the amount of added sugars, lipids, saturated fatty acids, trans-fatty acids, and sodium.

5.3 Bioaccessibility and Bioavailability of Nutritional Compounds

The evaluation of products in terms of their health benefits includes the measurement of bioavailability, bioaccessibility, and/or bioactivity, which are defined differently in pharmacology, nutrition, and food science (Fernández-García et al. 2009; Galanakis 2017).

Bioaccessibility is defined as the quantity of a compound that is released from the food matrix into the gastrointestinal (GI) tract that is then available for absorption or assimilation. It is highly influenced by the composition of the food matrix and by the synergistic and antagonistic effects present in the GI tract. It is usually determined by applying *in vitro* assays that simulate digestion in the GI tract using static or dynamic digestion models, and optionally is followed by measurement of the Caco-2 cell uptake (Galanakis 2017).

The *bioavailability* of a nutrient from a food product is defined as the fraction of the nutrient that reaches the systemic circulation. It can be determined using *in vivo* experiments, such as by determining the plasma concentration of the compound after food intake (Galanakis 2017).

Bioactivity is defined as the specific effect caused by a specific substance, and it includes tissue uptake, followed by the physiological response, and how the bioactive substance interacts with other biomolecules. Bioactive compounds present in food can modulate metabolic processes and thus cause beneficial health effects. Bioactivity is measured using *in vivo*, *ex vivo*, and *in vitro* assays (Galanakis 2017).

It is important to mention that numerous external and internal factors affect the bioaccessibility and bioavailability of bioactive compounds, while food processing and storage conditions may change the content of bioactive compounds in the food product and consequently influence the health benefits. This section gives an overview of the bioaccessibility and bioavailability research on bioactive compounds in cereal grains, such as polyphenols, fibers, and minerals.

5.3.1 Bioaccessibility and Bioavailability of Polyphenols

The structure and bioavailability of dietary polyphenols, the evidence of their protective effects against chronic diseases, and the role of the small intestine, colon, and microbiota in the determination of the metabolic fate of polyphenols are comprehensively reviewed by del Rio et al. (2013) and Williamson and Clifford (2017). Moreover, Bohn et al. (2015) stressed that the co-digestion of polyphenols from foods eaten in combinations can result in very different bioaccessibilities. For instance, a wheat-based breakfast cereal in combination with raspberry juice has no influence, while blueberries with oatmeal reduce the recovery of anthocyanins in total and polyphenols in total (Cebeci and Sahin-Yesilcubuk 2014; McDougall et al. 2005).

Polyphenols are the main source of antioxidants in cereals, but most of them are covalently bound to polysaccharides and proteins, and hence not accessible for enzymatic digestion in the GI tract, which in turn results in their low bioavailability. Processing technologies such as fermentation, mechanical, and/or thermal treatment facilitate the release of the bound polyphenols present in cereal grains when the cereal matrices are disintegrated through particle size reduction or fiber structure degradation. Wang et al. (2014) reviewed the impact of different mechanical treatments (milling, grinding, and microfuidization), thermal processing technologies (extrusion cooking), and bioprocessing (germination and fermentation) in the enhancement of the bioaccessibility and bioavailability of polyphenols in cereal grains. Mechanical treatments and bioprocessing have positive effects on the bioaccessibility and/or bioavailability of polyphenols from almost all cereal grains, while thermal treatments and extrusion cooking can show a positive or negative effect on the bioavailability of polyphenols depending on the processing parameters used and the type of cereal (Wang et al. 2014).

A study of the bioaccessibility of phenolic compounds from two common millets, finger millet (*Eleusine coracana*) and pearl millet (*Pennisetum glaucum*), which were raw, sprouted, and thermally treated, showed a 20% increase in the bioaccessibility of the polyphenols after sprouting and roasting. Pressure cooking, open-pan boiling and microwave heating reduced the bioaccessible polyphenol content by 30–35%. Out of 13 phenolic compounds identified in the native pearl millet, 10 were bioaccessible, of which around 80% of the activity was attributed to syringic acid and two unidentified compounds. The bioaccessible polyphenol content was 2.65 mg/g, which did not change significantly after sprouting and roasting (Hithamani and Srinivasan 2014a). A similar study by the same authors was performed on wheat (*Triticum aestivum*) and sorghum (*Sorghum bicolor*) (Hithamani and Srinivasan 2014b). The total polyphenol content of wheat and sorghum were 1.20 and 1.12 mg/g respectively, which were increased on roasting by 49 and 20% respectively.

The bioaccessibility of phenolic compounds in bread is thoroughly reviewed by Angelino et al. (2017). The results from the studies that were performed varied considerably, with regard to the type of flour, the applied bread-making processes, the methods of analysis that were used, and finally the ways in which the results were expressed. Generally, white wheat bread is characterized by a low bioaccessibility of phenolic acids, ranging from not detected to 10.2% (percentage of phenolic acid in the dialysate in relation to the original sample). In the studies where it was measured as the total phenolic content in the supernatant from static *in vitro* digestion, the range was 58–84%. When comparing the bioaccessibility of phenolic compounds from whole wheat bread with white wheat bread, the bioaccessibility was around fivefold higher in white bread compared with whole wheat, despite the fact that the content of phenolic compounds is around 10-fold higher in whole wheat bread (Hemery et al. 2010; Mateo Anson et al. 2009). Contrary to these findings, other studies reported 13.1% bioaccessibility of phenolic acids in whole-grain bread (Dall'Asta et al. 2016). When comparing the bioaccessibility results of *in vitro* digestion for the breads made from different grains, the following order can be established: wheat > buckwheat > rye > oat. The initial content of the phenolic compounds in bread made from one type of flour is: buckwheat > wheat > oat > rye. The high fiber and protein content in buckwheat and oat breads could be a cause for the lower bioaccessibility of the phenolic compounds (Angelino et al. 2017).

There are a few human studies that have investigated the bioavailability of polyphenols from bread, as single-dose or chronic dietary intervention studies, in which the conclusions were derived from an analysis of the concentrations of polyphenols in urine and/or serum. In all such studies, bioavailability was calculated as the ratio between the amount of phenolic compounds that were excreted and the amount of phenolic compounds that were ingested through the bread test sample. The urinary levels of phenolic acids and their metabolites were increased. The levels of phenolic acids were higher when bioprocessed bran was added to the bread formulation for some phenolic acids, like a twofold higher content of sinapic acid. However, no influence on the content of *p*-coumaric acid was noticed. Increased hippuric and hydroxyhippuric acid plasma levels were recorded after bread consumption, but since they may arrive from several different metabolic pathways, they cannot be attributed solely to the bread. Therefore, further research is needed to evaluate the effectiveness of innovative processes, including those involving the use of biofermentation in the bread-making process for better bioavailability of polyphenols from bread (Angelino et al. 2017). The beneficial health effects from the consumption of cereal-based products was demonstrated in intervention studies, such as the antihyperlipidemic effect in statin-treated patients after the consumption of buckwheat-enriched wheat bread (Stokić et al. 2015), decreased cardiovascular disease (CVD) risk markers after buckwheat consumption (Li et al. 2018), and improved lipid profiles and reduced inflammation after the consumption of buckwheat instant porridge (Mišan et al. 2017). A slight modification of the phenolic compounds occurs in the upper GI tract. In the small intestine they are metabolized by enzymes, and the aglycones are released, followed by further glucuronidation, sulfation, and/or methylation by specific enzymes. At the colonic level, the gut microbiota affect the metabolism of polyphenols by the rearrangement of their structure, hydroxylation/dehydroxylation, and methoxylation/

demetoxylation. Those metabolites rapidly reach the liver via the portal vein, where they are further metabolized. Lipophilic and polymeric polyphenolics can be transported back to the small intestine through the bile duct, metabolized, and delivered again to the liver, over several cycles. Therefore, very few aglycones are usually detected in the bloodstream, while secondary metabolites rarely exceed nanomolar concentrations (Angelino et al. 2013; del Rio et al. 2013; Terao et al. 2011; Walle et al. 2005; Williamson 2002; Zhang et al. 2007).

5.3.2 Bioaccessibility and Bioavailability of Fibers

Oat bran is rich in soluble fiber and β-glucan, and numerous studies have linked the consumption of oats with a reduction of lipids, anti-obesity activity, regulation of hypertension, hyperlipidemia and hyperglycemia, and lowering the risk of metabolic disorder and cardiovascular diseases (Lazaridou and Biliaderis 2007; Venkatakrishnan et al. 2019). β-glucans stimulate the immune system, modulating humoral and cellular immunity, and therefore have beneficial effects in fighting infectious diseases. The clinical and physiological importance of β-glucans from different sources is reviewed in Bashir and Choi (2017).

5.3.3 Bioaccessibility and Bioavailability of Minerals

The absorption of iron, zinc, and calcium from cereals (and plant food in general) is inhibited by the presence of phytic acid, thus lowering the content of phytic acid could improve the absorption of these minerals. At the same time, the removal of the bran reduces both the phytic acid and the iron, zinc, and calcium content, since the minerals could be found in the form of complexes of phytic acid and minerals. The potential for increasing the content and bioavailability of iron, zinc, and calcium in plants for human nutrition is reviewed in Frossard et al. (2000). Since iron deficiency is considered the most common human nutritional deficiency in developing and industrialized countries, the fortification of food with iron has been recommended as one of the preferred approaches. The bioavailability of micronized dispersible ferric pyrophosphate using anemic weaning pigs as a model showed that this form of iron can be recommended as the most suitable for the fortification of infant cereals (Caballero Valcárcel et al. 2019). Research on the thermal processing of cereals showed different results for mineral bioaccessibility, even when the same methods were used, like pressure or microwave cooking (Cilla et al. 2018; Hemalatha et al. 2007; Khanam and Platel 2016). A study on the mineral and phytic acid content of cooked buckwheat-enriched tagliatelle showed that the content of magnesium, zinc, manganese, and iron was significantly higher when the tagliatelle was made from unprocessed buckwheat flour, in comparison with autoclaved flour. On the other hand, the content of phytic acid was significantly reduced in the tagliatelle made from autoclaved flour (Jambrec et al. 2016). Winiarska-Mieczan et al. (2019) gave an overview of the copper, manganese, iron, and zinc content of different types of cereal products and their contribution to the recommended daily intake, with the conclusion that the cereal products should not be considered as the fundamental source of those microelements in the diet.

Taking into account all the findings on the prediction of the digestion *in vivo* using *in vitro* studies, there are still many questions related to the physiology of human digestion (Bohn et al. 2018). Food processing is one of the more successful ways to assure better bioavailability and bioaccessibility of the bioactive compounds from cereal foods. There is no single model for enhancing all the benefits from all of the nutrients and bioactive compounds present in cereals; rather, it is a compromise.

5.4 Rheological and Structural Properties of Healthy Grain Products

Whole-grain wheat flour or flour obtained from "healthy" cereals (barley, oat, rye, millet, teff, etc.) and pseudocereals (buckwheat, amaranth, quinoa) is used to replace completely or partially refined wheat flour in the preparation of bakery products, either: (i) due to economic reasons in wheat-import-dependent countries, or (ii) to improve the nutritional profile of the product and thus satisfy health-conscious consumers (Gómez et al. 2011; Rajiv et al. 2011). A lot of studies have been carried out on the use of whole-grain wheat flour or nonwheat flour to produce bread, cakes, cookies, biscuits, snack products, and other products. However, a strategy for the development of nutritionally enhanced grain-based products by incorporating healthy cereals results in changes to the structure of the product, thus affecting its processability and product quality characteristics (Rieder et al. 2012). A special challenge is presented by the utilization of whole-grain gluten-free flours in the development of leavened products, since their proteins do not possess the viscoelastic properties typical of the gluten complex (Vallons et al. 2011). However, the incorporation of gluten-containing cereals other than wheat, such as rye, triticale, barley, tritordeum, etc., or whole wheat flour in the formulation of leavened products in most cases requires process modifications in order to get a product of comparable quality to the product prepared from refined wheat flour. The majority of the processing challenges in the manufacturing of *healthy* grain bakery products arise from the incorporation of a bran or germ component in the formulation. This significantly improves the nutritional profile of the product in terms of a high content of dietary fiber and other whole grain components such as phenolics (Rieder et al. 2012; Slavin 2004). However, the use of bran and whole-grain flour in leavened products has a tremendous impact on the dough's rheology, and provokes difficulties in their manufacturing.

5.4.1 Properties of Bakery Products

In the production of bread and other leavened products, the replacement of refined flour with wholegrain flour results in an increase in the WA of the flour, a decrease in the stability of the dough, an increase in resistance to extension, and decreases in the dough's extensibility and peak pasting viscosity. The presence of whole-grain flour increases bread crumb moisture content, firmness, and hardness, and decreases the specific volume of the bread (Schmiele et al. 2012; Zhang and Moore 1997). The possible reasons are different mechanisms for WA and structure formation in refined and whole-grain flours. In refined gluten-containing flours, the main components responsible for WA are the gluten network proteins, which can absorb water up to three times their weight (Md Zaidul et al. 2004). In whole-grain products, the presence of fiber causes an increase in WA due to the water binding to the hydroxyl groups of the fiber (Nedeljković et al. 2017). A study performed by Rieder et al. (2012) has shown that, besides the content of fiber, the type of fiber and its MW also play important roles in the WA of the flour. Although their results indicated that partial substitution of wheat flour with 40% oat or barley flour significantly increased WA compared to wheat flour alone, the samples containing oat bran showed a significantly higher WA than those containing barley flour, even though the total amount of fiber was lower in oat bran. This phenomenon was explained by the fact that the content and MW of β-glucan is more important for WA than the content of other fibers, and that oat bran has a much higher content of β-glucan than barley flour (Rieder et al. 2012). Banu et al. (2010) have reported that the addition of whole buckwheat flour to whole wheat flour increased WA to a greater extent than the replacement of whole wheat flour with whole rye flour. Therefore, in order to make machinable dough and to obtain a good quality product from whole-grain flours, it is important to compensate for the high water-binding capacity of the fibers that are present (Barros et al. 2010; Holtekjølen et al. 2008). The aforementioned higher water requirement of whole-grain flours in comparison with refined flours may also be partly due to the differences in their milling techniques. The hammer mill, used to produce whole wheat flour, compared with the roller mill for refined flour, results in a higher content of damaged starch granules, which absorb more water than intact starch granules (Barros et al. 2010; Saxena and Rao 1997). Hüttner et al. (2010) identified the small flour particle size, damaged starch granules and high protein content as the key factors causing the increased water hydration capacity. The increase in the flour extraction rate results in an increase in protein and damaged starch content and a corresponding increase in WA (Barros et al. 2010; Ramírez-Wong et al. 2007).

The kinetics of WA influences the dough development time (DDT), i.e. the energy needed for dough structure formation. When added to refined wheat flour, whole-grain flours exhibit the combined effects of gluten dilution and decreased water availability for gluten network formation. The former leads to DDT shortening, while the latter prolongs DDT. The final influence on DDT depends on the prevalence of one effect over another. According to Schmiele et al. (2012), when whole-grain flour is added in lower amounts to refined wheat flour, a decrease in DDT is noticed, due to the dilution of the gluten proteins. Consequently, the lower amount of protein in the system requires less time to develop the gluten network (Schmiele et al. 2012). However, when flours containing insoluble fibers are added in higher amounts, the value of DDT increases, due to the lower hydration speed of insoluble fibers (cellulose or hemicellulose). In the latter case, the effects of hydration kinetics and water availability surpass the gluten dilution effect (Rieder et al. 2012; Schmiele et al. 2012). According to Barros et al. (2010), a longer time and more water is needed for whole wheat flour to hydrate and develop the gluten than for refined wheat flour, because the proteins have to compete with the fiber for water. Rieder et al. (2012) and Schmiele et al. (2012) reported that the addition of barley flour and oat bran, or whole wheat flour and bran, to refined wheat flour decreases the dough stability time (DST). Due to the ability of fiber to tightly bind high amounts of water, there is less available water in the composite flour system to develop the gluten network. Therefore, whole-grain flours yield stiffer and harder doughs relative to refined flours, which makes the former less machineable and more difficult to divide and round (Barros et al. 2010; Schmiele et al. 2012). The increased dough strength with the addition of whole-grain flour was also observed for the dough system in crackers (Li et al. 2014) and layer cake batters (Gómez et al. 2010).

Most of the studies reported that the incorporation of barley, oat, rye, triticale, tritordeum, finger millet flour, whole-grain wheat flour, or bran in bread, layer cake and muffin formulations diminishes product quality, reflected in a decrease in volume accompanied by an increase in firmness and increase in crumbliness compared to products baked with refined wheat flour (Gómez et al. 2010; Rajiv et al. 2011; Rieder et al. 2012; Schmiele et al. 2012). A lower specific volume of the healthy grain bread is mostly ascribed to the dilution of gluten and weakening of the gluten network formed in the presence of fiber (Schmiele et al. 2012). This deteriorating effect of the application of healthy cereals in bread making is, however, more than the effect of wheat gluten dilution. Fibers, especially insoluble ones, may mechanically interfere with gluten network formation (Gill et al. 2002; Salmenkallio-Marttila et al. 2001) and cause the rupture of gas cells (Courtin and Delcour 2002). Zhang and Moore (1999) suggested that the loaf-volume-depressing effect of fibrous materials was the result of reduced gas retention rather than reduced gas formation. According to Barros et al. (2010), whole wheat tortillas, despite the increased amount of water needed during dough making compared with that needed for refined wheat tortillas, had less moisture content, owing to their thinner and more porous structure, which allowed more steam to escape during baking. The nuclear magnetic resonance (NMR) measurements on the dough system of whole wheat crackers indicated that the water migrated from the gluten network into the arabinoxylan matrix, thus inhibiting the hydration of gluten proteins, which produced excessively firm dough that was unable to expand and retain gas during fermentation and baking (Li et al. 2014). Consequently, the lower specific bread volume mostly results in greater crumb firmness, due to its more compact and less porous structure (Schmiele et al. 2012). This was confirmed for buckwheat-enriched gluten-free bread (Torbica et al. 2010) as well as wheat bread containing amaranth flour (Sanz-Penella et al. 2013), where the increase in buckwheat and amaranth flour fractions led to an increase in crumb firmness.

However, there are a few studies reporting the positive effect of fibers, such as β-glucans and arabinoxylans, on the volume of composite breads. According to Skendi et al. (2010), the addition of high MW β-glucan to wheat flours with weak gluten quality improved the volume of the breads made with the optimal addition of water. A similar effect was observed for wheat breads containing water-extractable arabinoxylans, due to the contribution of arabinoxylans to increasing the viscosity of the dough in liquid phase and thereby stabilizing the gas cells (Courtin and Delcour 2002).

5.4.2 Properties of Pasta Products

Pasta is the second most consumed food after bread (Khan et al. 2014). The term "pasta" is used to describe a food group that comprises noodles, spaghetti, and similar commodities (Brennan 2013). Traditionally, pasta is produced by mixing and forming durum wheat semolina and water with the aid of cold extrusion, followed by drying (Heiniö et al. 2016). As opposed to pasta products, noodles are usually made from common wheat flour by a process of sheeting and cutting (Aydin and Gocmen 2011).

Pasta quality is dependent upon the hydration of the starch and protein fractions during the mixing stage and the formation of a cohesive protein–starch matrix (Brennan 2013; Heiniö et al. 2016). The incorporation of whole-grain cereals mostly negatively affects pasta quality due to the following effects: (i) whole-grain ingredients have poor homogeneity in size, thus influencing the formation of isolated pockets of materials, which interfere with the protein–starch matrix development; and (ii) the high water-binding capacity of the fiber present in whole-grain cereals leads to the uneven hydration of the semolina (Brennan 2013; Tudorica et al. 2002). For instance, a study by Vignola et al. (2018) illustrated that whole-grain flour pasta did not show the same technological quality as white flour pasta, i.e. the whole-grain flour pasta showed a shorter optimal cooking time and higher hardness than the white flour pasta samples. Research by Gauthier et al. (2006) and Tudorica et al. (2002) has shown that the incorporation of bran into durum wheat semolina led to pasta products that were brittle and sticky, and exhibited high cooking losses. In most cases, it was reported that the application of whole-grain cereals in pasta making led to a high level of cooking loss from the pasta. The increased cooking loss of durum wheat pasta and noodles with the addition of the β-glucan fiber fraction from barley and oat flour were also noticed by Cleary and Brennan (2006) and Aydin and Gocmen (2011), respectively. This represents a great technological problem, since high cooking losses indicate an increase in the amount of starch and protein solubilized in the cooking water, a weakening of the pasta structure, an increase in pasta surface stickiness, and disintegration on cooking (Brennan 2013). The inclusion of healthy cereals also alters the cooked pasta texture. Vignola et al. (2018) reported that the hardness of whole wheat pasta was higher than that of refined wheat pasta. However, the studies by Cleary and Brennan (2006) and Khan et al. (2014) demonstrated a decrease in the hardness of durum wheat pasta upon the addition of β-glucan from barley as well as whole-grain red sorghum flour and white sorghum flour respectively. The observed decrease in pasta hardness was associated with the disruption of the structural integrity of the pasta and a reduction in gluten content due to the addition of high fiber ingredients (Khan et al. 2014).

5.4.3 Properties of Extruded Products

Extrusion is a high-temperature short-time process during which the material is subjected to a combination of moisture, pressure, temperature, and mechanical shear (Gómez et al. 2011). Replacing refined flours by whole grains in extruded products, such as breakfast cereals or savory snacks, results in a reduced product size due to a lower expansion at the die exit (Camire 2004; Robin et al. 2012), followed by an increase in hardness and density along with a reduction in crispiness (Sozer and Poutanen 2013). The changes in the elastic properties of the starchy melt during the extrusion cooking, due to the low water-holding capacity of the bran fraction compared to starch, may partially explain the reduced expansion and hard/crunchy texture of extruded whole-grain flours compared to refined flours (Robin et al. 2011; Schaffer-Lequart et al. 2017). Sibakov et al. (2015) showed that the lower expansions of whole-grain and bran-enriched oat flour extrudates were related to the higher content of insoluble fiber in these fractions. In contrast, the addition of water-soluble oat bran concentrates enhanced expansion and resulted in a less hard texture. Beside insoluble fibers, lipids are also identified as components that restrict expansion. Several studies showed that a highly expanded oat-based extruded product is hard to produce, due to a high content of lipids and fiber (Yao et al. 2006, 2011).

5.4.4 Properties of Flour Confectionery Products

Compared with bread dough, cookie dough contains a lower amount of water and high amounts of sugar and fat. In this dough system, the gluten is in underdeveloped state (Heiniö et al. 2016). This means that gluten development is not considered favorable in cookie making since it would result in tougher cookies with reduced spread (Coleman et al. 2013). The cookie spread ratio is the main quality attribute of cookies. The incorporation of wheat bran or whole-grain flour into the cookie formulation mostly leads to an increase in flour WA capacity, which consequently leads to a decrease in cookie spread (Coleman et al. 2013; Haque et al. 2002; Heiniö et al. 2016). However, it was shown that the replacement of wheat flour with teff, amaranth, whole-grain finger millet, or sorghum resulted in an increase in cookie spread (Chauhan et al. 2016; Coleman et al. 2013; Handa et al. 2010). Dapčević Hadnađev et al. (2013) related the increase in the spread of buckwheat-enriched gluten-free cookie dough with the decrease in dough elasticity arising from the addition of buckwheat flour to refined rice flour. Moreover, most of the studies reported decreases in cookie hardness linked with increases in the amount of healthy cereals in the cookie formulations (Chauhan et al. 2016; Dapčević Hadnađev et al. 2013; Handa et al. 2010).

According to the aforementioned, the incorporation of healthy grain material into bread, cake, cookies, pasta, and extruded products mostly has a detrimental impact on the rheological and structural indices of the cereal food matrices. This indicates that, despite being nutritionally sound, the manufacturing of products containing healthy grains requires ingredient and process modifications in order to get products of comparable quality to the products prepared from refined wheat flour.

5.5 Technological Challenges in the Production of Healthy Grain Products

The strategies to overcome the technological challenges related to the incorporation of healthy cereals into grain-based products can be divided into (i) those related to the modification of ingredients and (ii) those related to the modification of processes. Grains can be modified by germination, which, beside the effects on the nutritional properties, influences the physical properties of the final product. According to Singkhornart et al. (2014), germination combined with CO_2 extrusion altered the structure, texture, and nutritional properties of whole-grain wheat extrudates, due to the degradation of the starch molecule as well as other chemical components by endoenzymatic action during the germination process. The modification of flour milling fractions in terms of particle size reduction or enzymatic modification was also employed to improve the techno-functionality of a product. It was shown that particle size reduction of whole-grain wheat or bran: (i) resulted in extrudates with better texture and higher radial expansion than coarse bran (Santala et al. 2014), and the ones with thinner cell walls and more expanded having crisp structure (Alam et al. 2014), (ii) reduced differences in rheological and baking properties between refined wheat flour cracker and whole-grain wheat cracker (Wang et al. 2016), (iii) decreased cooking loss, increased hardness, cohesiveness, and resilience of cooked noodles (Niu et al. 2014). The modification of bran by hydrolytic enzymes in a low-moisture process increased the crispiness and reduced the hardness and density of the bran-supplemented expanded extrudates, due to increased level of water extractable arabinoxylans and decreased water holding capacity of the bran (Santala et al. 2014). Transglutaminase treatment has also been applied in the gluten-free formulations in order to improve the quality of gluten-free breads by promoting a protein network formation (Gujral and Rosell 2004; Renzetti et al. 2008). The improving effect of xylanase in nonwheat bread formulation was also shown and attributed to the conversion of water-insoluble into soluble arabinoxylans with water binding capacity due to which dough firmness is decreased, bread volume increased and bread is characterized by finer and more uniform crumb (Pojić et al. 2017).

Innovative processing technologies such as high-pressure (HP) treatment have been also widely employed to improve the functional properties of gluten-free cereals. An increase in viscoelastic properties of HP-treated buckwheat, white rice and teff batters was observed by Vallons et al. (2011), and explained by the modifications occurring in the structure of biopolymers such as proteins and starch. It was also shown that HP treatment of sorghum flour increased the batter consistency, while the replacement of untreated sorghum flour with up to 2% HP treated (at 600 MPa) flour resulted in delayed bread staling (Vallons et al. 2010). According to Gómez et al. (2011) the application of extrusion process to bran led to modified dough rheology and did not negatively affect bread quality. On the contrary, it even improved the quality of breads with bran when improvers are added.

Process modification, such as application of sourdough fermentation, can also be employed to improve the quality of healthy cereal products. Sourdough fermentation resulted in improved dough structure, loaf volume and texture of whole-grain barley breads (Rieder et al. 2012), while fermentation of bran with dextran producing *Weissella confusa* decreased hardness and increased crispiness of extruded snacks (Nikinmaa et al. 2017).

5.6 Conclusion

Cereal science abounds with versatile solutions to enhance and preserve the content of functional ingredients in cereal-based products. Given that the cereal-based food represent a global dietary staple and the fact that the cereal and pseudocereal nutrients have proven its potential in health promotion and disease prevention, the aspiration of cereal and food scientists to offer new cereal/pseudocereal-based food solutions is justified. However, the application of the majority of the proposed solutions and technologies in the food industry is still limited and lacks the cost-effectiveness and the consumer acceptability assessment. In order to overcome those challenges and improve the marketability of the scientifically offered solutions, both scientific community and industry should strive for the wider engagement of consumers in mutual value creation and product innovation.

Acknowledgment

This work was financially supported by the Ministry of Education, Science and Technological Development of the Republic of Serbia (No. 451-03-68/2020-14/200222).

References

Abdel-Aala, E.-S.M., Huclb, P., and Rabalskia, I. (2018). Compositional and antioxidant properties of anthocyanin-rich products prepared from purple wheat. *Food Chemistry* 254: 13–19.

Adlercreutz, H. (2007). Lignans and human health. *Critical Reviews in Clinical Laboratory Sciences* 44: 483–525.

Adom, K.K. and Liu, R.H. (2002). Antioxidant activity of grains. *Journal of Agricultural and Food Chemistry* 50: 6182–6187.

Akhtar, S., Anjum, F.M., and Anjum, A. (2011). Micronutrient fortification of wheat flour: recent development and strategies. *Food Research International* 44 (3): 652–659.

Akter, S., Sasaki, H., Ikeda, Y. et al. (2018). γ-Oryzanol ameliorates the acute stress induced by behavioral anxiety testing in mice. *Journal of Pharmacological Sciences* 138 (2): 155–159.

Alam, S.A., Järvinen, J., Kirjoranta, S. et al. (2014). Influence of particle size reduction on structural and mechanical properties of extruded rye bran. *Food and Bioprocess Technology* 7 (7): 2121–2133.

Alongi, M., Melchior, S., and Anese, M. (2019). Reducing the glycemic index of short dough biscuits by using apple pomace as a functional ingredient. *LWT - Food Science and Technology* 100: 300–305.

Alvarez-Jubete, L., Arendt, E.K., and Gallagher, E. (2009). Nutritive value and chemical composition of pseudocereals as gluten-free ingredients. *International Journal of Food Sciences and Nutrition* 60 (sup4): 240–257.

Andersson, A.A.M., Kamal-Eldin, A., Fras, A. et al. (2008). Alkylresorcinols in wheat varieties in the HEALTHGRAIN diversity screen. *Journal of Agricultural and Food Chemistry* 56: 9722–9725.

Andreasen, M.F., Kroon, P.A., Williamson, G. et al. (2001). Intestinal release and uptake of phenolic antioxidant diferulic acids. *Free Radical Biology & Medicine* 31: 304–314.

Angelino, D., Berhow, M., Ninfali, P. et al. (2013). Caecal absorption of vitexin-2-O-xyloside and its aglycone apigenin, in the rat. *Food & Function* 4: 1339–1345.

Angelino, D., Cossu, M., Marti, A. et al. (2017). Bioaccessibility and bioavailability of phenolic compounds in bread: a review. *Food & Function* 8: 2368–2393.

Anson, N.M., Hemery, Y.M., Basta, A. et al. (2012). Optimizing the bioactive potential of wheat bran by processing. *Food & Function* 3: 362–375.

Aydin, E. and Gocmen, D. (2011). Cooking quality and sensorial properties of noodle supplemented with oat flour. *Food Science and Biotechnology* 20 (2): 507–511.

Banu, I., Vasilean, I., and Aprodu, I. (2010). Evaluation of rheological behaviour of whole rye and buckwheat blends with whole wheat flour using Mixolab. *Italian Journal of Food Science* 22 (1): 83–89.

Barros, F., Alviola, J.N., and Rooney, L.W. (2010). Comparison of quality of refined and whole wheat tortillas. *Journal of Cereal Science* 51 (1): 50–56.

Bashir, K.M.I. and Choi, J.S. (2017). Clinical and physiological perspectives of β-glucans: the past, present, and future. *International Journal of Molecular Sciences* 18 (9): 1906.

Bastos, R., Coelho, E., and Coimbra, M.A. (2018). Arabinoxylans from cereal by-products: insights into structural features, recovery, and applications. In: *Sustainable Recovery and Reutilization of Cereal Processing By-Products* (ed. C.M. Galanakis), 227–251. Woodhead Publishing.

Bell, S., Goldman, V.M., Bistrian, B.R. et al. (1999). Effect of beta-glucan from oats and yeast on serum lipids. *Critical Reviews in Food Science and Nutrition* 39: 189–202.

Benítez, V., Esteban, R.M., Moniz, E. et al. (2018). Breads fortified with wholegrain cereals and seeds as source of antioxidant dietary fibre and other bioactive compounds. *Journal of Cereal Science* 82: 113–120.

Bhatty, R.S. (1993). Nonmalting uses of barley. In: *Barley Chemistry and Technology* (eds. A.W. MacGregor and R.S. Bhatty), 355–417. St. Paul, MN: American Association of Cereal Chemists.

Biró, B., Fodor, R., Szedljak, I. et al. (2019). Buckwheat-pasta enriched with silkworm powder: technological analysis and sensory evaluation. *LWT - Food Science and Technology* 116: 108542.

Bohn, T., McDougall, G.J., Alegría, A. et al. (2015). Mind the gap-deficits in our knowledge of aspects impacting the bioavailability of phytochemicals and their metabolites – a position paper focusing on carotenoids and polyphenols. *Molecular Nutrition & Food Research* 59: 1307–1323.

Bohn, T., Carriere, F., Day, L. et al. (2018). Correlation between in vitro and in vivo data on food digestion. What can we predict with static in vitro digestion models? *Critical Reviews in Food Science and Nutrition* 58 (13): 2239–2261.

Bolarinwa, I.F., Aruna, T.E., and Raji, A.O. (2019). Nutritive value and acceptability of bread fortified with moringa seed powder. *Journal of the Saudi Society of Agricultural Sciences* 18 (2): 195–200.

Bondia-Pons, I., Aura, A.-M., Vuorela, S. et al. (2009). Rye phenolics in nutrition and health. *Journal of Cereal Science* 49: 323–336.

Brennan, C.S. (2013). Fibre-enriched and wholewheat pasta. In: *Fibre-Rich and Wholegrain Foods* (eds. J.A. Delcour and K. Poutanen), 273–290. Cambridge: Woodhead Publishing Ltd.

Brennan, C.S. and Cleary, L.J. (2005). The potential use of cereal (1-3, 1-4)-β-D-glucans as functional food ingredients. *Journal of Cereal Science* 42: 1–13.

Bryszewska, M.A., Tomás-Cobos, L., Gallego, E. et al. (2019). In vitro bioaccessibility and bioavailability of iron from breads fortified with microencapsulated iron. *LWT - Food Science and Technology* 99: 431–437.

Caballero Valcárcel, A.M., Martínez Graciá, C., Martínez Miró, S. et al. (2019). Iron bioavailability of four iron sources used to fortify infant cereals, using anemic weaning pigs as a model. *European Journal of Nutrition* 58 (5): 1911–1922.

Camire, M.E. (2004). Technological challenges of whole grains. *Cereal Foods World* 49 (1): 22–22.

Cauvain, S.P. (2019). Reduced salt and sodium in bread and other baked products. In: *Reducing Salt in Foods*, 2e (eds. C. Beeren, K. Groves and P.M. Titoria), 213–229. Woodhead Publishing.

Cebeci, F. and Sahin-Yesilcubuk, N. (2014). The matrix effect of blueberry, oat meal and milk on polyphenols, antioxidant activity and potential bioavailability. *International Journal of Food Sciences and Nutrition* 65: 69–78.

Cedola, A., Cardinali, A., D'Antuono, I. et al. (2020). Cereal foods fortified with by-products from the olive oil industry. *Food Bioscience* 33: 100490.

Charalampopoulos, D., Wang, R., Pandiella, S.S. et al. (2002). Application of cereals and cereal components in functional foods: a review. *International Journal of Food Microbiology* 79: 131–141.

Chauhan, A., Saxena, D.C., and Singh, S. (2016). Physical, textural, and sensory characteristics of wheat and amaranth flour blend cookies. *Cogent Food & Agriculture* 2 (1): 1125773.

Cilla, A., Bosch, L., Barberá, R. et al. (2018). Effect of processing on the bioaccessibility of bioactive compounds – a review focusing on carotenoids, minerals, ascorbic acid, tocopherols and polyphenols. *Journal of Food Composition and Analysis* 68: 3–15.

Cleary, L. and Brennan, C. (2006). The influence of a (1→ 3)(1→ 4)-β-D-glucan rich fraction from barley on the physico-chemical properties and in vitro reducing sugars release of durum wheat pasta. *International Journal of Food Science & Technology* 41 (8): 910–918.

Coda, R., Rizzello, C.G., and Gobbetti, M. (2010). Use of sourdough fermentation and pseudo-cereals and leguminous flours for the making of a functional bread enriched of γ-aminobutyric acid (GABA). *International Journal of Food Microbiology* 137 (2–3): 236–245.

Coleman, J., Abaye, A.O., Barbeau, W. et al. (2013). The suitability of teff flour in bread, layer cakes, cookies and biscuits. *International Journal of Food Sciences and Nutrition* 64 (7): 877–881.

Courtin, C.M. and Delcour, J.A. (2002). Arabinoxylans and endoxylanases in wheat flour bread-making. *Journal of Cereal Science* 35 (3): 225–243.

Cox, D.N., Evans, G., and Lease, H.J. (2011). The influence of product attributes, consumer attitudes and characteristics on the acceptance of: (1) novel bread and milk, and dietary supplements and (2) fish and novel meats as dietary vehicles of long chain omega 3 fatty acids. *Food Quality and Preference* 22 (2): 205–212. https://doi.org/10.1016/j.foodqual.2010.10.003.

Dall'Asta, M., Bresciani, L., Calani, L. et al. (2016). In vitro bioaccessibility of phenolic acids from a commercial aleurone-enriched bread compared to a whole grain bread. *Nutrients* 8 (1): 42.

Dapčević Hadnađev, T.R., Torbica, A.M., and Hadnađev, M.S. (2013). Influence of buckwheat flour and carboxymethyl cellulose on rheological behaviour and baking performance of gluten-free cookie dough. *Food and Bioprocess Technology* 6 (7): 1770–1781.

Dapčević Hadnađev, T., Hadnađev, M., Pojić, M. et al. (2015). Functionality of OSA starch stabilized emulsions as fat replacers in cookies. *Journal of Food Engineering* 167, Part B: 133–138.

Dapčević-Hadnađev, T., Hadnađev, M., and Pojić, M. (2018). The healthy components of cereal by-products and their functional properties. In: *Sustainable Recovery and Reutilization of Cereal Processing By-Products* (ed. C.M. Galanakis), 27–61. Woodhead Publishing.

Delcour, L.A., Rouau, X., Courtin, C.M. et al. (2012). Technologies for enhanced exploitation of the health-promoting potential of cereals. *Trends in Food Science & Technology* 25: 78–86.

Diana, M., Rafecas, M., and Quílez, J. (2014). Free amino acids, acrylamide and biogenic amines in gamma-aminobutyric acid enriched sourdough and commercial breads. *Journal of Cereal Science* 60 (3): 639–644.

Dornez, E., Gebruers, K., Delcour, J. et al. (2009). Grain-associated xylanases: occurrence, variability, and implications for cereal processing. *Trends in Food Science & Technology* 20: 495–510.

Đurović, S., Vujanović, M., Radojković, M. et al. (2020). The functional food production: application of stinging nettle leaves and its extracts in the baking of a bread. *Food Chemistry* 312: 126091.

Dziki, D., Rózyło, R., Gawlik-Dziki, U. et al. (2014). Current trends in the enhancement of antioxidant activity of wheat bread by the addition of plant materials rich in phenolic compounds. *Trends in Food Science and Technology* 40: 48–61.

Fardet, A. (2014). Grain-based products, food structure and health potential: holism vs reductionism. *Journal of Nutritional Health & Food Engineering* 1 (1): 00001.

Fendri, L.B., Chaari, F., Maaloul, M. et al. (2016). Wheat bread enrichment by pea and broad bean pods fibers: effect on dough rheology and bread quality. *LWT - Food Science and Technology* 73: 584–591.

Fernandes, S.S. and de las Mercedes Salas-Mellado, M. (2017). Addition of chia seed mucilage for reduction of fat content in bread and cakes. *Food Chemistry* 227: 237–244.

Fernández-García, E., Carvajal-Lérida, I., and Pérez-Gálvez, A. (2009). In vitro bioaccessibility assessment as a prediction tool of nutritional efficiency. *Nutrition Research* 29: 751–760.

Filipčev, B., Šimurina, O., Sakač, M. et al. (2011). Feasibility of use of buckwheat flour as an ingredient in ginger nut biscuit formulation. *Food Chemistry* 125 (1): 164–170.

Foschia, M., Peressini, D., Sensidoni, A. et al. (2013). The effects of dietary fibre addition on the quality of common cereal products. *Journal of Cereal Science* 58: 216–227.

Frossard, E., Bucher, M., Mächler, F. et al. (2000). Potential for increasing the content and bioavailability of Fe, Zn and Ca in plants for human nutrition. *Journal of the Science of Food and Agriculture* 80: 861–879.

Fu, J.-T., Chang, Y.-H., and Shiau, S.-Y. (2015). Rheological, antioxidative and sensory properties of dough and Mantou (steamed bread) enriched with lemon fiber. *LWT - Food Science and Technology* 61 (1): 56–62.

Galanakis, C. (2017). Introduction. In: *Nutraceutical and Functional Food Components* (ed. C.M. Galanakis), 1–14. Academic Press.

Gallagher, E., O'Brien, C.M., Schanell, A.G.H., and Arendt, E.A. (2003). Evaluation of sugar replacers in short dough biscuit production. *Journal of Food Engineering* 56: 261–263.

Gani, A., Wani, S.M., Masoodi, F.A. et al. (2012). Whole-grain cereal bioactive compounds and their health benefits: a review. *Journal of Food Processing & Technology* 3: 146.

Garrod, M.G., Buchholz, B.A., Miller, J.W. et al. (2019). Vitamin B12 added as a fortificant to flour retains high bioavailability when baked in bread. *Nuclear Instruments and Methods in Physics Research Section B: Beam Interactions with Materials and Atoms* 438: 136–140.

Gauthier, J., Gélinas, P., and Beauchemin, R. (2006). Effect of stone-milled semolina granulation on the quality of bran-rich pasta made from khorasan (Kamut®) and durum wheat. *International Journal of Food Science & Technology* 41: 596–599.

Gill, S., Vasanthan, T., Ooraikul, B. et al. (2002). Wheat bread quality as influenced by the substitution of waxy and regular barley flours in their native and extruded forms. *Journal of Cereal Science* 36 (2): 219–237.

Gobetti, M., Rizzello, C.G., Di Cagno, R. et al. (2014). How the sourdough may affect the functional features of leavened baked goods. *Food Microbiology* 37: 30–40.

Gökmen, V., Mogol, B.A., Lumaga, R.B., and Shimoni, E. (2011). Development of functional bread containing nanoencapsulated omega-3 fatty acids. *Journal of Food Engineering* 105 (4): 585–591. https://doi.org/10.1016/J.Jfoodeng.2011.03.021.

Goloran, J.B., Johnson-Beebout, S.E., Morete, M.J. et al. (2019). Grain Zn concentrations and yield of Zn-biofortified versus Zn-efficient rice genotypes under contrasting growth conditions. *Field Crops Research* 234: 26–32.

Gómez, M., Manchón, L., Oliete, B. et al. (2010). Adequacy of wholegrain non-wheat flours for layer cake elaboration. *LWT - Food Science and Technology* 43 (3): 507–513.

Gómez, M., Jiménez, S., Ruiz, E. et al. (2011). Effect of extruded wheat bran on dough rheology and bread quality. *LWT - Food Science and Technology* 44 (10): 2231–2237.

Gujral, H.S. and Rosell, C.M. (2004). Functionality of rice flour modified with a microbial transglutaminase. *Journal of Cereal Science* 39 (2): 225–230.

Handa, C., Goomer, S., and Siddhu, A. (2010). Effects of whole-multigrain and fructoligosaccharide incorporation on the quality and sensory attributes of cookies. *Food Science and Technology Research* 17 (1): 45–54.

Haque, M., Shams-Ud-Din, M., and Haque, A. (2002). The effect of aqueous extracted wheat bran on the baking quality of biscuit. *International Journal of Food Science & Technology* 37 (4): 453–462.

Hassan, O.M.S., Di Folco, U., Nardone, M.R. et al. (2020). Fiber enrichment of pasta: metabolic effects and diet adherence in obese subjects. *Mediterranean Journal of Nutrition and Metabolism* 13 (1): 53–62.

Heiniö, R.L., Noort, M.W.J., Katina, K. et al. (2016). Sensory characteristics of wholegrain and bran-rich cereal foods – a review. *Trends in Food Science & Technology* 47: 25–38.

Hemalatha, S., Platel, K., and Srinivasan, K. (2007). Influence of heat processing on the bioaccessibility of zinc and iron from cereals and pulses consumed in India. *Journal of Trace Elements in Medicine and Biology* 21 (1): 1–7.

Hemery, Y.M., Anson, N.M., Havenaar, R. et al. (2010). Dry-fractionation of wheat bran increases the bioaccessibility of phenolic acids in breads made from processed bran fractions. *Food Research International* 43: 1429–1438.

Hithamani, G. and Srinivasan, K. (2014a). Effect of domestic processing on the polyphenol content and bioaccessibility in finger millet (*Eleusine coracana*) and pearl millet (*Pennisetum glaucum*). *Food Chemistry* 164: 55–62.

Hithamani, G. and Srinivasan, K. (2014b). Bioaccessibility of polyphenols from wheat (*Triticum aestivum*), sorghum (*Sorghum bicolor*), green gram (*Vigna radiate*), and chickpea (*Cicer arietinum*) as influenced by domestic food processing. *Journal of Agricultural and Food Chemistry* 62: 11170–11179.

Hoehnel, A., Axel, C., Bez, J. et al. (2019). Comparative analysis of plant-based high-protein ingredients and their impact on quality of high-protein bread. *Journal of Cereal Science* 89: 102816.

Holtekjølen, A.K., Olsen, H.H.R., Færgestad et al. (2008). Variations in water absorption capacity and baking performance of barley varieties with different polysaccharide content and composition. *LWT - Food Science and Technology* 41 (10): 2085–2091.

Hüttner, E.K., Dal Bello, F., and Arendt, E.K. (2010). Rheological properties and bread making performance of commercial wholegrain oat flours. *Journal of Cereal Science* 52 (1): 65–71.

Jacobs, D.R. Jr., Marquart, L., Slavin, J. et al. (1998). Whole-grain intake and cancer: an expanded review and meta-analysis. *Nutrition and Cancer* 30: 85–96.

Jacobs, D.R. Jr., Pereira, M.A., Stumpf, K. et al. (2002). Whole grain food intake elevates serum enterolactone. *The British Journal of Nutrition* 88: 111–116.

Jambrec, D., Sakač, M., Jovanov, P. et al. (2016). Effect of processing and cooking on mineral and phytic acid content of buckwheat-enriched tagliatelle. *Chemical Industry and Chemical Engineering Quarterly* 22: 319–326.

Katina, K., Juvonen, R., Laitila, A. et al. (2012). Fermented wheat bran as a functional ingredient in baking. *Cereal Chemistry* 89: 126–134.

Kaur, N., Kaur, G., and Mavi, G.S. (2020). Assessment of nutritional and quality traits in biofortified bread wheat genotypes. *Food Chemistry* 302: 125342.

Khan, I., Yousif, A.M., Johnson, S.K. et al. (2014). Effect of sorghum flour addition on in vitro starch digestibility, cooking quality, and consumer acceptability of durum wheat pasta. *Journal of Food Science* 79 (8): S1560.

Khanam, A. and Platel, K. (2016). Bioaccessibility of selenium: selenomethionine and selenocysteine from foods and influence of heat processing on the same. *Food Chemistry* 194: 1293–1299.

Lachman, J., Martinek, P., Kotíková, Z. et al. (2017). Genetics and chemistry of pigments in wheat grain – a review. *Journal of Cereal ScienceVolume* 74: 145–154.

Laguna, L., Vallons, K.J.R., Jurgens, A., and Sanz, T. (2013). Understanding the effect of sugar and sugar replacement in short dough biscuit. *Food and Bioprocess Technology* 6: 3143–3154.

Landberg, R., Kamal-Eldin, A., Andersson, R. et al. (2006). Alkylresorcinol content and homologue composition in durum wheat (*Triticum durum*) kernels and pasta products. *The Journal of Agricultural and Food Chemistry* 54: 3012–3014.

Lazaridou, A. and Biliaderis, C.G. (2007). Molecular aspects of cereal β-glucan functionality: physical properties, technological applications, and physiological effects. *Journal of Cereal Science* 46: 101–118.

Li, J., Hou, G.G., Chen, Z. et al. (2014). Studying the effects of whole-wheat flour on the rheological properties and the quality attributes of whole-wheat saltine cracker using SRC, alveograph, rheometer, and NMR technique. *LWT - Food Science and Technology* 55 (1): 43–50.

Li, L., Lietz, G., and Seal, C. (2018). Buckwheat and CVD risk markers: a systematic review and meta-analysis. *Nutrients* 10 (5): 619.

Lin, J., Teo, L.M., Leong, L.P., and Zhou, W. (2019). In vitro bioaccessibility and bioavailability of quercetin from the quercetin-fortified bread products with reduced glycemic potential. *Food Chemistry* 286: 629–635.

Littardi, P., Paciulli, M., Carini, E. et al. (2020). Quality evaluation of chestnut flour addition on fresh pasta. *LWT - Food Science and Technology* 126: 109303.

Lloyd, B.J., Siebenmorgen, T.J., and Beers, K.W. (2000). Effects of commercial processing on antioxidants in rice bran. *Cereal Chemistry* 77: 551–555.

Mandić, A., Sakač, M., and Mišan, A. (2011). Antinutrients in food. In: *Nutritional Insights and Food Safety* (ed. J. Švarc-Gajić), 1–37. New York: Nova Science Publishers.

Mateo Anson, N., Selinheimo, E., Havenaar, R. et al. (2009). Bioprocessing of wheat bran improves in vitro bioaccessibility and colonic metabolism of phenolic compounds. *Journal of Agricultural and Food Chemistry* 57: 6148–6155.

McDougall, G.J., Dobson, P., Smith, P. et al. (2005). Assessing potential bioavailability of raspberry anthocyanins using an in vitro digestion system. *Journal of Agricultural and Food Chemistry* 53: 5896–5904.

McRae, M.P. (2017). Health benefits of dietary whole grains: an umbrella review of meta-analyses. *Journal of Chiropractic Medicine* 16: 10–18.

Md Zaidul, I.S., Abd Karim, A., Manan, D.M.A. et al. (2004). A farinograph study on the viscoelastic properties of sago/wheat flour dough systems. *Journal of the Science of Food and Agriculture* 84 (7): 616–622.

Melini, V., Panfili, G., Fratianni, A., and Acquistucci, R. (2019). Bioactive compounds in rice on Italian market: pigmented varieties as a source of carotenoids, total phenolic compounds and anthocyanins, before and after cooking. *Food Chemistry* 277: 119–127.

Mendis, M. and Simsek, S. (2014). Arabinoxylans and human health. *Food Hydrocolloids* 42: 239–243.

Millar, K.A., Barry-Ryan, C., Burke, R. et al. (2019). Dough properties and baking characteristics of white bread, as affected by addition of raw, germinated and toasted pea flour. *Innovative Food Science & Emerging Technologies* 56: 102189.

Millen, B.E., Abrams, S., Adams-Campbell, L. et al. (2016). The 2015 dietary guidelines advisory committee scientific report: development and major conclusions. *Advances in Nutrition* 7: 438–444.

Miñarro, B., Albanell, E., Aguilar, N. et al. (2012). Effect of legume flours on baking characteristics of gluten-free bread. *Journal of Cereal Science* 56 (2): 476–481.

Mišan, A., Sakač, M., Medić, Đ. et al. (2016). Antioxidant and physicochemical properties of hydrogen peroxide-treated sugar beet dietary fibre. *Phytotherapy Research* 30: 855–860.

Mišan, A., Petelin, A., Stubelj, M. et al. (2017). Buckwheat – enriched instant porridge improves lipid profile and reduces inflammation in participants with mild to moderate hypercholesterolemia. *Journal of Functional Foods* 36: 186–194.

Monnet, A.-F., Laleg, K., Michon, C., and Micard, V. (2019). Legume enriched cereal products: a generic approach derived from material science to predict their structuring by the process and their final properties. *Trends in Food Science & Technology* 86: 131–143.

Moriano, M.E., Cappa, C., Casiraghi, M.C. et al. (2019). Reduced-fat biscuits: interplay among structure, nutritional properties and sensory acceptability. *LWT - Food Science and Technology* 109: 467–474.

Nedeljković, N., Hadnađev, M., Dapčević Hadnađev, T. et al. (2017). Partial replacement of fat with oat and wheat bran gels: optimization study based on rheological and textural properties. *LWT - Food Science and Technology* 86: 377–384.

Neuhouser, M.L. (2019). The importance of healthy dietary patterns in chronic disease prevention. *Nutrition Research* 70: 3–6.

Nikinmaa, M., Alam, S.A., Raulio, M. et al. (2017). Bioprocessing of bran with exopolysaccharide producing microorganisms as a tool to improve expansion and textural properties of extruded cereal foams with high dietary fibre content. *LWT - Food Science and Technology* 77: 170–177.

Niu, M., Hou, G.G., Wang, L., and Chen, Z. (2014). Effects of superfine grinding on the quality characteristics of whole-wheat flour and its raw noodle product. *Journal of Cereal Science* 60 (2): 382–388.

Olojede, A.O., Sanni, A.I., and Banwo, K. (2020). Effect of legume addition on the physiochemical and sensorial attributes of sorghum-based sourdough bread. *LWT - Food Science and Technology* 118: 108769.

Ötles, S. and Cagindi, Ö. (2006). Cereal based functional foods and nutraceuticals. *Acta Scientiarum Polonorum Technologica Alimentaria* 5 (1): 107–112.

Pasqualone, A., Caponio, F., Ambrogina Pagani, M. et al. (2019). Effect of salt reduction on quality and acceptability of durum wheat bread. *Food Chemistry* 289: 575–581.

Peressini, D., Cavarape, A., Brennan, M.A. et al. (2020). Viscoelastic properties of durum wheat doughs enriched with soluble dietary fibres in relation to pasta-making performance and glycaemic response of spaghetti. *Food Hydrocolloids* 102: 105613.

Perez-Ternero, C., de Sotomayor, M.A., and Herrera, M.D. (2017). Contribution of ferulic acid, γ-oryzanol and tocotrienols to the cardiometabolic protective effects of rice bran. *Journal of Functional Foods* 32: 58–71.

Pestorić, M., Filipčev, B., Mišan, A. et al. (2017). Food applications of medicinal plants: the use of "Vitalpant" herbal mixture in cereal-based products. In: *Medicinal Plants: Production, Cultivation and Uses* (eds. A. Matthias and N. Laisné), 135–210. New York: Nova Science Publishers, Inc.

Pojić, M., Dapčević Hadnađev, T., Hadnađev, M. et al. (2015). Bread supplementation with hemp seed cake: a by-product of hemp oil processing. *Journal of Food Quality* 38 (6): 431–440.

Pojić, M., Dapčević Hadnađev, T., Hadnađev, M. et al. (2017). Optimization of additive content and their combination to improve the quality of pure barley bread. *Journal of Food Science and Technology* 54 (3): 579–590.

Pontonio, E., Lorusso, A., Gobbetti, M., and Rizzello, C.G. (2017). Use of fermented milling by-products as functional ingredient to develop a low-glycaemic index bread. *Journal of Cereal Science* 77: 235–242.

Pontonio, E., Dingeo, C., Di Cagno, R. et al. (2020). Brans from hull-less barley, emmer and pigmented wheat varieties: from by-products to bread nutritional improvers using selected lactic acid bacteria and xylanase. *International Journal of Food Microbiology* 313: 108384.

Poutanen, K., Soze, N., and Della Valle, G. (2014). How can technology help to deliver more of grain in cereal foods for a healthy diet. *Journal of Cereal Science* 59: 327–336.

Purić, M., Rabrenović, B., Rac, V. et al. (2020). Application of defatted apple seed cakes as a by-product for the enrichment of wheat bread. *LWT - Food Science and Technology* 130: 109391. https://doi.org/10.1016/j.lwt.2020.109391.

Rajiv, J., Soumya, C., Indrani, D. et al. (2011). Effect of replacement of wheat flour with finger millet flour (*Eleusine Corcana*) on the batter microscopy, rheology and quality characteristics of muffins. *Journal of Texture Studies* 42 (6): 478–489.

Ramírez-Wong, B., Walker, C.E., Ledesma-Osuna, A.I. et al. (2007). Effect of flour extraction rate on white and red winter wheat flour compositions and tortilla texture. *Cereal Chemistry* 84 (3): 207–213.

Rebellato, A.P., Lima, J.C., Silva, J.G.S. et al. (2017). Mineral bioaccessibility in French breads fortified with different forms iron and its effects on rheological and technological parameters. *Journal of Cereal Science* 74: 56–63.

Renzetti, S., Dal Bello, F., and Arendt, E.K. (2008). Microstructure, fundamental rheology and baking characteristics of batters and breads from different gluten-free flours treated with a microbial transglutaminase. *Journal of Cereal Science* 48 (1): 33–45.

Rieder, A., Holtekjølen, A.K., Sahlstrøm, S. et al. (2012). Effect of barley and oat flour types and sourdoughs on dough rheology and bread quality of composite wheat bread. *Journal of Cereal Science* 55 (1): 44–52.

del Rio, D., Rodriguez-Mateos, A., Spencer, J.P. et al. (2013). Dietary (poly)phenolics in human health: structures, bioavailability, and evidence of protective effects against chronic diseases. *Antioxidants & Redox Signaling* 18: 1818–1892.

Rizzello, C.G., Nionelli, L., Coda, R. et al. (2010). Use of sourdough fermented wheat germ for enhancing the nutritional, texture and sensory characteristics of the white bread. *European Food Research and Technology* 230: 645–654.

Robin, F., Dubois, C., Pineau, N. et al. (2011). Expansion mechanism of extruded foams supplemented with wheat bran. *Journal of Food Engineering* 107: 80–89.

Robin, F., Dubois, C., Pineau, N. et al. (2012). Process, structure and texture of extruded whole wheat. *Journal of Cereal Science* 56 (2): 358–366.

da Rosa Machado, C. and Thys, R.C.S. (2019). Cricket powder (Gryllus assimilis) as a new alternative protein source for gluten-free breads. *Innovative Food Science & Emerging Technologies* 56: 102180.

Rosell, C.M. (2003). The nutritional enhancement of wheat flour. In: *Bread Making – Improving Quality* (ed. S.P. Cauvain), 253–269. Cambridge, England: CRC Press, Woodhead Publishing Ltd.

Ross, A.B., Kamal-Eldin, A., Lundin, E.A. et al. (2003a). Cereal alkylresorcinols are absorbed by humans. *Journal of Nutrition* 133: 2222–2224.

Ross, A.B., Shepherd, M.J., Schupphaus, M. et al. (2003b). Alkylresorcinols in cereals and cereal products. *The Journal of Agricultural and Food Chemistry* 51: 4111–4118.

Rubel, I.A., Pérez, E.E., Manrique, G.D., and Genovese, D.B. (2015). Fibre enrichment of wheat bread with Jerusalem artichoke inulin: effect on dough rheology and bread quality. *Food Structure* 3: 21–29.

Saikia, D. and Deka, S.C. (2011). Cereals: from staple food to nutraceuticals. *International Food Research Journal* 18: 21–30.

Sakač, M.B., Čanadanovic-Brunet, J.M., Mišan, A.Č. et al. (2010). Antioxidant activity of phytic acid in lipid model system. *Food Technology and Biotechnology* 48: 524–529.

Salmenkallio-Marttila, M., Katina, K., and Autio, K. (2001). Effects of bran fermentation on quality and microstructure of high-fiber wheat bread. *Cereal Chemistry* 78 (4): 429–435.

Santala, O., Kiran, A., Sozer, N. et al. (2014). Enzymatic modification and particle size reduction of wheat bran improves the mechanical properties and structure of bran-enriched expanded extrudates. *Journal of Cereal Science* 60 (2): 448–456.

Sanz-Penella, J.M., Wronkowska, M., Soral-Smietana, M. et al. (2013). Effect of whole amaranth flour on bread properties and nutritive value. *LWT - Food Science and Technology* 50 (2): 679–685.

Šarić, B., Mišan, A., Mandić, A. et al. (2016). Valorisation of raspberry and blueberry pomace through the formulation of value-added gluten-free cookies. *Journal of Food Science and Technology* 53 (2): 1140–1150. https://doi.org/10.1007/s13197-015-2128-1.

Saxena, D.C. and Rao, P.H. (1997). Influence of grinding method on some physicochemical characteristics of whole wheat flour and its suitability for the preparation of "tandoori roti". *Sciences des Aliments* 17: 183–191.

Schaffer-Lequart, C., Lehmann, U., Ross, A.B. et al. (2017). Whole grain in manufactured foods: current use, challenges and the way forward. *Critical Reviews in Food Science and Nutrition* 57 (8): 1562–1568.

Schmiele, M., Jaekel, L.Z., Patricio, S.M.C. et al. (2012). Rheological properties of wheat flour and quality characteristics of pan bread as modified by partial additions of wheat bran or whole grain wheat flour. *International Journal of Food Science & Technology* 47 (10): 2141–2150.

Sedej, I., Sakač, M., Mandić, A. et al. (2011). Assessment of antioxidant activity and rheological properties of wheat and buckwheat milling fractions. *Journal of Cereal Science* 54: 347–353.

Sibakov, J.K., Kirjoranta, S.J., Alam, S.A. et al. (2015). Effect of oat bran fractions on extrudates made of defatted oats. *Food and Bioprocess Technology* 8 (2): 445–458.

Sinesio, F., Raffo, A., Peparaio, M. et al. (2019). Impact of sodium reduction strategies on volatile compounds, sensory properties and consumer perception in commercial wheat bread. *Food Chemistry* 301: 125252.

Singkhornart, S., Edou-ondo, S., and Ryu, G.H. (2014). Influence of germination and extrusion with CO2 injection on physicochemical properties of wheat extrudates. *Food Chemistry* 143: 122–131.

Siró, I., Kápolna, E., Kápolna, B. et al. (2008). Functional food. Product development, marketing and consumer acceptance – a review. *Appetite* 51: 456–467.

Skendi, A., Biliaderis, C.G., Papageorgiou, M. et al. (2010). Effects of two barley β-glucan isolates on wheat flour dough and bread properties. *Food Chemistry* 119 (3): 1159–1167.

Slavin, J. (2004). Whole grains and human health. *Nutrition Research Reviews* 17 (1): 99–110.

Sozer, N. and Poutanen, K. (2013). Fibre in extruded food products. In: *Fibre-Rich and Wholegrain Foods* (eds. J.A. Delcour and K. Poutanen), 256–268. Cambridge: Woodhead Publishing Ltd.

Stokić, E., Mandić, A., Sakač, M. et al. (2015). Quality of buckwheat-enriched wheat bread and its antihyperlipidemic effect in statin treated patients. *LWT - Food Science and Technology* 63: 556–561.

Terao, J., Murota, K., and Kawai, Y. (2011). Conjugated quercetin glucuronides as bioactive metabolites and precursors of aglycone in vivo. *Food & Function* 2: 11–17.

Tiong, S.A., Chandra-Hioe, M.V., and Arcot, J. (2015). Thiamin fortification of bread-making flour: retention in bread and levels in Australian commercial fortified bread varieties. *Journal of Food Composition and Analysis* 38: 27–31.

Torbica, A., Hadnađev, M., and Dapčević, T. (2010). Rheological, textural and sensory properties of gluten-free bread formulations based on rice and buckwheat flour. *Food Hydrocolloids* 24 (6–7): 626–632.

Tudorica, C.M., Kuri, V., and Brennan, C.S. (2002). Nutritional and physicochemical characteristics of dietary fiber enriched pasta. *Journal of Agricultural and Food Chemistry* 50 (2): 347–356.

Valcárcel-Yamani, B. and Lannes, S.D.S. (2012). Applications of quinoa (*Chenopodium quinoa* Willd.) and amaranth (*Amaranthus* spp.) and their influence in the nutritional value of cereal based foods. *Food and Public Health* 2 (6): 265–275.

Vallons, K.J., Ryan, L.A., Koehler, P. et al. (2010). High pressure–treated sorghum flour as a functional ingredient in the production of sorghum bread. *European Food Research and Technology* 231 (5): 711–717.

Vallons, K.J., Ryan, L.A., and Arendt, E.K. (2011). Promoting structure formation by high pressure in gluten-free flours. *LWT - Food Science and Technology* 44 (7): 1672–1680.

Venkatakrishnan, K., Chiub, H.-F., and Wang, C.-K. (2019). Extensive review of popular functional foods and nutraceuticals against obesity and its related complications with a special focus on randomized clinical trials. *Food & Function* 10 (5): 2313–2329.

Vignola, M.B., Bustos, M.C., and Pérez, G.T. (2018). Comparison of quality attributes of refined and whole wheat extruded pasta. *LWT - Food Science and Technology* 89: 329–335.

Vinardell, P.M. and Mitjans, M. (2017). Lignins and their derivatives with beneficial effects on human health. *The International Journal of Molecular Sciences* 18: 1219.

Virág, D., Kiss, A., Forgó, P. et al. (2013). Study on Maillard-reaction driven transformations and increase of antioxidant activity in lysine fortified biscuits. *Microchemical Journal* 107: 172–177. https://doi.org/10.1016/j.microc.2012.06.018.

Vishwas, T., Anjana, S., and Tashfeen, A.M. (2018). Avenanthramides of oats: medicinal importance and future perspectives. *Pharmacognosy Reviews* 12: 66–71.

Walle, T., Browning, A.M., Steed, L.L. et al. (2005). Flavonoid glucosides are hydrolyzed and thus activated in the oral cavity in humans. *Journal of Nutrition* 135: 48–52.

Wang, T., He, F., and Chen, G. (2014). Improving bioaccessibility and bioavailability of phenolic compounds in cereal grains through processing technologies: a concise review. *Journal of Functional Foods* 7: 101–111.

Wang, N., Hou, G.G., Kweon, M. et al. (2016). Effects of particle size on the properties of whole-grain soft wheat flour and its cracker baking performance. *Journal of Cereal Science* 69: 187–193.

WHO (2018). Fact sheets: Healthy diet. https://www.who.int/en/news-room/fact-sheets/detail/healthy-diet

Whole Grains Council (2018). Definition of a whole grain. https://www.wholegrainscouncil.org/definition-whole-grain

Williamson, G. (2002). The use of flavonoid aglycones in in vitro systems to test biological activities: based on bioavailability data, is this a valid approach? *Phytochemistry Reviews* 1: 215–222.

Williamson, G. and Clifford, M.N. (2017). Role of the small intestine, colon and microbiota in determining the metabolic fate of polyphenols. *Biochemical Pharmacology* 139: 24–39.

Winiarska-Mieczan, A., Kowalczuk-Vasilev, E., Kwiatkowska, K. et al. (2019). Dietary intake and content of Cu, Mn, Fe, and Zn in selected cereal products marketed in Poland. *Biological Trace Element Research* 187 (2): 568–578.

Wood, P.J. (1992). Aspects of the chemistry and nutritional effects of non-starch polysaccharides of cereals. In: *Developments in Carbohydrate Chemistry* (eds. R.J. Alexander and H.F. Zobel), 293–314. St. Paul: American Association of Cereal Chemistry.

Xu, Z. and Godber, J.S. (1999). Purification and identification of components of γ-oryzanol in rice bran oil. *Journal of Agricultural and Food Chemistry* 47: 2724–2728.

Yao, N., Jannink, J., Alavi, S. et al. (2006). Physical and sensory characteristics of extruded products made from two oat lines with different β-glucan concentrations. *Cereal Chemistry* 83 (6): 692–699.

Yao, N., White, P.J., and Alavi, S. (2011). Impact of β-glucan and other oat flour components on physic-chemical and sensory properties of extruded oat cereals. *International Journal of Food Science & Technology* 46 (3): 651–660.

Zarzycki, P., Sykut-Domańska, E., Sobota, A. et al. (2020). Flaxseed enriched pasta-chemical composition and cooking quality. *Foods* 9 (4): 404.

Zhang, D. and Moore, W.R. (1997). Effect of wheat bran particle size on dough rheological properties. *Journal of the Science of Food and Agriculture* 74 (4): 490–496.

Zhang, D. and Moore, W.R. (1999). Wheat bran particle size effects on bread baking performance and quality. *Journal of the Science of Food and Agriculture* 79 (6): 805–809.

Zhang, L., Zuo, Z., and Lin, G. (2007). Intestinal and hepatic glucuronidation of flavonoids. *Molecular Pharmacology* 4: 833–845.

Zhao, J., Liu, X., Bai, X., and Wang, F. (2019). Production of biscuits by substitution with different ratios of yellow pea flour. *Grain & Oil Science and Technology* 2 (4): 91–96.

6
Sprouted Cereal Grains and Products

Alessandra Marti, Gaetano Cardone, and Maria Ambrogina Pagani

Department of Food, Environmental and Nutritional Sciences (DeFENS), Università degli Studi di Milano, Milan, Italy

6.1 Introduction

Cereals, pseudocereals, and pulses are good sources of energy as well as micro- and macronutrients. The uneven distribution of nutrients in the different parts of the grain, together with bioactive compounds, vitamins, and minerals in the external layers, make them natural choices for wholesome consumption (www.healthgrain.eu). In fact, whole-grains are unanimously considered important ingredients of a healthy diet, preventing various human pathologies, as reviewed by Jonnalagadda et al. (2011). Moreover, whole grains satisfy consumer demand for health-promoting foods that come directly from primary production or have been exposed to minimal processing. Therefore, it is worth considering the present and ever-growing interest in sprouted grains in Western countries. Indeed, in the past few decades, in developed and industrialized countries, germination or sprouting has been limited to producing malted cereals for the brewing industry (Finney 1982). However, in many African and Asian countries, germination has always been considered an economic as well as excellent process to improve the nutritional value, flavor, and taste of grains while, at the same time, decreasing the amount of anti-nutrients present in raw cereals and pulses. Moreover, until a few years ago, the sprouting of cereals in the field, prior to harvesting (i.e. pre-harvest sprouting, PHS), had been exclusively associated with negative consequences, such as decreasing the commercial value of cereals and making them too difficult to process for human food products.

Food manufacturers' growing interest in germination is accompanied by a proliferation of research on sprouted grains from various perspectives, including nutrition and food production. A review of the scientific literature of the last 20 years, using "sprouting/germination of pulses or cereals" as search terms, resulted in the identification of about 1285 scientific papers (Figure 6.1).

Innovative Processing Technologies for Healthy Grains, First Edition.
Edited by Milica Pojić and Uma Tiwari.

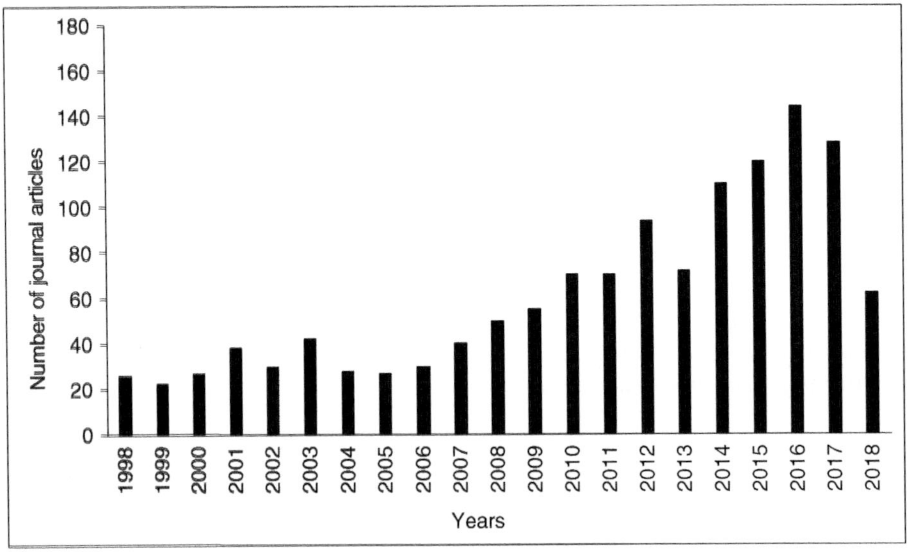

Figure 6.1 Number of scientific articles on sprouting (1998–2018). *Source*: FSTA (July 2018).

On conducting a systematic review, articles dealing with pre-harvest sprouting were excluded, as pre-harvest sprouting negatively affects product functionality. The number of contributions has doubled in the last four years, highlighting the growing interest in this topic ("sprouting/germination").

Awareness of the health benefits of sprouted grains might account for the increase, about 60%, in the number of products containing sprouted grains, which were launched on the market from 2013 to 2016 (Pagand et al. 2017). In particular, North America followed by Europe and Australia, influence these trends in the market. From January 2015 to April 2017, snacks represented 22% of all the products launched, followed by meals (19%) and bakery products (15%) (Pagand et al. 2017). Health benefits and improvements in flavor and texture that sprouted grains bring to food products constitute the driving forces behind the growing interest in this product category.

6.2 Definition

In scientific literature, the term "germination" is often used loosely and sometimes incorrectly, so it is important to clarify its meaning. The intent of this section is not to propose a definition, but to stimulate discussion with respect to the increasing interest in this topic.

Providing a generic definition of *germination* is difficult because the different actors that make up the chain (from physiologists to agronomists to seed testers, grain elevator technicians, food scientists and consumers) use different terms and meanings. The most common definition for this phenomenon is: "the process by which a plant grows from a seed" (Bewley and Black 1994). According to physiologists, *germination* begins with water uptake by the seed (imbibition) and ends with the start of the elongation of the embryonic axis, usually the radicle. Therefore, germination promotes the *de novo* synthesis of several hydrolytic enzymes necessary to depolymerize the cell–wall polysaccharides and mobilize the depolymerized storage macromolecules (Mares et al. 2016). Thus, germination does not include seedling growth, which commences when germination finishes (Bewley and Black 1994). According to this definition, processes occurring in the nascent seedling, such as the mobilization of the major storage reserves, are also not part of germination; rather they are post-germination events. However, since breeders and agronomists are interested in monitoring the establishment of a vigorous plant of agronomic value, they refer to germination as seedling emergence from soil, even if germination ends sometime before the seedling is visible (Bewley and Black 1994).

It is more difficult to define germinated grains as the last link in the value chain of grains, i.e. food industries and consumers. Indeed, currently no regulated definition of "sprouted grain" is available. The Whole Grains Council (www.wholegrainscouncil.org) suggests to consumers who want to understand what they are eating, and companies who are considering manufacturing or marketing sprouted grains, to start by reading how the American Association of Cereal Chemists International (AACCI) defines sprouted grains. In early 2008, the AACC International Board approved the following statement regarding sprouted grains: "*Malted or sprouted grains containing all of the original bran, germ, and endosperm shall be considered whole-grains as long as sprout growth does not exceed kernel length and nutrient values have not diminished. These grains should be labeled as malted or sprouted whole-grains*" (https://www.aaccnet.org/initiatives/definitions/Pages/WholeGrain.aspx).

In addition to the lack of definition, the terms "sprouting" "germination" and "malting" are often used as synonyms, generating confusion in some cases. From a botanical standpoint, there is no difference between sprouting and germination, but for most people sprouting is the practice of germinating seeds to obtain sprouts to be eaten raw or cooked. The germination or sprouting process is similar to malting, which is used extensively in the brewing and distilling industries. However, in the malting process the rootlet emergence, called "chitting" occurs prior to the end of steeping (Pyler and Thomas 2000). The germination phase is typically allowed to proceed for three to five days, during which time approximately 40–50% of the proteins should have been solubilized and high levels of starch-degrading enzymes should have been released (Izydorczyk and Dexter 2004).

Finally, the term "pre-harvest germination" or "pre-harvest sprouting" defines the biochemical changes that take place when cereals (mainly wheat) are exposed to prolonged wet or foggy conditions during their growth in the field. The consequent huge accumulation of hydrolytic enzymes (above all amylases) can impair the quantity and quality of wheat grains. For further details about the effects of pre-harvest sprouting on wheat kernel quality and bread-making performance, see the recent review by Olaerts and Courtin (2018).

In this chapter, the terms "sprouting" and "germination" will be interchangeably used for the process occurring after seed harvesting and under controlled processing conditions.

6.3 Mechanisms of Grain Germination

After harvesting, sound grains are characterized by moisture values of no higher than 14%, corresponding to water activity values of less than 0.60 (at 25 °C). Under these conditions, grains are in a dormant/quiescent state as no depolymerization activities by hydrolytic enzymes have yet occurred (Delcour and Hoseney 2010).

As mentioned in Section 6.2, the germination process begins with the soaking or steeping of the dry grains in water. During this phase, the water uptake is influenced by seed size, seed-coat permeability, the quantity of available water, the chemical composition of seeds, and concentration of solutes in solution. For germination at a faster rate, most cereal seeds require 20–30 °C as the optimal temperature range (Porter and Gawith 1999). This depends on the species, genetic differences, varietal variation, source of seeds, and age. During this phase, an extensive physiological and biochemical process begins to support plant growth (Figure 6.2). Briefly, the gibberellin hormone – that produces gibberellic acid – translocates from the embryo to the aleurone layer (1,2), promoting the synthesis and secretion of enzymes such as

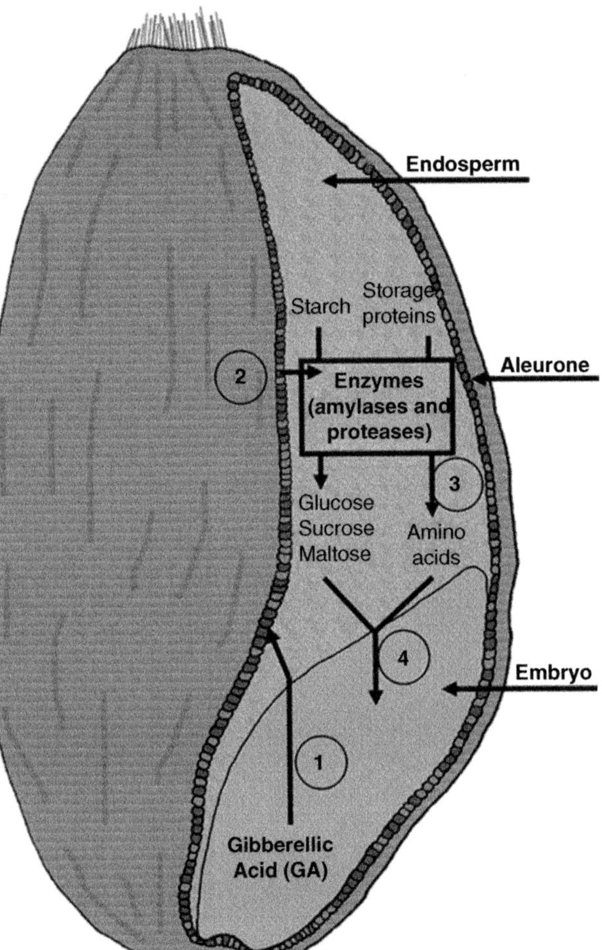

Figure 6.2 Sequence of main metabolic events during the sprouting process in the wheat kernel. (1),(2) Release of gibberellic acid from the embryo triggers the secretion of hydrolytic enzymes from the aleurone layer and the scutellum; (3) amylases and proteases hydrolyze starch – to glucose, sucrose, and maltose – and proteins to amino acids; and (4) these products are used by the embryo to support growth. *Source*: adapted from Nelson et al. (2013).

amylases and proteases to act on storage molecules in the endosperm, releasing simple sugars and amino acids, respectively (3). These products are used by the embryo to support growth (4).

6.3.1 Effect of Germination on the Carbohydrate Complex of Cereal Grains

α-amylases are endo-amylases that hydrolyse the α-(1,4)-linkages of starch, yielding soluble sugars and low molecular weight α-dextrins. β-amylases hydrolyze the α-(1,4)-linkages at the non-reducing ends of amylose and amylopectin molecules to produce β-maltose and β-limited dextrin (from amylopectin molecules). Sound wheat contains significant amounts of β-amylase, but little α-amylase. However, the β-amylases have little effect on undamaged, native starch granules, while their hydrolytic action is enhanced by that of α-amylase (Olaerts and Courtin 2018).

Germination induces starch hydrolysis to yield simple sugars by the increased activity of amylases (Van Hung et al. 2011a), while the extent of starch degradation depends upon the length of sprouting (Lorenz and Valvano 1981). Differences in the degree of starch degradation throughout the kernel were observed, with starch granules being more degraded near the aleurone layer and germ region, than in the inner endosperm (Faltermaier et al. 2015). Indeed, amylases are found principally in the pericarp layer and are responsible for the breakdown of starch during the early phases of development (Dedio et al. 1975). During germination, α-amylases are *de novo* synthesized in the scutellum and aleurone. Linked to other seed proteins, β-amylases initially are only partially present in a free or soluble form, while during germination, they are progressively released in a soluble form, presumably due to proteases secreted by the aleurone and/or scutellum (Ziegler 1995). As the starch is degraded by amylases, increased sucrose occurs during early wheat germination, with glucose and maltose predominating during later germination stages (Aoki et al. 2006).

6.3.2 Effect of Germination on the Protein Complex of Cereal Grains

Germination also induces the accumulation of proteolytic enzymes that catalyze the hydrolysis of peptide bonds in proteins. Proteases can be subdivided into two major groups according to their site of action: exopeptidases and endopeptidases. Exopeptidases, also referred to as proteinase, cleave the peptide bond proximal to the amino or carboxyl termini of the substrate, whereas endopeptidases cleave peptide bonds distant from the end of the substrate (Miguel et al. 2013). Storage proteins such as globulins, prolamins, and glutelins, after two to three days from the beginning of imbibition, undergo varying degrees of proteolysis that frees stored nitrogen and carbon for the growing plant during sprouting (Müntz 1996). In addition, endopeptidases might induce conformational changes that subsequently facilitate further breakdown by both endo- and exopeptidases (Müntz 1996).

Conflicting results have been reported for the effect of germination on protein content (Nelson et al. 2013). Some studies reported the decrease in protein content associated with the hydrolytic action of proteases (Koehler et al. 2007), while others reported the increase in protein content in whole-grain flours during germination associated with the loss of dry matter, mainly in the form of carbohydrates (Lorenz et al. 1981) or to the synthesis of enzymes during germination (Bau et al. 1997). However, a turnover of protein and non-protein nitrogen resulting in equilibrium of the degradation and synthesis processes during germination may account for the lack of changes (Bau et al. 1997). Thus, it is clear that the effects of germination depend on seeds from different plant species, varieties, or cultivars, as well as variations in germination conditions (temperature, light, moisture, and time of germination) (Yang et al. 2001).

The products released by starch and protein hydrolysis (i.e. soluble sugars and amino acids) are then absorbed by the germ, thereby transitioning the grain from dormancy to active metabolism (Figure 6.2).

Such biochemical changes deeply affect the native grain's composition to support the growth of a young plant, thus altering the nutritional (Danisova et al. 1994) and physicochemical (Noda et al. 2004) properties of the grain.

6.4 Nutritional Profile of Germinated Cereal Grains and Their Health Benefits

Germination has been carried out for millennia to improve the nutritional properties of cereals and pulses. Most studies dealing with germination have focused on how germination affects the specific cereal components having positive or negative nutritional effects, as recently reviewed by Hübner and Arendt (2013) and Singh et al. (2015). A comprehensive review of current research leads to a general conclusion: caution must be applied when comparing the results of different studies, since the types and varieties of grains, soaking conditions (e.g. water quality, soaking time, and temperature), germination conditions (e.g. duration and temperature), and measurement methods differ from one study to another. In addition, optimal germination conditions may vary with grain type and the compound of interest. Indeed, it has been reported that the concentration of the same compound may be ascribed as increased, decreased, or unchanged within the same type of grain or for different grains (Nelson et al. 2013). Moreover, many nutrients and bioactive compounds increase and then decrease as they are utilized by the growing plant (Yang et al. 2001), making the germination time an important variable. In some cases, the increase in nutrients could be due to the loss of starch during germination, which results in a decrease with respect to the weight of the grain, and thus in an increase of the other macromolecules (Van Hung et al. 2015).

Nelson et al. (2013) recently provided a comprehensive review of the phytochemical and health effects of germination on both cereals and pseudocereals. Most of the literature selected by the authors (about 60% of the papers) focused on wheat, which is the most common cereal for bread and baked products. The great attention paid to the nutritional effects of germinated barley (about 26% of the papers) is probably due to the brewing industry's interest in malt (Hübner and Arendt 2013). Likewise, germinated brown rice also gained interest due to its high level of γ-aminobutyric acid (GABA) content (Patil and Khan 2011). According to the authors, the amount of GABA in germinated brown rice was doubled when compared to that of unsprouted brown rice. They also observed that soaking for 3 hours and sprouting for 21 hours proved to be the optimum method for reaching the highest GABA content in germinated brown rice. The numerous health benefits of GABA are the main reason for the popularity of germinated brown rice. In this context, Okada et al. (2000) reported that a diet including GABA lowered blood pressure and decreased insomnia, and autonomic disorders observed during the menopausal or presenile periods. Similarly,

Kayahara and Tsukahara (2000) also showed that brown rice sprouts contain a potent inhibitor of the enzyme called protylendopetidase, which is involved in Alzheimer's disease.

Unexpectedly, very few studies addressed the health benefits of germinated sorghum and millet, although germination is commonly used in African countries to improve the protein digestibility of these less common cereals, due to the degradation of storage proteins that become more easily available for pepsin hydrolysis (Sehgal and Kawatra 2001). Among the various mechanisms proposed to explain the above-mentioned health benefits of germinated sorghum and millet (Annor et al. 2017), a decrease in anti-nutrients such as phytic acid, tannins, and other phenolics, as well as protease inhibitors, are also occurring (Sehgal and Kawatra 2001). The main phytochemical effects of cereal germination, its mechanism effect, potential health benefits, and the impact on product quality, are summarized in Table 6.1. As previously discussed in Section 5.3, starch and protein are hydrolyzed during germination by the action of amylases and proteases, respectively.

The occurrence of starch degradation and the release of oligosaccharides and simple sugars suggest an increase in both starch digestibility and postprandial levels of blood glucose. However, Marti et al. (2018) showed that the amount of rapid (RDS) and slow digestible starch (SDS) in bread decreased and increased, as the percentage of sprouted wheat flour increased from 50 to 100%, respectively. The data of this study were obtained by *in vitro* studies and are in partial agreement with those reported by Świeca et al. (2017), who attributed the decrease in starch digestibility of sprouted wheat-enriched bread to an increase in the aliquot of resistant starch and/or to the high phenolic content of sprouted wheat.

Considering that the glycemic response appears to be directly related to the amount of RDS, while insulin demand is inversely correlated to the SDS fraction (Garsetti et al. 2005), data on sprouted wheat bread favor the use of sprouted wheat in new bread formulations aiming to reduce postprandial levels of blood glucose. Indeed, *in vivo* studies on bread enriched with germinated wheat reported favorable blood glucose effects compared with non-germinated wheat bread (Andersen et al. 2008), by reducing the glycemic response to sprouted grain bread in both healthy subjects (Andersen et al. 2008) and in overweight or obese men (Mofidi et al. 2012). However, the above studies show that health benefits are not solely dependent on the germination process, but other factors as well. Therefore, further research is required to assess the beneficial health effects of germination and clarify whether the potential health benefits of bread enriched with sprouted wheat can be attributed to sprouting. Indeed, none of the above-cited studies investigated the characteristics of bread made from the same wheat variety before and after sprouting.

In protein, protease activities result in an increase in peptides and/or free amino acids, including the essential amino acids. For example, the lysine content – limiting amino acid in wheat – increased by 10% after germination (Sibian et al. 2017). An increase in digestibility of proteins has been observed in sprouted wheat, in particular after 48 hours of germination at 20 °C, whereas after 96 hours at 25 °C it decreased (Świeca and Dziki 2015). This decrease might be due to the accumulation of phenolic compounds. Indeed, there is evidence that the interactions between the phenolic compounds and the digestive enzymes and/or the food matrix could decrease, directly or indirectly, the bioavailability of nutrients (Świeca et al. 2013). Studies show that the gluten decreases in germinated wheat (Koehler et al. 2007), likewise the proteases are responsible for the degradation of gliadin peptides (Hartmann et al. 2006). Thus, people suffering from gluten sensitivity may benefit from these

Table 6.1 Effect, cause, and potential health benefits of cereal germination on selected nutrients.

Component		Cereal	Effect and sprouting conditions	Cause	Potential health effects
Carbohydrates	Starch	Wheat	Decreased (>1 day @ 25 °C)[1] (3 days @ 20 °C)[2] (2 days @ 24 °C)[3]	Hydrolysis by α-amylases	Decrease in starch digestibility[1] Increase in slowly digestible starch fraction[3]
		Brown rice	Decreased (3 days @ 35 °C)[4]		
		Barley	Decreased (4 days @ 17 °C)[5]		
		Oat	Decreased (6 days @ 16 °C)[6]		
	Sugars	Wheat	Increased (4 days @ 28 °C)[7] (3 days @ 20 °C)[2] (2 days @ 24 °C)[3]	Starch hydrolysis by α-amylases with production of simple sugars and oligosaccharides	Decrease in postprandial glucose[8,9] Prebiotic effect[10]
	Total Fiber	Wheat	Decreased (4 days @ 15–20 °C)[11]	Loss of starch	Delay in the increase of glucose in the blood[13]
			Increased (2 days @ 15–20 °C)[11]	Hydrolysis by pentosanases and xylanases	
		Barley	Not changed (3 days @ 15 °C)[12]	–	
	Soluble Fiber	Wheat	Increased (> 4 days @ 20 °C)[11] (> 6 h @ 30 °C)[14]	Hydrolysis by pentosanases and xylanases	
	Insoluble Fiber	Wheat	Decreased (> 7 days @ 20 °C)[11] (> 6 h @ 30 °C)[14]	–	

Proteins	Wheat	Decreased (> 2 days @ 25 °C)[11] (> 2 days @ 20–25 °C)[15]	Hydrolysis by proteases or leaching of water-soluble peptides during steeping	Decrease in gluten sensitivity[16]
	Sorghum	Decreased (2 days @ 25 °C)[17]		
	Brown rice	(5 days @ 20 °C)[18]		
	Wheat	Increased (5 days @ 16.5 °C)[19] (> 1 days @ 30 °C)[14] (> 4 days @ 28 °C)[20]	Decrease in starch content[14]	–
	Barley	Increased (5 days @ 16.5 °C)[19]		
	Oat	Increased (5 days @ 16.5 °C)[19] (3 days @ 17 °C)[12]		
	Brown rice	Increased (5 days @ 16.5 °C)[19]		
	Wheat	Not changed (3 days @ 25 °C)[21] (< 1 day @ 30 °C)[14]	–	–
Lipids	Wheat	Decreased (5 days @ 16.5 °C)[19] (2 days @ 30 °C)[22]	Hydrolysis by lipases	Improvement of plasma free fatty acids[8]
	Barley	Decreased (5 days @ 22 °C)[23]		
	Oat	Decreased (6 days @ 16 °C)[24]		
	Brown rice	Decreased (1–5 days @ 25–30 °C)[25]		
	Wheat	Increased (> 12 h @ 30 °C)[14]	Following the decrease in the content of starch[26,14]	–
	Wheat	Not changed (2 days @ 30 °C)[22]	–	–

(*continued overleaf*)

Table 6.1 *(continued)*

Component	Cereal	Effect and sprouting conditions	Cause	Potential health effects
Total polyphenols	Wheat	Increased (2 days @ 30 °C)[22]	Hydrolysis by phenol oxidases and peroxidases[28]	Decrease in plasma polyphenols and antioxidants measured *in vivo*[8]
	Barley	Increased (>1 day @ 28 °C)[27]		
	Sorghum	Increased (2 days @ 20 °C)[29]		
	Brown rice	Increased (4 days @ 34 °C)[30]		
Phytate	Wheat	Decreased (3 days @ 20 °C)[31]	Hydrolysis by phytases[32]	Increase in bio-accessibility of vitamins and minerals
	Barley	Decreased (3 days @ 20 °C)[31] (3 days @ 22 °C)[33]		
	Brown rice	Decreased (4 days @ 34 °C)[30]		
	Oat	Decreased (6 days @ 10 °C)[34]		

Vitamin	Wheat	Decreased (vit. E) (4 days @ 28 °C)[35]	–	–
		Increased (vit. E) (7 days @ 16.5 °C)	–	–
	Barley	No change (vit. E) (4 days @ 15 °C)[37]	–	–
	Brown rice	No change (vit. E) (1 days @ 28 °C)[25]	–	–
		Increased (vit. E) (3 days @ 30 °C)[38]	–	–
	Wheat	Increased (vit. A, B_1, B_2, B_6, C) (4 days @ 28 °C)[35]	–	–
	Sorghum	Increased (vit. B_1, B_2, C) (3–4 days @ 25 °C)[39]	–	–
	Rice	Increased (vit. B1, B2) (3–4 days @ 25 °C)[40]	–	–
		Decreased or no change (B_1, B_2, B_3, B_9) (1 day @ 28 °C)[25]	–	–

(*continued overleaf*)

Table 6.1 (*continued*)

Component	Cereal	Effect and sprouting conditions	Cause	Potential health effects
Minerals	Wheat	Decreased (Fe; 3 days @ 25 °C)[21] (Ca, Mn, Na; 4 days @ 28 °C)[35] (Mn)[41]	–	–
	Sorghum	Decreased (Fe, Zn; 3 days @ 20 °C)[42]	–	–
	Wheat	Increased (Na, P, S, Cl, K, Ca, Fe, Cu, Zn)[41] (Cu, K, Mg, Z; 4 days @ 28 °C)[35]	Hydrolysis by phytases[32]	–
	Rice	Increased (Fe, Zn; 2 days @ 20 °C)[43]		
	Barley	Increased (Fe, Zn; 2 days @ 20 °C)[43]		
	Wheat	No change (Fe; 4 days @ 28 °C)[35]	–	–
	Brown rice	No change (Zn; 3 days @ 30 °C)[44]		

(*Source*: [1]Świeca et al. 2017; [2]Marti et al. 2017; [3]Marti et al. 2018; [4]Xu et al. 2012; [5]Vinje et al. 2015; [6]Peterson 1998; [7]Sibian et al. 2017; [8]Andersen et al. 2008; [9]Mofidi et al. 2012; [10]Hübner and Arendt 2013; [11]Koehler et al. 2007; [12]Teixeira et al. 2016; [13]Anderson et al. 2004; [14]Van Hung et al. 2015; [15]Świeca and Dziki 2015; [16]Hartmann et al. 2006; [17]Elmaki et al. 1999; [18]Mohan et al. 2010; [19]Donkor et al. 2012; [20]Sibian et al. 2017; [21]Zambiazi da Silva et al. 2014; [22]Van Hung et al. 2011b; [23]Chung et al. 1989; [24]Peterson 1998; [25]Watanabe et al. 2004; [26]Lorenz et al. 1981; [27]Ha et al. 2016; [28]Barron et al. 2007; [29]Hithamani and Srinivasan 2014; [30]Cáceres et al. 2014; [31]Bartnik and Szafrańska 1987; [32]Larsson and Sandberg 1992; [33]Centeno et al. 2001; [34]Hübner et al. 2010; [35]Plaza et al. 2003; [36]Yang et al. 2001; [37]Haraldsson et al. 2004; [38]Esa et al. 2011; [39]Asiedu et al. 1993; [40]Trachoo et al. 2006; [41]Pongrac et al. 2016; [42]Afify et al. 2011; [43]Platel et al. 2010; [44]Liang et al. 2008) kinds of products. The germination process also affects the total lipids, resulting in an overall increase because of free lipids (Van Hung et al. 2015). However, the bound lipids did not change, regardless of the sprouting time. Furthermore, the germination process does not seem to cause any changes in the fatty acid composition (Van Hung et al. 2011b, 2015). Lipid hydrolysis during wheat germination should be further investigated due to its potential role in affecting starch digestibility in other cereals (Annor et al. 2017).

The effect of germination on total fiber content is still unclear, with conflicting results available in the literature. Fiber changes are more pronounced when germination occurs at 15–20 °C than at higher temperatures (25–30 °C) (Koehler et al. 2007). At higher temperatures, a distinct increase in total dietary fiber concentration (> 25%) was found at prolonged time points (e.g. 102 hours) (Koehler et al. 2007). Concerning fiber solubility, insoluble fiber generally decreased in wheat by 50% (Koehler et al. 2007), whereas soluble fiber values reported for wheat remained constant up to 96 hours, and then increased steadily to 168 hours (Koehler et al. 2007), likely due to the increase in cell–wall degrading enzymes, including endoxylanases (Olaerts and Courtin 2018). This aspect is of great interest from a nutritional point of view as soluble dietary fiber has been associated with certain health benefits such as the maintenance of normal blood cholesterol levels due to its ability to form viscous solutions in the intestine (Kumar et al. 2012).

Increased polyphenol content and antioxidant activity have been reported in several *in vitro* studies (Van Hung et al. 2011a, 2015; Žilić et al. 2014; Świeca and Dziki 2015), because of increased polyphenol oxidase activity compared with non-germinated grain. Indeed, polyphenol oxidase activity increased in wheat following 8–16 hours of soaking (Demeke et al. 2001). Furthermore, an increase in free phenolics at the expense of bound phenolics has also been reported (Van Hung et al. 2011a). Besides serving as antioxidants and protecting against lipid oxidation and the development of off-flavors (Maillard et al. 1996), increasing the intake of phenolic compounds can have positive health effects due to anti-carcinogenic, antioxidant, and anti-inflammatory effects for some of the compounds investigated (Santos-Buelga and Scalbert 2000). Further studies are warranted to explore the transition from *in vitro* findings to *in vivo* effects (Nelson et al. 2013).

Results on the effects of the germination process on vitamins and minerals are conflicting. Most studies found an improvement of vitamin values as a result of sprouting (Yang et al. 2001; Plaza et al. 2003). However, drying conditions, especially if high drying temperatures are applied, may cause the loss of some vitamins.

Mineral content in sprouted cereals also showed inconsistent results, depending on the applied methods of soaking and germination. During the process, some minerals are leached or absorbed by the hydrating and germinating seeds (Finney 1982). In general, if high mineral-containing water is used to steep and germinate, sprouted seeds will increase their uptake of total minerals and ash, resulting in increased germination. Conversely, if distilled water is used, minerals will invariably leach out (Omary et al. 2012).

The accessibility of minerals is the result of the interactions of many factors – such as the type of mineral, the composition and structure of grains, and the processes used (Erba et al. 2018). Germination is generally reported as a process that improves mineral accessibility in grains by reducing anti-nutritional factors, including phytates, which form insoluble complexes, not only with minerals, but also with proteins, carbohydrates, and lipids, thus reducing their bioavailability (Kumar et al. 2010).

Additionally, Hübner and Arendt (2013) suggested a potential prebiotic effect of sprouted grains as substrates for the growth of probiotic bacteria, providing high levels of fermentable sugars.

Despite the nutritional enhancement associated with sprouting, up to now, sprouted grains have typically been used in relatively small amounts when incorporated into bakery products. Thus, both industries and consumers should be aware that they are not going to reap all of the nutritional benefits of sprouting when so little is used.

6.5 From Traditional to Industrial Germination Processes

Germination of seeds for consumption has been practiced for thousands of years in order to improve their technological and nutritional properties. For example, sorghum and millet are commonly germinated to improve their protein digestibility (Elmaki et al. 1999; Annor et al. 2017); whereas pulses are germinated both to decrease anti-nutritional factors (Soetan and Oyewole 2009) and to facilitate the dehulling and cooking process (Bellaio et al. 2011; Zamprogna Rosenfeld et al. 2011). The basic germination process (Figure 6.3) consists of steeping the grains in water for 8–24 hours, until they reach the moisture content needed to start the growth of the seedling (30%) (Bewley and Black 1985), after which the steeping water is drained and the grains are left to germinate under controlled conditions. Aeration and mixing are advisable in this phase to allow the grain to use the nutrients to germinate and sprout. On the other hand, time and temperature determine the retention of enzymatic activity and

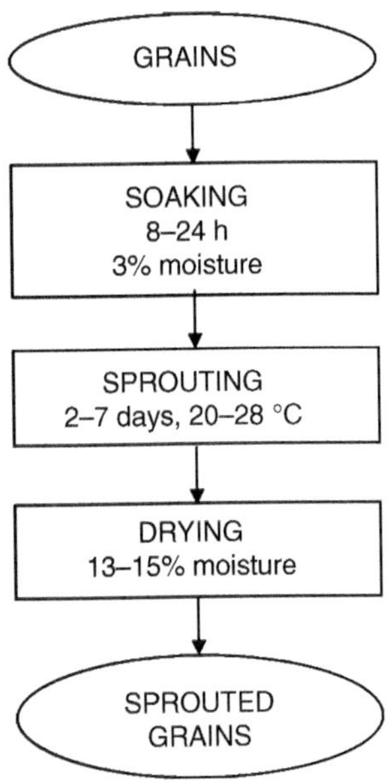

Figure 6.3 Flow-chart of the controlled sprouting process.

the development of color and flavor compounds. The germinated grains are then consumed in the form of sprouts or further processed (i.e. dried or roasted).

Traditionally, the germination process has been conducted at home, neglecting the potential grain safety risk raised with regard to microbial growth that may be associated with the uncontrolled process. Controlling the process seems the unique way of decreasing the safety risk while preserving the nutritional and technological benefits of the product.

With the exception of the malting process, most of the studies are conducted at the laboratory scale, where the germination is carried out in thermostatic cells for a variable number of days, often without controlling relative humidity. Since both the mechanisms of grain germination and the related changes in phytochemicals are strongly dependent on temperature and relative humidity (apart from time and type of grains), an understanding of the phenomenon, a comparison of results, and the repeatability and reproducibility of the experiments is difficult to obtain.

After germination, seeds are stabilized by drying with hot air to extend their shelf-life. The thermal treatment stops the sprouting and reduces the moisture content of 30 to 50% to less than 12%. The drying temperature must be chosen carefully in order to keep the enzymatic pool developed during germination active and to improve the technological performance of flours. Nevertheless, the sprouted grains can be toasted to create additional flavors. Besides guaranteeing safety, the monitoring and controlling of the sprouting conditions standardize the nutritional and sensory properties of the final product, thus assuring consistency in quality.

6.6 Utilization of Germinated Cereal Grains in Different Food Products

Germination is of fundamental importance for the food industry. It has been applied for millennia to pulses to reduce their anti-nutritional components, such as trypsin inhibitors and phytic acid. At the same time, germination decreases the digestive discomfort caused by ROFs (raffinose family of oligosaccharides), while developing sweet taste notes. In addition to the nutritional aspects listed above, germination affects the technological performance of grains and related flours. Surprisingly, germination facilitates the dehulling and cooking process of grains (Bellaio et al. 2011; Zamprogna Rosenfeld et al. 2011). This is of great interest since it would facilitate the consumption of whole grains. Indeed, even while reducing the risk of cardiovascular disease and inflammation, the eating of grains is not so common in many countries, due to their long cooking times and bitter and pungent flavors (Heiniö et al. 2016). In the case of brown rice, germinated kernels require less time for cooking and from the sensory standpoint taste sweeter and softer than regular brown rice (Patil and Khan 2011).

6.6.1 Malting for Brewing Products

Malt from barley is another example of using germination in the food industry. It is a special form of limited germination aimed at producing enzymes, which hydrolyze starch to make sugars available for fermentation. Although amylolytic enzymes are of prime importance, other enzymes play a key role in the production of flavor compounds, contributing to the quality of the malt (Mäkinen and Arendt 2015). However, reserve mobilization should not proceed to completion, to avoid losing potentially fermentable materials. This characteristic can be described by the diastatic power that measures the combination effect of starch-degrading enzymes. Low levels of hydrolytic enzymes create problems for fermentability and high wort viscosities (Mäkinen and Arendt 2015). Other factors, such as complex cell–wall polysaccharides, which are not readily hydrolysed, may also contribute to poor brewing performances (Mäkinen and Arendt 2015). Although barley is the main malting cereal worldwide, other cereals, including sorghum, millets, and pseudocereals, can also be used (Hager et al. 2014). Malts produced from pseudocereals and some tropical grains can have low levels of amylotic activity and often these malts have poor brewing performance, thus requiring exogenous enzymes when brewing (Hager et al. 2014).

In addition to brewing, barley and wheat malt are used to optimize and standardize the α-amylase levels in wheat flour, or as sources of color and flavor (Kulp 1993). The amylase content in wheat flour can be variable and low α-amylase activity results in low bread volume and quality. Thus, adding α-amylases or malt to the flour increases loaf volume and crumb softness during storage (Kulp 1993). On the other hand, an excessive concentration of amylases results in a sticky dough, which is difficult to handle, and in bread with an irregular crumb structure (McCleary and Sturgeon 2002).

6.6.2 Bakery Products

Marti et al. (2017) explored the enzymatic activities developed during sprouting in bread-making, with the aim of decreasing or substituting the use of commercially enzymes, such as flour improvers that are commonly present in the formulation of baked products. Their study incorporated 0.5% of sprouted wheat to stiff refined flour, as an alternative to conventional flour improvers (i.e. malt, proteases, and an enzymatic improver based on xylanases). Small amounts of sprouted wheat flour were effective in increasing bread specific volume and crumb softness. Moreover, for up to three days of storage, sprouted wheat flour showed an effect similar to malt in lowering the staling process in the bread. The authors concluded that flour from sprouted wheat is a promising and interesting ingredient for formulating baked products, as it eliminates the need for enzymatic improvers, which is a plus for consumer acceptance and facilitates the adoption of clean labels (Marti et al. 2017). It is logical that such small amounts of sprouted wheat only play a technological role, with no nutritional benefits. At the same time, the activity of amylases and proteases induced by germination – if excessive – negatively affects the technological performances of wheat (Morris and Rose 1996), which then becomes unsuitable for baked foods, as shown in Figure 6.4.

This might occur directly in the field (i.e. pre-harvest sprouting) or when the germination is carried out under severe and uncontrolled conditions, whereby starch and protein are excessively hydrolyzed. This seems to be the case in most of the studies that highlighted decreases in bread volume when up to 20% of sprouted brown rice was added to wheat flour (Watanabe et al. 2004). However, beneficial effects were more obvious for sprouted grain-enriched products than for unsprouted ones. Although the enzymatic activities of sprouted grains were not assessed by Watanabe et al. (2004), it is reasonable to assume that the processing conditions

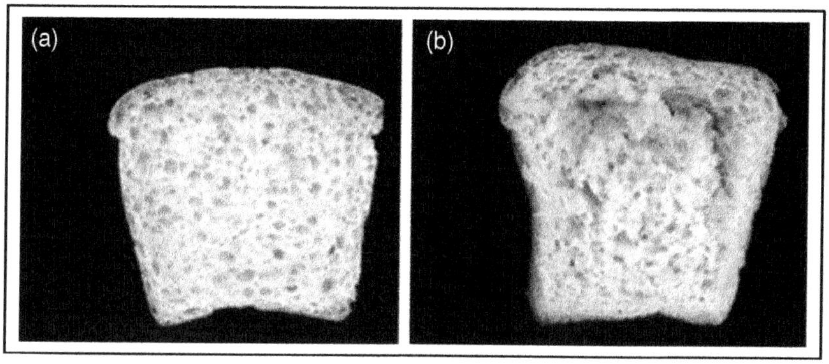

Figure 6.4 Slices of bread made with flour from (a) unsprouted wheat and (b) sprouted wheat characterized by excessive amylase and protease activities. *Source*: courtesy of Alessandra Marti, Gaetano Cardone and Maria Ambrogina Pagani.

adopted by previous studies allowed high concentrations of hydrolytic enzymes to accumulate, negatively affecting the technological performance of flour and making it unsuitable for baked foods. Both amylases and xylanases are responsible for dough stickiness. On the one hand, high levels of α-amylases extensively degrade damaged starch during dough mixing and fermentation, generating high levels of sugars and dextrins, and releasing water that was previously bound by the starch (MacArthur et al. 1981). On the other hand, an excess of endoxylanases can cause extensive degradation of the arabinoxylans, releasing the water that was previously bound to them (Courtin et al. 2001). In protein, a marked decrease in insoluble residue protein in sprouted wheat samples has already been shown to compromise the baking performance of sprouted wheat (Koehler et al. 2007; Simsek et al. 2014). Moreover, the breakdown of gluten proteins by peptidases occurs, even during dough processing. Experiments with the Farinograph® (Brabender GmbH & Co. KG, Duisburg, Germany) and Mixograph® (National Manufacturing, Lincoln, NE) showed a decrease in dough development time and stability and in water absorption of dough as controlled germination time increased (Sekhon et al. 1995). On the other hand, wheat germination carried out for 48 hours at 20 °C promoted a limited accumulation of proteases, so that gluten was still able to aggregate and form a network with good bread-making performance (Marti et al. 2017, 2018). At the same time, both pasting and gelation properties of starch were not affected when sprouting was carried out for up to 72 hours (Grassi et al. 2018). However, there is the risk that the amylases synthetized during sprouting could be activated during baking, thus promoting strong starch degradation that might negatively affect bread crumb structure and crust color. By limiting starch and gluten degradation, bread from 100% sprouted wheat can be made (Richter et al. 2014; Marti et al. 2018). The resulting dough was not sticky and required less time for leavening, resulting in a final product characterized by a higher volume than the control bread (Richter et al. 2014; Marti et al. 2018), as shown in Figure 6.5. Higher loaf volume is mostly explained by the greater CO_2 production due to increased amounts of fermentable sugars released by higher α-amylase activity (Van der Maarel 2009). In addition, bread from sprouted wheat was able to retain its soft crumb texture for up to six days of storage (Marti et al. 2018).

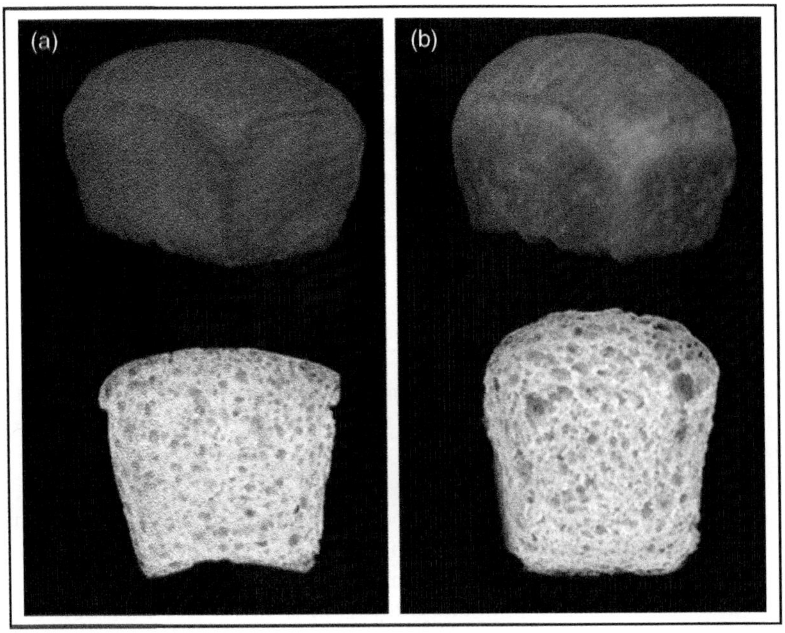

Figure 6.5 Bread loaves and slices of bread prepared with flour from (a) unsprouted and (b) sprouted wheat. *Source*: courtesy of Alessandra Marti, Gaetano Cardone and Maria Ambrogina Pagani.

In addition to bread-making performance, germination influences the sensory properties of grains, giving them a typical flavor and odor generally perceived as pleasant. During germination, reducing sugars and amino acids are released, which subsequently react during heating, giving rise to Maillard reaction products (Hefni and Witthöft 2011). Moreover, both germination and drying decrease the musty and earthy odor notes, favoring the perception of roasted, nutty, and intense flavors (Heiniö et al. 2001) and masking the typical bitterness of quinoa seeds (Suárez-Estrella et al. 2018) and whole-grain bread (Richter et al. 2014).

More recently, Marti et al. (2018) investigated the effect of high enrichment levels of sprouted wheat flour (from 15 to 100%) on the rheological properties of dough and baking performance. Regardless of the amount added, sprouted wheat improved dough development and gas production during leavening. The best results – in terms of bread volume and crumb porosity – were obtained with 50% of sprouted wheat instead of using sprouted wheat flour alone. This enrichment level was certainly suitable for obtaining a product with enhanced sensory and nutritional benefits, without compromising bread-making performance and *in vitro* starch digestibility. In this context, controlling the germination process seems the only way of enhancing nutritional and sensory benefits while optimizing flour performance to ensure satisfactory and consistent products.

6.7 Monitoring of Seed Germination

The attempt to optimize the germination time leads to another issue, since no universally useful biochemical marker of the progress of germination has been found (Bewley and Black 1994). The only stage of germination that we can determine precisely is its termination. The emergence of the radicle from the seed is normally used to define when germination has been completed. In this context, some companies and/or researchers use the radicle emergence as a marker, but this approach is often based on their own experience rather than on a scientific and systematic approach. Moreover, the evidence of the germination process may occur in a seed where radical emergence does not occur. In cases where the radicle is not evident, different methods have been proposed, and their strengths and weaknesses are summarized in Table 6.2. These methods are based on the direct or indirect quantification of enzymes (mostly α-amylases) present in cereals such as wheat and barley. The presence of high α-amylase activity in wheat is generally associated with pre-harvest sprouting that also promotes the *de novo* synthesis of proteolytic enzymes, which critically impair the commercial quality of grains (Olaerts and Courtin 2018).

6.7.1 Falling and Stirring Number

Among the listed methods, the Falling Number (FN) and the Stirring Number (SN) are the most commonly used to measure the effect of amylase activity on flour quality. Neither test measures α-amylase activity directly, but indirectly by quantifying the viscosity of the starch hydrolyzed by the enzymes during the test. They have been proposed as simple and rapid techniques and are performed according to international standards (AACC 56–81.03 and ICC 107–1 for the FN; AACC 22–08.01 and ICC 161 for the SN) (AACCI 2011, ICC 1968). The optimal value for baking purposes is 250 seconds; with FN < 250 seconds, the dough looks sticky while a FN > 300 seconds corresponds to a flour with almost no amylase activity (www.perten.com). Generally, the increasing levels of α-amylase result in a decrease in FN down to 60 seconds, beyond which further increases in activity cannot be measured (MacArthur et al. 1981). A low FN value is generally associated with pre-harvest sprouting.

The SN is defined as the apparent viscosity in Rapid Visco Units (RVU) after 180 seconds of stirring a hot aqueous flour suspension undergoing liquefaction in the Rapid Visco Analyzer® (RVA, Newport Scientific Pvt. Ltd., Warriewood, Australia). Due to the action of the hydrolytic α-amylase, viscosity decreases, and the SN decreases as well (Figure 6.6).

However, the limitation of the FN and SN methods has turned out to be a reduced sensitivity to low levels of α-amylase activity, due to rapid heating during the analysis (Ross and Bettge 2009). However, this is a widely used method for wheat grading. Also, although widely used to detect pre-harvest sprouting in wheat kernels, the FN and SN seem to overestimate the extent of starch hydrolysis in sprouted wheat under controlled conditions (Grassi et al. 2018).

Indeed, running the test in the presence of an amylase inhibitor (i.e. $AgNO_3$) highlighted that changes in viscosity were caused by α-amylase activities during analysis and not by inherent changes in starch swelling, pasting, and gelation properties (Grassi et al. 2018). Hence, starch in sprouted wheat with an FN lower than 250 seconds is still of good quality (Grassi et al. 2018).

6.7.2 Amylograph

Similar to RVA, the presence of sprouting can also be detected using the Amylograph® (Brabender GmbH & Co. KG, Duisburg, Germany), by measuring the activity of α-amylase in an aqueous suspension of flour while it gelatinizes during heating. The

Table 6.2 Main useful approaches for characterizing flour from sprouted wheat.

Index	Official method	Device	Principle of the method	Advantage	Disadvantage
Falling number	AACC 56-81B; ICC-standard 107/01	Falling number system 1500 (Perten Instruments)	Indirect evaluation of α-amylases by evaluating the time of a plunger falling into a flour and water gel	Rapid test (5–10 minutes); Low influence of the analyst; Low sample size	Not effective in determining the effect of low levels of α-amylase
Stirring number	AACC 22–08; ICC-standard 161	Rapid Viscoanalyzer (RVA-4500, Perten Instruments)	Indirect evaluation of α-amylases by measuring the viscosity of a flour/water suspension during rapid gelatinization	Rapid test (3 minutes); Low influence of the analyst; Low sample size	Not effective in determining the effect of low levels of α-amylase
α-amylase	AACC 22–02; ICC-standard 303	Enzymatic kit (Megazyme K-CERA 02/17)	Extraction of α-amylases from the flour made to act on a p-nitrophenyl-malto-heptoside substrate (BPNPG7), which possesses the non-reducing end blocked by the p-nitrophenol reactive chromophore	Very low amount of sample (100 mg)	Possibility of non-homogeneous sampling; Long analysis times (2 hours); Need for expert analysts
Viscosity	AACC 22–10.01; ICC standard 126/1	Visco-Amylograph (Brabender)	Measurement of the viscosity of a suspension of mixing flour during programmed heating	Sensitive to low levels of amylasic activity	Use of a non-water solvent (aqueous solution consisting of anhydrous disodium phosphate and citric acid monohydrate); Large amount of sample (65 g)

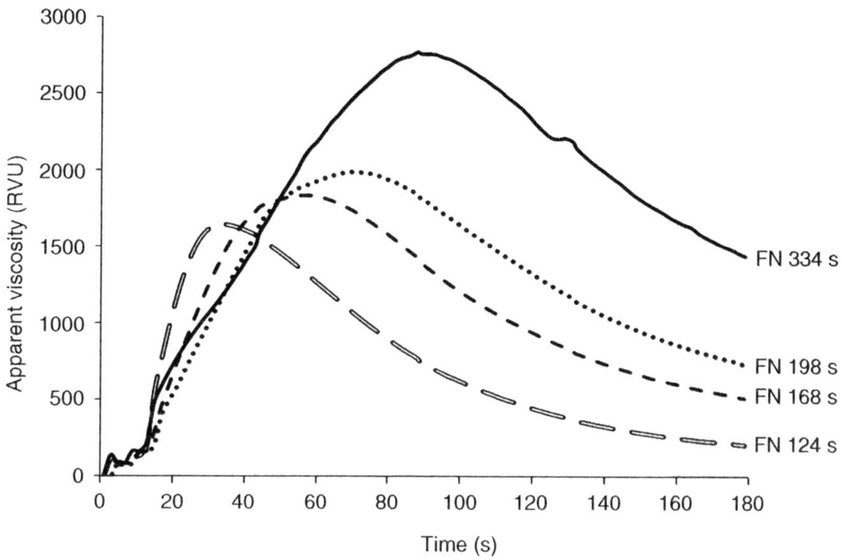

Figure 6.6 Examples of RVA plots of wheat flours with different values of Falling Number. *Source*: based on Grassi et al. (2018).

peak viscosity is inversely correlated to the integrity of the starch granules. When enzymatically weakened, starch granules lose their resistance to swelling; these structural changes result in the lowering of the paste viscosity of the sprouted grains (Simsek et al. 2014).

In addition, the above-mentioned methods use the starch as substrate, neglecting the effect of other hydrolytic enzymes such as protease on gluten proteins. Indeed, changes in protein aggregation properties during germination are worth investigating (Marti et al. 2018), since samples with similar FN or SN values can be different in composition and functionality (Kruger 1994). As reported in Figure 6.7, semolina samples having the same FN value (about 62 seconds) showed different gluten aggregation kinetics that were measured by the GlutoPeak® Test (Brabender GmbH & Co. KG, Duisburg, Germany).

6.7.3 Alpha-Amylase Activity

The enzymatic activities developed during germination can be directly quantified using standard methods such as those for α-amylase (AACC Method 22–02) and protease (AACC Method 22–62) (AACCI 2011). These tests require laboratory equipment and operator expertise. Furthermore, the development of these approaches is a laborious process, involving the collaboration of different laboratories before proposing an official and internationally recognized method (Bridges and Wrigley 2016). In fact, industries ideally need a rapid, non-destructive, and in-line approach to monitor the germination process. Spectroscopic techniques seem to answer this need. Infrared spectroscopy was applied to flour to detect grain germination and the starting time of the sprouting process in wheat and barley (Burke et al. 2016). Burke et al. (2016) obtained good Partial Least Squares models for Fourier-transform infrared (FT-IR) data (cross-validated coefficient

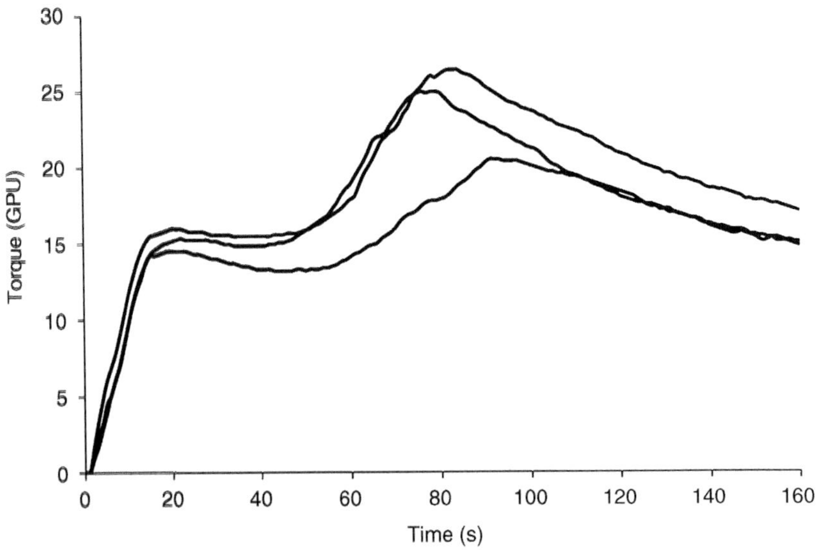

Figure 6.7 Flours with similar Falling Number (about 62 seconds) but with different gluten aggregation properties as measured with the GlutoPeak (R) (Brabender GmbH & Co. KG, Duisburg, Germany).

of determination of 0.75), but poor validation results by Fourier-transform near-infrared (FT-NIR) modeling due to high overfitting. As for the Near-Infrared Hyperspectral Imaging, this technique combined with different discriminant classification techniques proved to be a useful tool to distinguish unsprouted from sprouted kernels (Singh et al. 2009). However, lack of information regarding the relation between wheat functionality and hyperspectral images makes it difficult to draw any conclusions useful for the technological development of robust simplified and cost-effective spectroscopic systems. More recently, Grassi et al. (2018) collected spectra – in the spectral range of 950–1650 nm – of both wet and dried kernels at different time intervals (from 24 hours up to 72 hours of sprouting) using a MicroNIR OnSite (VIAVI, Santa Rosa, California), equipped with a shaker probe. The spectral profiles of sprouted grains differed greatly when compared to those of unsprouted samples. The multivariate analysis of the spectra highlighted that differences between sprouted and unsprouted samples are associated with the different absorptions in the range 1500–1626 nm, related to starch absorption signals (Juhász et al. 2005). Likewise, the differences between early stage germination (up to 36 hours) and late-stage germination are influenced by the absorption occurring in the range of 1360–1440 nm, probably related to C—H combination bands (Workman and Weyer 2008). In addition, data showed that the most interesting changes occurred in the first 48 hours, whereas longer germination times generated no further relevant changes. Thus, a near-infrared (NIR) portable device can predict the progress of controlled sprouting processes directly on wet kernels, thus skipping both the drying and refinement steps, providing information similar to that obtained by the complex and time-consuming analyses on refined flour (Grassi et al. 2018). The development of this approach could help food companies to standardize and monitor the sprouting process, as well as to produce novel cereal-based foods with guaranteed and consistent characteristics. In addition, the monitoring of the sprouting process and defining when the process begins and when it ends it is of special interest when discussing the potential health benefits of sprouted grains. Thus, as stated above, the nutritional benefits of sprouting depend on many factors, including the type of grain, the variety, processing conditions, and processing time.

6.8 Conclusion and Further Remarks

The transformation of a grain into a new plant is based on complex and interdependent phenomena that, starting from the hydrolysis of storage macromolecules into soluble substances, permits the growth of the first sprout. Although the effects of germination on the nutritional quality of grains have been shown, most of the nutritional benefits have been assessed in *in vitro* and raw materials. Further *in vivo* studies on the final products as eaten are needed to determine the fate of phytochemical components during processing.

The nutritional improvement of sprouted grains was practically neglected by Western consumers until recently and the consumption of whole sprouted grains was almost non-existent in Western diet. In fact, germination was judged negatively based on the poor technological characteristics of the resultant flour. Indeed, the loss of baking properties is normally observed in pre-harvest sprouted wheat. It is therefore easy to understand the efforts to develop and set up fast and reliable tests capable of recognizing so-called "sprout damage". Indices related to this defect are included in the grading procedures of all countries, both exporters and importers, to check raw materials unsuitable for industrial transformation.

Conflicting results about the effects of germination suggest the need for more research to optimize cultivars and germination conditions of grains. Recent research carried out on wheat seems to indicate interesting developments that were unforeseeable a few years ago. Indeed, flour from wheat germinated under controlled conditions, although rich in hydrolytic enzymes (amylases and proteases), could nonetheless be successfully transformed into bread with good quality characteristics. It is therefore essential to understand the molecular and structural differences between germination carried out under uncontrolled (pre-harvest germination) and controlled conditions. Only in this way, can the parameters directly related to the maintenance of good cross-linking properties of proteins be identified, thus replacing the present indirect indices, based on amylase activity.

Besides the positive effects of germination on nutritional and technological features, the conditions applied (i.e. high relative humidity) might favor the growth of pathogens, making the safety risk a critical point of the process. In this context, the treatment of seeds with either ozone or non-thermal technologies (i.e. cold plasma) needs to be further explored to potentially improve the microbiological quality of sprouted grains.

References

AACC International. (2011) Approved methods of analysis (*11th Edition*). St. Paul: American Association of Cereal Chemists. methods.aaccnet.org

Afify, A.E.M.R., El-Beltagi, H.S., El-Salam, S.M.A. et al. (2011). Bioavailability of iron, zinc, phytate and phytase activity during soaking and germination of white sorghum varieties. *PLoS One* 6 (10): 1–7.

Andersen, G., Koehler, P., and Somoza, V. (2008). Postprandial glucose and free fatty acid response is improved by wheat bread fortified with germinated wheat seedlings. *Current Topics in Nutraceutical Research* 6 (1): 15–22.

Anderson, J.W., Randles, K.M., Kendall, C.W.C. et al. (2004). Carbohydrate and fiber recommendations for individuals with diabetes: a quantitative assessment and meta-analysis of the evidence. *Journal of the American College of Nutrition* 23 (1): 5–17.

Annor, G.A., Tyl, C., Marcone, M. et al. (2017). Why do millets have slower starch and protein digestibility than other cereals? *Trends in Food Science and Technology* 66: 73–83.

Aoki, K., Taniyama, M., Nagayama, C. et al. (2006). T cell immunity to glutamic acid decarboxylase in fulminant type 1 diabetes without significant elevation of serum amylase. *Annals of the New York Academy of Sciences* 1079 (1): 181–185.

Asiedu, M., Lied, E., Nilsen, R. et al. (1993). Effect of processing (sprouting and/or fermentation). On sorghum and maize: II. Vitamins and amino acid composition. Biological utilization of maize protein. *Food Chemistry* 48 (2): 201–204.

Barron, C., Surget, A., and Rouau, X. (2007). Relative amounts of tissues in mature wheat (*Triticum aestivum* L.). grain and their carbohydrate and phenolic acid composition. *Journal of Cereal Science* 45 (1): 88–96.

Bartnik, M. and Szafrańska, I. (1987). Changes in phytate content and phytase activity during the germination of some cereals. *Journal of Cereal Science* 5 (1): 23–28.

Bau, H.M., Villaume, C., Nicolas, J.P. et al. (1997). Effect of germination on chemical composition, biochemical constituents and antinutritional factors of soya bean (*Glycine max*). Seeds. *Journal of the Science of Food and Agriculture* 73 (1): 1–9.

Bellaio, S., Zamprogna Rosenfeld, E., and Jacobs, M. (2011). Novel process based on partial germination to enhance milling yield and nutritional properties of pulses. *Cereal Foods World* 56 (S4): A30.

Bewley, J.D. and Black, M. (1985). Storage, imbibition, and germination. In: *Seeds: Physiology of Development and Germination* (ed. e.J. Bewley), 89–134. Boston: Springer.

Bewley, J.D. and Black, M. (1994). *Seeds: Physiology of Development and Germination*. Boston: Springer.

Bridges, A.R. and Wrigley, C.W. (2016). Standardized test methods for grains and grain-based products. In: *Encyclopedia of Food Grains* (eds. C.W. Wrigley, H. Corke, K. Seetharaman, et al.), 300–307. Oxford: Elsevier.

Burke, M., Small, D.M., Antolasic, F. et al. (2016). Infrared spectroscopy-based metabolomic analysis for the detection of preharvest sprouting in grain. *Cereal Chemistry* 93 (5): 444–449.

Cáceres, P.J., Martínez-Villaluenga, C., Amigo, L. et al. (2014). Assessment on proximate composition, dietary fiber, phytic acid and protein hydrolysis of germinated ecuatorian brown rice. *Plant Foods for Human Nutrition* 69 (3): 261–267.

Centeno, C., Viveros, A., Brenes, A. et al. (2001). Effect of several germination conditions on total p, phytate p, phytase, and acid phosphatase activities and inositol phosphate esters in rye and barley. *Journal of Agricultural and Food Chemistry* 49 (7): 3208–3215.

Chung, T.Y., Nwokolo, E.N., and Sim, J.S. (1989). Compositional and digestibility changes in sprouted barley and canola seeds. *Plant Foods for Human Nutrition* 39 (3): 267–278.

Courtin, C.M., Gelders, G.G., and Delcour, J.A. (2001). Use of two endoxylanases with different substrate selectivity for understanding arabinoxylan functionality in wheat flour breadmaking. *Cereal Chemistry* 78 (5): 564–571.

Danisova, C., Holotnakova, E., Hozova, B. et al. (1994). Effect of germination on a range of nutrients of selected grains and legumes. *Acta Alimentaria* 23 (3): 287–298.

Dedio, W., Simmonds, D.H., Hill, R.D. et al. (1975). Distribution of α-amylase in the triticale kernel during development. *Canadian Journal of Plant Science* 55 (1): 29–36.

Delcour, J.A. and Hoseney, R.C. (eds.) (2010). Storage of cereals. In: *Principles of Cereal Science and Technology*, 107–120. St. Paul: American Association of Cereal Chemists (AACC).

Donkor, O.N., Stojanovska, L., Ginn, P. et al. (2012). Germinated grains – sources of bioactive compounds. *Food Chemistry* 135 (3): 950–959.

Elmaki, H.B., Babiker, E., and El Tinay, A.H. (1999). Changes in chemical composition, grain malting, starch and tannin contents and protein digestibility during germination of sorghum cultivars. *Food Chemistry* 64: 331–336.

Erba, D., Angelino, D., Marti, A. et al. (2018). Effect of sprouting on nutritional quality of pulses. *International Journal of Food Sciences and Nutrition* 70 (1): 30–40.

Esa, N.M., Kadir, K.A., Amom, Z. et al. (2011). Improving the lipid profile in hypercholesterolemia-induced rabbit by supplementation of germinated brown rice. *Journal of Agricultural and Food Chemistry* 59 (14): 7985–7991.

Faltermaier, A., Zarnkow, M., Becker, T. et al. (2015). Common wheat (*Triticum aestivum* L.): evaluating microstructural changes during the malting process by using confocal laser scanning microscopy and scanning electron microscopy. *European Food Research and Technology* 241 (2): 239–252.

Finney, P.L. (1982). Effect of germination on cereal and legume nutrient changes and food or feed value: a comprehensive review. In: *Mobilization of Reserves in Germination* (eds. C. Nozzolillo, P.J. Lea and F.A. Loewus), 229–305. New York: Plenum Press.

Garsetti, M., Vinoy, S., Lang, V. et al. (2005). The glycemic and insulinemic index of plain sweet biscuits: relationships to in vitro starch digestibility. *Journal of the American College of Nutrition* 24 (6): 441–447.

Grassi, S., Cardone, G., Bigagnoli, D. et al. (2018). Monitoring the sprouting process of wheat by non-conventional approaches. *Journal of Cereal Science* 3: 180–187.

Ha, K.S., Jo, S.H., Mannam, V. et al. (2016). Stimulation of phenolics, antioxidant and α-glucosidase inhibitory activities during barley (*Hordeum vulgare* L.) seed germination. *Plant Foods for Human Nutrition* 71 (2): 211–217.

Hager, A.S., Taylor, J.P., Waters, D.M. et al. (2014). Gluten free beer–a review. *Trends in Food Science & Technology* 36 (1): 44–54.

Haraldsson, A.K., Rimsten, L., Alminger, M.L. et al. (2004). Phytate content is reduced and β-glucanase activity suppressed in malted barley steeped with lactic acid at high temperature. *Journal of the Science of Food and Agriculture* 84 (7): 653–662.

Hartmann, G., Koehler, P., and Wieser, H. (2006). Rapid degradation of gliadin peptides toxic for coeliac disease patients by proteases from germinating cereals. *Journal of Cereal Science* 44 (3): 368–371.

Hefni, M. and Witthöft, C.M. (2011). Increasing the folate content in Egyptian baladi bread using germinated wheat flour. *LWT-Food Science and Technology* 44 (3): 706–712.

Heiniö, R.L., Oksman-Caldentey, K.M., Latva-Kala, K. et al. (2001). Effect of drying treatment conditions on sensory profile of germinated oat. *Cereal Chemistry* 78 (6): 707–714.

Heiniö, R.L., Noort, M.W.J., Katina, K. et al. (2016). Sensory characteristics of wholegrain and bran-rich cereal foods–a review. *Trends in Food Science & Technology* 47: 25–38.

Hithamani, G. and Srinivasan, K. (2014). Effect of domestic processing on the polyphenol content and bioaccessibility in finger millet (*Eleusine coracana*) and pearl millet (*Pennisetum glaucum*). *Food Chemistry* 164: 55–62.

Hübner, F. and Arendt, E.K. (2013). Germination of cereal grains as a way to improve the nutritional value: a review. *Critical Reviews in Food Science and Nutrition* 53 (8): 853–861.

ICC (1968). International Association of Cereal Science and Technology, Standard Methods. Verlag Moritz Schafer.

Izydorczyk, M.S. and Dexter, J.E. (2004). Barley: milling and processing. In: *Encyclopedia of Food Grains* (eds. C. Wrigley, H. Corke, K. Seetharaman, et al.), 57–68. Oxford: Elsevier.

Jonnalagadda, S.S., Harnack, L., Hai Liu, R. et al. (2011). Putting the whole grain puzzle together: health benefits associated with whole grains – summary of American Society for Nutrition 2010 satellite symposium. *The Journal of Nutrition* 141 (5): 1011S–1022S.

Juhász, R., Gergely, S., Gelencsér, T. et al. (2005). Relationship between NIR spectra and RVA parameters during wheat germination. *Cereal Chemistry* 82 (5): 488–493.

Kayahara, H. and Tsukahara, K. (eds.) (2000). *Flavor, Health and Nutritional Quality of Pre-Germinated Brown Rice*. Hawaii: Proceedings of the International Chemical Congress of Pacific Basin Societies (December 2000).

Koehler, P., Hartmann, G., Wieser, H. et al. (2007). Changes of folates, dietary fiber, and proteins in wheat as affected by germination. *Journal of Agricultural and Food Chemistry* 55 (12): 4678–4683.

Kruger, J.E. (1994). Enzymes of sprouted wheat and their possible technological significance. In: *Wheat* (eds. W. Bushuk and V.F. Rasper), 143–153. Boston: Springer.

Kulp, K. (1993). Enzymes as dough improvers. In: *Advances in Baking Technology* (eds. B.S. Kamel and C.E. Stauffer), 152–178. Boston: Springer.

Kumar, V., Sinha, A.K., Makkar, H.P.S. et al. (2010). Dietary roles of phytate and phytase in human nutrition: a review. *Food Chemistry* 20 (4): 945–959.

Larsson, M. and Sandberg, A.S. (1992). Phytate reduction in oats during malting. *Journal of Food Science* 57 (4): 994–997.

Liang, J., Han, B.Z., Nout, M.J.R. et al. (2008). Effects of soaking, germination and fermentation on phytic acid, total and in vitro soluble zinc in brown rice. *Food Chemistry* 110 (4): 821–828.

Lorenz, K. and Valvano, R. (1981). Functional characteristics of sprout-damaged soft white wheat flours. *Journal of Food Science* 46 (4): 1018–1020.

Lorenz, K., Collins, F., and Kulp, K. (1981). Sprouting of cereal grains-effects on starch characteristics. *Starch-Stärke* 33 (6): 183–187.

MacArthur, L.A., D'Appolonia, B.L., and Banasik, O.J. (1981). The falling number test—what is it and how does it work? North Dakota Agricultural Experiment Station Farm Research 38: 15–22.

Maillard, M.N., Soum, M.H., Boivin, P. et al. (1996). Antioxidant activity of barley and malt: relationship with phenolic content. *LWT-Food Science and Technology* 29 (3): 238–244.

Mäkinen, O.E. and Arendt, E.K. (2015). Nonbrewing applications of malted cereals, pseudocereals, and legumes: a review. *Journal of the American Society of Brewing Chemists* 73 (3): 223–227.

Mares, D.J., Mrva, K., and Fincher, G.B. (2016). The enzymes associated with food grains and their functions in processing. In: *Encyclopedia of Food Grains* (eds. C. Wrigley, H. Corke, K. Seetharaman, et al.), 138–145. Oxford: Elsevier.

Marti, A., Cardone, G., Nicolodi, A. et al. (2017). Sprouted wheat as an alternative to conventional flour improvers in bread-making. *LWT-Food Science and Technology* 80: 230–236.

Marti, A., Cardone, G., Pagani, M.A. et al. (2018). Flour from sprouted wheat as a new ingredient in bread-making. *LWT-Food Science and Technology* 89: 237–243.

McCleary, B.V. and Sturgeon, R. (2002). Measurement of alpha-amylase in cereal, food, and fermentation products. *Cereal Foods World* 47 (7): 299.

Miguel, Â.S.M., Martins-Meyer, T.S., Veríssimo da Figueiredo, É. et al. (2013). Enzymes in bakery: current and future trends. In: *Food Industry* (ed. I. Muzzalupo), 287–321. Rijeka: InTech.

Mofidi, A., Ferraro, Z.M., Stewart, K.A. et al. (2012). The acute impact of ingestion of sourdough and whole-grain breads on blood glucose, insulin, and incretins in overweight and obese men. *Journal of Nutrition and Metabolism* 2012: 1–9.

Mohan, B.H., Malleshi, N.G., and Koseki, T. (2010). Physico-chemical characteristics and non-starch polysaccharide contents of *Indica* and *Japonica* brown rice and their malts. *LWT-Food Science and Technology* 43 (5): 784–791.

Morris, C.F. and Rose, S.P. (1996). Wheat. In: *Cereal Grain Quality* (eds. R.J. Henry and P.S. Kettlewell), 3–54. Dordrecht: Springer.

Müntz, K. (1996). Proteases and proteolytic cleavage of storage proteins in developing and germinating dicotyledonous seeds. *Journal of Experimental Botany* 47 (5): 605–622.

Nelson, K., Stojanovska, L., Vasiljevic, T. et al. (2013). Germinated grains: a superior whole grain functional food? *Canadian Journal of Physiology and Pharmacology* 91 (6): 429–441.

Noda, T., Takigawa, S., Matsuura-Endo, C. et al. (2004). The physicochemical properties of partially digested starch from sprouted wheat grain. *Carbohydrate Polymers* 56 (3): 271–277.

Okada, T., Sugishita, T., Murakami, T. et al. (2000). Effect of the defatted rice germ enriched with GABA for sleeplessness, depression, autonomic disorder by oral administration. *Nippon Shokuhin Kagaku Kogaku Kaishi/Journal of the Japanese Society for Food Science and Technology* 47: 596–603.

Olaerts, H. and Courtin, C.M. (2018). Impact of preharvest sprouting on endogenous hydrolases and technological quality of wheat and bread: a review. *Comprehensive Reviews in Food Science and Food Safety* 17 (3): 698–713.

Omary, M.B., Fong, C., Rothschild, J. et al. (2012). Effects of germination on the nutritional profile of gluten-free cereals and pseudocereals: a review. *Cereal Chemistry* 89 (1): 1–14.

Pagand, J., Heirbaut, P., Pierre, A. et al. (2017). The magic and challenges of sprouted grains. *Cereal Foods World* 62 (5): 221–226.

Patil, S.B. and Khan, M.K. (2011). Germinated brown rice as a value added rice product: a review. *LWT-Journal of Food Science and Technology* 48 (6): 661–667.

Peterson, D.M. (1998). Malting oats: effects on chemical composition of hull-less and hulled genotypes. *Cereal Chemistry Journal* 75 (2): 230–234.

Platel, K., Eipeson, S.W., and Srinivasan, K. (2010). Bioaccessible mineral content of malted finger millet (*Eleusine coracana*), wheat (*Triticum aestivum*), and barley (*Hordeum vulgare*). *Journal of Agricultural and Food Chemistry* 58 (13): 8100–8103.

Plaza, L., de Ancos, B., and Cano, P.M. (2003). Nutritional and health-related compounds in sprouts and seeds of soybean (*Glycine max*), wheat (*Triticum aestivum* L). And alfalfa (*Medicago sativa*). Treated by a new drying method. *European Food Research and Technology* 216 (2): 138–144.

Pongrac, P., Potisek, M., Fraś, A. et al. (2016). Composition of mineral elements and bioactive compounds in tartary buckwheat and wheat sprouts as affected by natural mineral-rich water. *Journal of Cereal Science* 69: 9–16.

Porter, J.R. and Gawith, M. (1999). Temperatures and the growth and development of wheat: a review. *European Journal of Agronomy* 10 (1): 23–36.

Pyler, R.E. and Thomas, D.A. (2000). Malted cereals: their production and use. In: *Handbook of Cereal Science and Technology* (eds. K. Kulp and J.G. Ponte), 685–696. New York: Marcel Dekker Inc.

Richter, K., Christiansen, K., and Guo, G. (2014). Wheat sprouting enhances bread baking performance. *Cereal Foods World* 59 (5): 231–233.

Ross, A.S. and Bettge, A.D. (2009). Passing the test on wheat end-use quality. In: *Wheat: Science and Trade* (ed. B.F. Carver), 468–470. New York: Wiley.

Santos-Buelga, C. and Scalbert, A. (2000). Proanthocyanidins and tannin-like compounds-nature, occurrence, dietary intake, and effects on nutrition and health. *Journal of the Science of Food and Agriculture* 80: 1094–1117.

Sehgal, S. and Kawatra, A. (2001). In vitro protein and starch digestibility of pearl millet (*Pennisetum gluacum* L.). as affected by processing techniques. *Molecular Nutrition & Food Research* 45 (1): 25–27.

Sekhon, K., Singh, N., Kaur, H. et al. (1995). Improving the functional and bread making properties of sprouted Indian wheat. *Journal of Food Processing and Preservation* 19 (2): 147–160.

Sibian, M.S., Saxena, D.C., and Riar, C.S. (2017). Effect of germination on chemical, functional and nutritional characteristics of wheat, brown rice and triticale: a comparative study. *Journal of the Science of Food and Agriculture* 97 (13): 4643–4651.

da Silva, S.Z., da Rosa, T.C.M., Tessaro, D. et al. (2014). Technological quality of non-germinated and germinated whole wheat flour. *Journal of Food, Agriculture and Environment* 12 (2): 132–134.

Simsek, S., Ohm, J.B., Lu, H. et al. (2014). Effect of pre-harvest sprouting on physicochemical changes of proteins in wheat. *Journal of the Science of Food and Agriculture* 94 (2): 205–212.

Singh, C.B., Jayas, D.S., Paliwal, J. et al. (2009). Detection of sprouted and midge-damaged wheat kernels using near-infrared hyperspectral imaging. *Cereal Chemistry* 86 (3): 256–260.

Singh, A.K., Rehal, J., Kaur, A. et al. (2015). Enhancement of attributes of cereals by germination and fermentation: a review. *Critical Reviews in Food Science and Nutrition* 55 (11): 1575–1589.

Soetan, K.O. and Oyewole, O.E. (2009). The need for adequate processing to reduce the anti-nutritional factors in plants used as human foods and animal feeds: a review. *African Journal of Food Science* 3 (9): 223–232.

Suárez-Estrella, D., Torri, L., Pagani, M.A. et al. (2018). Quinoa bitterness: causes and solutions for improving product acceptability. *Journal of the Science of Food and Agriculture* 98 (11): 4033–4041.

Świeca, M. and Dziki, D. (2015). Improvement in sprouted wheat flour functionality: effect of time, temperature and elicitation. *International Journal of Food Science & Technology* 50 (9): 2135–2142.

Świeca, M., Baraniak, B., and Gawlik-Dziki, U. (2013). In vitro digestibility and starch content, predicted glycemic index and potential in vitro antidiabetic effect of lentil sprouts obtained by different germination techniques. *Food Chemistry* 138 (2–3): 1414–1420.

Świeca, M., Dziki, D., and Gawlik-Dziki, U. (2017). Starch and protein analysis of wheat bread enriched with phenolics-rich sprouted wheat flour. *Food Chemistry* 228: 643–648.

Teixeira, C., Nyman, M., Andersson, R. et al. (2016). Effects of variety and steeping conditions on some barley components associated with colonic health. *Journal of the Science of Food and Agriculture* 96 (14): 4821–4827.

Trachoo, N., Boudreaux, C., Moongngarm, A. et al. (2006). Effect of geminated rough rice media on growth of selected probiotic bacteria. *Pakistan Journal of Biological Sciences* 9 (14): 2657–2661.

Van der Maarel, M.J. (2009). Starch-processing enzymes. In: *Enzymes in Food Technology* (eds. R.J. Whitehurst and M. van Oort), 320–331. Blackwell Publishing Ltd.

Van Hung, P., Hatcher, D.W., and Barker, W. (2011a). Phenolic acid composition of sprouted wheats by ultra-performance liquid chromatography (UPLC) and their antioxidant activities. *Food Chemistry* 126 (4): 1896–1901.

Van Hung, P., Maeda, T., Yamamoto, S. et al. (2011b). Effects of germination on nutritional composition of waxy wheat. *Journal of the Science of Food and Agriculture* 92: 667–672.

Van Hung, P., Maeda, T., and Morita, N. (2015). Improvement of nutritional composition and antioxidant capacity of high-amylose wheat during germination. *LWT-Journal of Food Science and Technology* 52 (10): 6756–6762.

Vinje, M.A., Duke, S.H., and Henson, C.A. (2015). Comparison of factors involved in starch degradation in barley germination under laboratory and malting conditions. *Journal of the American Society of Brewing Chemists* 73 (2): 195–205.

Watanabe, M., Maeda, T., Tsukahara, K. et al. (2004). Application of pregerminated brown rice for breadmaking. *Cereal Chemistry* 81 (4): 450–455.

Workman, J.J. and Weyer, L. (2008). *Practical Guide to Interpretive near-Infrared Spectroscopy*. Boca Raton: CRC Press Inc.

Xu, J., Zhang, H., Guo, X. et al. (2012). The impact of germination on the characteristics of brown rice flour and starch. *Journal of the Science of Food and Agriculture* 92 (2): 380–387.

Yang, T.K., Basu, B., and Ooraikul, F. (2001). Studies on germination conditions and antioxidant contents of wheat grain. *International Journal of Food Sciences and Nutrition* 52 (4): 319–330.

Zamprogna Rosenfeld, E., Bellaio, S., Mane, D. et al. (2011). A new family of healthy, safe, and convenient food products based on partial germination of pulses. *Cereal Foods World* 56 (S4): A71.

Ziegler, P. (1995). Carbohydrate degradation during seed development. In: *Seed Development and Germination* (eds. J. Kigel and G. Galili), 447–474. New York: Marcel Dekker Inc.

Žilić, S., Basić, Z., Hadži-Tašković Šukalović, V. et al. (2014). Can the sprouting process applied to wheat improve the contents of vitamins and phenolic compounds and antioxidant capacity of the flour? *International Journal of Food Science and Technology* 49: 1040–1047.

7
Novel Ingredients from Cereals

Dominic Agyei[1], Jaison Jeevanandam[2], Christian Kwesi Ofotsu Dzuvor[3], Sharadwata Pan[4], Michael Kobina Danquah[5], Caleb Acquah[6], and Chibuike C. Udenigwe[6]

[1] *Department of Food Science, University of Otago, Dunedin, New Zealand*
[2] *CQM – Centro de Química da Madeira, Universidade da Madeira, Funchal, Portugal*
[3] *Bioengineering Laboratory, Department of Chemical Engineering, Monash University, Clayton, Victoria, Australia*
[4] *School of Life Sciences, Weihenstephan, Technical University of Munich, Freising, Germany*
[5] *Department of Chemical Engineering, University of Tennessee, Chattanooga, TN, USA*
[6] *School of Nutrition Sciences, University of Ottawa, Ottawa, Ontario, Canada*

7.1 Introduction

Cereals are the most widely cultivated and consumed plant food known to humans, as they account for 50% of dietary calories and proteins worldwide (Yu and Tian 2018). Historically, cereals have been a significant component of food formulation used by humanitarian and disaster relief/aid agencies (Caiafa and Wrabel 2019). Also, cereals are the subject of many plant breeding programs aimed at improving the nutritional status of indigenous staple foods (Yu and Tian 2018). Moreover, the consumption of breakfast cereals is a growing trend in both developed and developing economies, and cereals are one of the most widely used base materials to which micronutrients are added in biofortification interventions (Wiemer 2018). Whole cereal grains are rich sources of essential nutrients such as vitamins, minerals, dietary fiber, proteins, lipids, and carbohydrates. Cereals are also rich in secondary metabolites such as polyphenols, anthocyanidins, anthocyanins, and avenanthramides (AVAs), with demonstrated bioactive and health applications. Since, the key macromolecules in cereals are generally fermentable, fermentation as a processing step can be used as an *in situ* approach for the production and/or accumulation of important nutrients such as the B vitamins, as well as to further improve bioaccesibiltiy/bioavailability (Coda et al. 2017; Chamlagain et al. 2018). Their rich biomolecule profile gives cereals the ability to be used in food formulations for nutritional, health, sensorial, and functional applications. Therefore, this chapter gives a broad overview of the functional ingredients in cereals.

Innovative Processing Technologies for Healthy Grains, First Edition.
Edited by Milica Pojić and Uma Tiwari.

7.2 Structure, Biochemistry, and Bioactivity of Cereal Ingredients

The economic value of a cereal is determined by the presence and quantity of active ingredients that are essential for applications in several branches of industry, apart from the food industry – pharmaceutical, nutraceutical, and cosmetic industry. The structure and biochemistry of the cereal ingredients help in enhancing their bioactivity, making them suitable active ingredients and essential nutrient sources for humans and animals. The active ingredients present in most of the cereals are carbohydrates, proteins, peptides, lipids, secondary metabolites, vitamins, and minerals.

7.2.1 Carbohydrates

Cereals, legumes, pseudocereals, unripe fruits, and tubers generally consist of 40–80% of dry matter as carbohydrates. Carbohydrates consist of carbon, hydrogen, and oxygen atoms, and based on these atomic compositions, they form monosaccharides, disaccharides, oligosaccharides, and polysaccharides (Doubet et al. 1989; Lütteke et al. 2005). The abundance and uncomplicated extraction of carbohydrates via inexpensive methods make them a significant nutritional entity and a principal energy source in the human diet. Storage or structural polysaccharides and oligosaccharides or simple sugars are the forms in which carbohydrates are present in plants (Repo-Carrasco-Valencia and Arana 2017). In cereals, carbohydrates can be categorized into starch and non-starch polysaccharides.

7.2.1.1 Starch The majority of cereals possess carbohydrates in the form of starch, which is a unique calorie source for human nutrition and are used in a wide variety of industrial sectors as renewable raw materials (Carciofi et al. 2012). Starch is a biopolymer that is formed as a plant photosynthetic carbon fixation product (Schwartz and Whistler 2009), and was reported to be used in the pre-dynastic period by the Egyptians for producing cemented papyrus strips in combination with wheat adhesives (Inglett 1974). Since then, starch from various cereals and plant materials has been used in several applications such as textiles (Teli et al. 2009a), paper production (Roux and Voigt 2014), color printing (Teli et al. 2009b), and cement additives (Reddy et al. 2013). In modern biomedical sciences, starches are widely used in pharmaceutical applications such as bone cements, drug delivery carriers (Pereira et al. 1998), tablet excipient (Garr and Bangudu 1991), thermoplastic implant materials (de Carvalho and Trovatti 2016), and biopackaging (Alvarez et al. 2017). High-amylose, waxy, chemically, and naturally modified starches are the most widely used in the food industry. Generally, starch polymers are composed of amylose molecules which were modified and increased from 65 to 85% to form high-amylose starch (Whistler and Doane 1961; Richardson et al. 2000). Modified starches are used in candy manufacturing (Sajilata and Singhal 2005), tomato paste (Bo-Linn et al. 1982), and as an apple sauce modifier (Bae 2014), as well as in degradable plastics (Woggum et al. 2015). Likewise, waxy starch is a unique starch formed with a single amylopectin molecule which is obtained from the endosperm of the cereal kernel (Franco et al. 1998). Physical methods, such as extrusion (Jiamjariyatam et al. 2015), gelatinization (Smrcková et al. 2013), annealing (Wang et al. 2014), gelation (Kong et al. 2015), heat moisture treatment (Lee et al. 2012), and biochemical methods, namely hydrolysis (Kittisuban et al. 2014), acetylation (Pietrzyk et al. 2018), and amylopectin debranching (Klaochanpong et al. 2015), are used for processing of waxy starch. Waxy starches from cereals, such as glutinous rice, maize, and barley, are useful in the food industry due to their stickiness and carbohydrate content (Šárka and Dvořáček 2017).

In recent years, resistant starch has gained importance among researchers due to its desired bioavailability and enhanced dietary fiber contents (Zhao et al. 2018). These starch types are modified by enzymatic processes (*in vitro* treatment of pullulanase and α-amylase) to make them resistant to hydrolysis (Englyst et al. 1992). Resistant starch possesses unique characteristics such as smaller particle size, bland flavor, white appearance, low water-holding capacity (Fausto et al. 1997), and necessary physicochemical properties such as increased viscosity, water-binding capability, swelling, and gel formation (Sajilata et al. 2006). These starch types have benefits such as improved dietary fiber components (Rideout et al. 2017), colon cancer prevention (Topping and Clifton 2001), hypoglycemic complication reduction (Sun et al. 2017), prebiotics (Zaman and Sarbini 2016), inhibition of gall stone formation (Fuentes-Zaragoza et al. 2010), fat accumulation (Keenan et al. 2006), hypocholesterolemia conditions (Anandharaj et al. 2014), and in mineral absorption (Zeng et al. 2017). There are several sub-classes of starches such as types A, B, and C (categorized based on X-ray diffraction), as well as rapidly digestible and slowly digestible starches (categorized based on rate of enzyme action). Resistant starches are classified into types I, II, III, and IV, based on their nutritional characteristics (Sajilata et al. 2006). Chemical and natural modifications of these starches are necessary to improve their quality and performance. The chemically modified starches help in yielding frozen, encapsulated, dehydrated, instant, and heat-and-serve food products with suitable shelf life, texture, and improved tolerance to processing conditions such as mechanical shear, improved heat, and acid stability (Włodarczyk-Stasiak et al. 2017). However, chemically modified starches may cause toxic reactions in humans, so they are often replaced with naturally modified starches. Developments in genetics and cross-breeding are widely utilized to modify the biosynthesis of cereal starches naturally, which eventually supports the starch modification. These modifications, through genes, are used to determine amylose and amylopectin molecule structure and ratio, produce intermediate and high amylose starch, or waxy starch forms (Gérard et al. 2001; Waterschoot et al. 2015).

7.2.1.2 Non-Starch Polysaccharides Non-starch polysaccharides include a wide variety of polysaccharides other than α-glucans (starch). Cellulose, non-cellulose polymer, and pectic polysaccharides are the important classifications of non-starch polysaccharides. Non-cellulose polymers are sub-classified into xyloglucan, β-glucans, arabinoxylans, and mannans, whereas arabinogalactan, galactan, and polygalacturonic acid replaced with arabinan are the sub-classes of pectic polysaccharides (Sinha et al. 2011). However, the majority of cereals contain cellulose, arabinoxylans, and β-glucans as non-starch polysaccharides, which contribute value-added nutrition to the consumers.

7.2.1.2.1 Cellulose Cellulose, a complex polysaccharide, is present in the plant cell wall as an elementary structural component. The digestion of cellulose in humans is not possible due to a lack of a cellulase enzyme needed to cleave its β (1–4) glycosidic bonds, unlike in animals (O'Sullivan 1997; Poletto et al. 2014). Cereals contain approximately 2.5% of cellulose, depending on the species (Koehler and Wieser 2013). Cellulose from cereals is used in paper products (Ververis et al. 2004), textile fibers

(Costa et al. 2013), thin film chromatography (Vomhof and Tucker 1965), and biofuels (Janssen et al. 2010). Moreover, the cellulose (nano-sized cellulose crystals) from cereals is used for drug delivery applications (Lin and Dufresne 2014).

7.2.1.2.2 Arabinoxylans The most important non-starch polysaccharide in cereals such as wheat, sorghum, barley, rye, oat, and rice, are arabinoxylans. Most cereal grains possess around 60–70% of arabinoxylans, whereas their amount varies in barley (20%) and rice (40%) (Koehler and Wieser 2013). It was reported that non-endospermic wheat tissues and pericarp have 64% of high arabinoxylans content (Selvendran and Du Pont 1980). On the basis of their solubility, arabinoxylans are classified into water extractable and unextractable forms. Wheat consists of 25–30%, while rye contains 15–25% of water extractable arabinoxylans with a predominant role in bread-making. This type of non-starch polysaccharide can absorb up to 15–20 times more water than their own weight, which helps them to form high viscous solutions and increases their gas holding capacity, especially in wheat (Jan et al. 1995). Two pentoses, xylose and arabinose, are the sugars in cereal arabinoxylan (Izydorczyk and Biliaderis 1995). Cereal arabinoxylans are insoluble in water due to their alkali-labile ester cross-links mediated anchoring of the cell wall (Izydorczyk 2017). Highly branched arabinoxylan peptides that are soluble in water are present in pentosan fractions which do not have any significant effect in the processing of cereals (Fausch et al. 1963; Huang et al. 2018). The intermolecular hydrogen bonds between the unsubstituted xylan backbone region enables these polysaccharides to form junction zones (Buksa et al. 2014). These zones are highly beneficial in determining the solubility and conformation of arabinoxylans and also in predicting their anti-nutritional activities (Buksa et al. 2014).

7.2.1.2.3 β-Glucans β-glucans are linear D-glucose chains with links of mixed β-(1,3)- and β-(1,4)-glycosides with higher water solubility than arabinoxylans. They are also called lichenins; and barley and oats are composed of 3–7% and 3.5–5% of β-glucans, respectively. Almost all cereals possess at least 2% of β-glucans (Klopfenstein 1988; Zielke et al. 2018). These β-glucans have been accepted as bioactive and functional ingredients for the last two decades (Cui and Wood 2000). These non-starch polysaccharides from cereals, present in the endospermic and sub-aleurone cell wall (Ebringerová 2005), are widely associated with the control of postprandial serum glucose levels and reduction of plasma cholesterol in humans and animals (Granfeldt et al. 2008; Sima et al. 2018). The ubiquitous β-glucans structure is of high importance, since it determines their functionality, physical properties, and physiological responses (Wood 2010). It should be noted that cellulose and β-glucans consists of similar β-linked glucose units, except the former is stiff, non-soluble, and highly crystalline, whereas the latter is flexible and soluble (Mikkelsen et al. 2015). Additionally, these polysaccharides form a gel under favorable conditions which is utilized to modify the structure and texture of different food systems, such as salad dressings and ice creams (Borchani et al. 2016).

7.2.2 Proteins, Peptides, and Amino Acids

Cereals contain varying amounts of proteins, but overall the levels of proteins are lower than carbohydrates. Cereal genotype, such as variety and species, their growing conditions, namely soil, climate, and fertilization, especially the quantity of nitrogen fertilization, are the factors that determine the amount of proteins present in each cereal (Belitz et al. 2009). Moreover, cereal proteins are classified into four major groups such as storage, structural, metabolic, and protective proteins, along with bioactive peptides (Wrigley and Bekes 2001). In addition, cereal seeds contain Osborne fractions in three different tissues of the grain (Shewry and Halford 2002).

7.2.2.1 Osborne Proteins Albumins, prolamins, glutelins, and globulins are the sub-classes of proteins in cereals, as categorized by Osborne (1907). Among these proteins, only albumins are soluble in water (Shewry 2017), whereas solubility was achieved with globulins using dilute salt solutions (Larkins et al. 2017), prolamins using 60–70% ethanol (Balindong et al. 2016), and glutelins using aqueous alcohol mixtures, disaggregating compounds, and reducing agents (Rosell et al. 2014). Enzyme and enzyme inhibitor-mediated metabolism is the main function of albumins and globulins (Singh and Skerritt 2001), while glutelins and prolamins are considered as storage proteins. These storage proteins possess a crucial function of supplying amino acid and nutrition for cereal seedlings during their germination (Shewry and Tatham 1990).

7.2.2.2 Storage Proteins Storage proteins differ from one another in terms of their molecular weight, intra-and interchain disulfide linkages, and amino acid sequence, as well as composition. Thus, wheat, barley, and rye are identified as containing storage proteins that are closely related, except for oats, which consist of structurally divergent glutelins. Based on sulfur content and molecular weight, prolamin storage proteins are categorized into sulfur-rich, sulfur-poor, and high molecular weight prolamins. Storage proteins are classified into low-, high-, and medium molecular weight group, based on their molecular masses and sequences of amino acids. Monomeric proteins, namely γ-40 k-secalins, γ-hordeins, avenins, α, β, and γ-gliadins, along with aggregative proteins such as glutenin subunits, B-hordeins, and γ-75 k-secalins, are present in the low molecular weight group, including the 28000–35000 range of molecular weight and 300 amino acid residues. Likewise, the high molecular weight group includes glutenin subunits, secalins, and D-hordeins in wheat, rye, and barley, respectively, with molecular weights of between 70000 and 90000 and residues of 600–800 amino acids. Similarly, the medium molecular weight group contains residues of amino acid in the range of 300–400 with molecular weight of 40000 along with ω-secalins, C-hordeins, and ω 1,2-gliadins. It should be noted that the properties and structure of storage proteins are predominantly determined by their disulfide bonds. Intra- and inter-chain cysteine residues of the sulfhydryl groups play a crucial role in the formation of disulfide bonds. These storage proteins are highly useful as nutrients for humans, and this determines the market value of a cereal. Likewise, structural (Guerrieri and Cavaletto 2018), metabolic (Shewry and Casey 1999), and protective proteins (Shewry 2000), are beneficial in the growth of cereal plants and are useful as sources of high-value nutrients for humans.

7.2.2.3 Bioactive Peptides Protein molecules that are smaller than 10 kDa are called peptides (Farrokhi et al. 2008), which can be obtained either naturally or derived from cryptic native protein sequences. Digestion mediated hydrolysis, plant, and microbial proteolytic enzymes are some of the methods that can lead to the release of peptides (Coda et al. 2012). Bioactive peptides possess biomedical properties such as antioxidant, antithrombotic, antimicrobial, antiproliferative properties, blood pressure lowering ability, opioid-like activities, and bioavailability (Zambrowicz et al. 2013). It has been reported in the literature that cereals such as wheat, rice, corn, barley, and pseudocereals, namely amaranth and buckwheat, contain huge quantities of bioactive peptides (Malaguti et al. 2014). Cereals, such as barley and rye contain enormous bioactive peptides, such as lunasin, xanthan oxidase, and angiotensin converting enzyme inhibitory enzymes (Jeong et al. 2010; Lee et al. 2010). These bioactive peptides present in cereals are proved to be extremely helpful in the treatment of chronic colitis, reducing postprandial glucose level in diabetic patients, reducing hypoglycemia risk, and also in inhibiting growth of cancer cells (Cavazos and Gonzalez de Mejia 2013).

7.2.2.4 Amino Acids On average, cereals contain varying levels of all dietary amino acids (Table 7.1). Broadly speaking, these cereal amino acids are classified into trimethyl glycine, essential amino acids, and branched chain amino acids (BCAAs).

7.2.2.4.1 Trimethyl Glycine Cereals are an important source of methyl donors such as choline, betaine, and folate in the diet (Bruce et al. 2010). Betaine, or glycine betaine, which is a form of N,N,N-trimethyl glycine, are crucial in the human body as a methyl donor and an osmolyte (Craig 2004). Betaine also helps to decrease homocysteine circulation, where elevated concentration of homocysteine in plasma indicates a cognitive impairment and vascular disease risk (Bates et al. 2010). Recent reports suggest that cereals such as wheat and pseudocereals, namely quinoa, amaranth, and buckwheat, possess a high quantity of betaine (Ross et al. 2014). Similarly, esterified choline and 5-methyltetrahydrofolate (5-CHO-H4 folate) are also present in cereals in considerable quantities (Hefni et al. 2018).

7.2.2.4.2 Essential Amino Acids Among amino acids, histidine, isoleucine, leucine, lysine, methionine, phenylalanine, threonine, tryptophan, and valine are grouped as essential amino acids (Shewry 2007). Several studies show that oats, millet, and wheat possess typical proteins with essential amino acids in elevated quantities (see Table 7.1) (Fontaine et al. 2002; Amadou et al. 2013). The inclusion of these essential amino acids from cereals in human and animal diets is highly essential to modulate adiposity, and enhance immune function and antioxidant activity. Thus, the presence of amino acids in cereals is utilized in promoting whole body and muscle protein synthesis for active lifestyles (Ha and Zemel 2003).

7.2.2.4.3 Branched Chain Amino Acids Animal and human metabolism can be directly or indirectly affected by branched amino acid mediated nutrient signals, which includes essential amino acids, namely leucine, valine, and isoleucine (Lu et al. 2013). These amino acids are involved in stimulating the mammalian target of rapamycin complex 1 (mTORC1), signaling a pathway for regulating mRNA translation (Vary and Lynch 2007), and eventually aiding protein synthesis. Cereals such as sorghum, millet, oats, triticale, and corn possess high quantities of BCAAs (see Table 7.1) (Wang et al. 2018; Kawaguchi et al. 2017). The BCAAs help in metabolic health improvement and glucose homeostasis in humans and animals (Lynch and Adams 2014; Yoon 2016). However, elevated levels of these BCAAs in plasma will lead to metabolic disorders, including insulin resistance and type 2 diabetes (Wang et al. 2011b; Rohini et al. 2018).

Table 7.1 Amino acid contents of some commonly-consumed cereals and pseudocereals.

Amino acids	Triticale	Quinoa	Rye	Sorghum	Millet	Oats	Barley	Buckwheat	Brown rice	Wheat	Yellow corn
Tryptophan	0.16	0.17	0.11	0.12	0.12	0.18	0.18	0.19	0.09	0.05	0.07
Threonine	0.41	0.42	0.29	0.35	0.35	0.38	0.36	0.51	0.27	0.17	0.35
Isoleucine	0.48	0.50	0.21	0.43	0.47	0.50	0.38	0.50	0.31	0.22	0.34
Leucine	0.91	0.84	0.56	1.49	1.40	0.98	0.71	0.83	0.60	0.43	1.16
Lysine	0.37	0.77	0.29	0.23	0.21	0.64	0.39	0.67	0.28	0.16	0.27
Methionine	0.20	0.31	0.15	0.17	0.22	0.21	0.20	0.17	0.16	0.10	0.20
Cystine	0.28	0.20	0.44	0.13	0.21	0.46	0.23	0.23	0.09	0.12	0.17
Phenylalanine	0.64	0.59	0.20	0.55	0.58	0.67	0.59	0.52	0.37	0.30	0.46
Tyrosine	0.38	0.27	0.32	0.32	0.34	0.40	0.30	0.24	0.27	0.14	0.38
Valine	0.61	0.59	0.45	0.56	0.58	0.69	0.52	0.68	0.42	0.27	0.48
Arginine	0.67	1.09	0.19	0.36	0.38	0.85	0.53	0.98	0.55	0.27	0.47
Histidine	0.31	0.41	0.41	0.25	0.24	0.28	0.24	0.31	0.18	0.15	0.29
Alanine	0.49	0.59	0.56	1.03	0.99	0.56	0.41	0.75	0.42	0.21	0.71
Aspartic acid	0.79	1.13	2.29	0.74	0.73	1.22	0.66	1.13	0.68	0.31	0.66
Glutamic acid	4.01	1.87	0.42	2.44	2.40	2.83	2.74	2.05	1.47	1.88	1.77
Glycine	0.56	0.69	0.80	0.35	0.29	0.64	0.38	1.03	0.36	0.23	0.39
Proline	1.18	0.77	0.46	0.85	0.88	0.45	1.25	0.51	0.34	0.62	0.82
Serine	0.59	0.57	0.00	0.46	0.64	0.71	0.44	0.69	0.37	0.30	0.45
[a]Total essential amino acids	4.08	4.60	2.67	4.15	4.16	4.52	3.56	4.38	2.68	1.85	3.60
[b]Total branch chain amino acids	2.00	1.94	1.23	2.49	2.44	2.17	1.61	2.01	1.33	0.92	1.97

[a] Calculated from the sum of the content of the essential amino acids: includes histidine, isoleucine, leucine, lysine, methionine, phenylalanine, threonine, tryptophan, and valine.

[b] Calculated from the sum of the content of the branched chain amino acids: leucine, isoleucine, and valine; g/100 g.

(*Source*: adapted from USDA 2018)

7.2.3 Lipids

There has been paramount interest in cereal lipids because of their nutritional and health benefits (Tong et al. 2014). According to Koehler and Wieser (2013), cereal lipids developed from membranes, organelles, and spherosomes exhibit distinct chemical structures. These lipids are largely distributed in the germ layer, followed by the aleurone layer and to a smaller degree in the endosperm (Morrison 1988; Koehler and Wieser 2013). For instance, in wheat seeds, the germ contains 25–30% lipid, whereas the endosperm contains only 1%. The majority of these lipids are primarily attached with protein bodies and starch. The latter is separated into three components; starch lipids, starch surface lipids, and non-starch lipids in the endosperm (Tong et al. 2014). In addition, non-starch lipids can be divided into nonpolar and polar forms. The nonpolar lipids are present in the free form, whereas the polar lipids are complexed with proteins (Koehler and Wieser 2013) and include glycolipids (GLs) and phospholipids. The starch lipids primarily consist of lysophospholipids (LPLs). Cereal starches from wheat, barley, rye, and triticale contain LPLs exclusively, while other cereals such as rice and maize have large quantities of nonpolar lipids in the form of free fatty acids (Morrison 1988; Koehler and Wieser 2013; Tong et al. 2014). Moreover, rice grain contains bran (5%), of which 12–18.5% is oil. Rice bran oil in turn contains 10–15% of high-quality proteins along with lipids in the form of monounsaturated (47%), polyunsaturated (33%), and saturated (20%) fats (Gul et al. 2015).

7.2.3.1 Lysophospholipids LPLs are a very prominent subclass of phospholipids, which show exceptional physical and biological characteristics not found in their parent phospholipids. LPLs merge with starch to form an amylose-lipid complex in the cereal endosperm (Koehler and Wieser 2013; Tong et al. 2014; Chuan et al. 2016). This inclusion complex affects the viscosity and swelling properties of starch (Tong et al. 2014), thus reducing the digestibility of amylose. In addition, LPLs have a consequential connection with pasting properties, viscosity, breakdown, and consistency of cereal such as rice (Tong et al. 2014; Chuan et al. 2016).

LPLs (also known as lysolecithin) are normal and natural ingredient cereal grains, which are formed during the ripening stages. An LPL is a phosphate-containing lipid characterized by a glycerol backbone which lacks one acyl chain and has only one hydroxyl group of the glycerol backbone acylated. Moreover, they play important roles in phospholipid metabolism and thus function as second messengers (D'Arrigo and Servi 2010). Although LPLs are available in small quantities in biological cell membranes, they can cause cell fusion and lysis at higher concentration (Tigyi 2013).

LPLs exist in three different forms: sphingosine-1-phosphate (S1P), lysophosphatidic acid (LPA), and sphingosylphosphorylcholine (SPC) (Figure 7.1) (Karliner 2002; Tigyi 2013). These lipids act as ligands for heptahelical membrane-spanning G-protein coupled receptors (Karliner 2002). S1P are ceramide derivatives found in the myelin sheath, which contains phosphocholine or phosphoethanelamine as a base group (Bourtsala and Galanopoulou 2018). This SIP is extracellularly generated through phosphorylation of sphingosine by sphingosine kinase (Karliner 2002; Bourtsala and Galanopoulou 2018). SIP engages five S1P receptors (S1PR1–S1PR5). All of these S1P belong to cell surface heterotrimeric G protein-coupled receptors (Brewer 2013; Bourtsala and Galanopoulou 2018). LPLs have been implicated in several physiological processes such as the reproduction, vascular, and nervous systems. For instance, S1P and LPA may contribute to angiogenesis, atherosclerosis, myocardial hypertrophy,

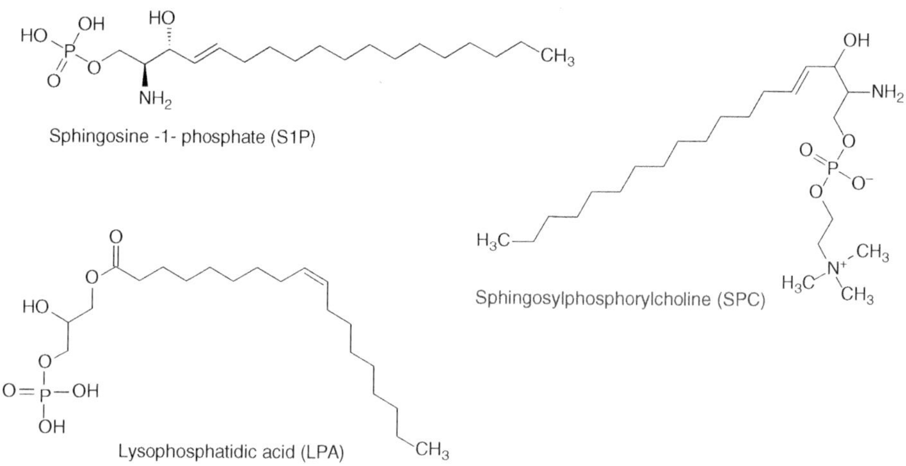

Figure 7.1 Chemical structures of some cereal-derived lysophospholipids. *Source*: Karliner (2002); Tigyi (2013).

obesity, and cancer (Karliner 2002; D'Arrigo and Servi 2010). In addition, LPAs play an important role in transmembrane signal transduction processes, and hence act as a platelet activating factor.

7.2.3.2 Glycolipids Glycolipds (GLs) are a class of polar, amphipathic lipids that has gained increasing interest because of their important nutritional and health benefits. GLs are also characterized by two fatty acid molecules bonded to a glycerol backbone at position sn1 and sn2, which combine with a sugar molecule to form the head. In wheat and most cereals, the sugar is mainly galactose (Rosentrater and Evers 2017). GLs are rich in 16- and 18-carbon saturated and unsaturated fatty acids and often contain polyunsaturated fats (PUFAs), which are crucial for the maintenance and improvement of cellular health (da Costa et al. 2016). GLs comprise of three major classes: monogalactosyldiacyl glycerolipids (MGDGs), digalactosyl diaglycerolipids (DGDGs), and sulfoquinovosyl diacylglycerolipids (SQDG) (Figure 7.2). DGDGs has a cylindrical shape, whereas MGDG is conical in shape (Rosentrater and Evers 2017). GLs possess biological activities such as antifungal, antiviral, antitumoral, and antimicrobial properties. In addition, they possess health benefits such as anti-inflammatories (da Costa et al. 2016).

7.2.4 Secondary Metabolites

Secondary metabolites have received tremendous attention over the past few decades due to their roles in human health. They are products of secondary metabolism in plants (Gani et al. 2012). Cereals are rich sources of secondary metabolites, which can be categorized into three broad group: phytoanticipins, signaling molecules, and phytoalexin (Bhanja Dey et al. 2016; Meyer et al. 2016). In addition, the secondary metabolites include polyphenols and AVAs. Polyphenols can be further classified into flavonoids, phenolic acid, stillbenes, and lignans. These substances are present in the bran or germ fraction of cereal grains and are specific to certain cereals (Fardet et al. 2008;

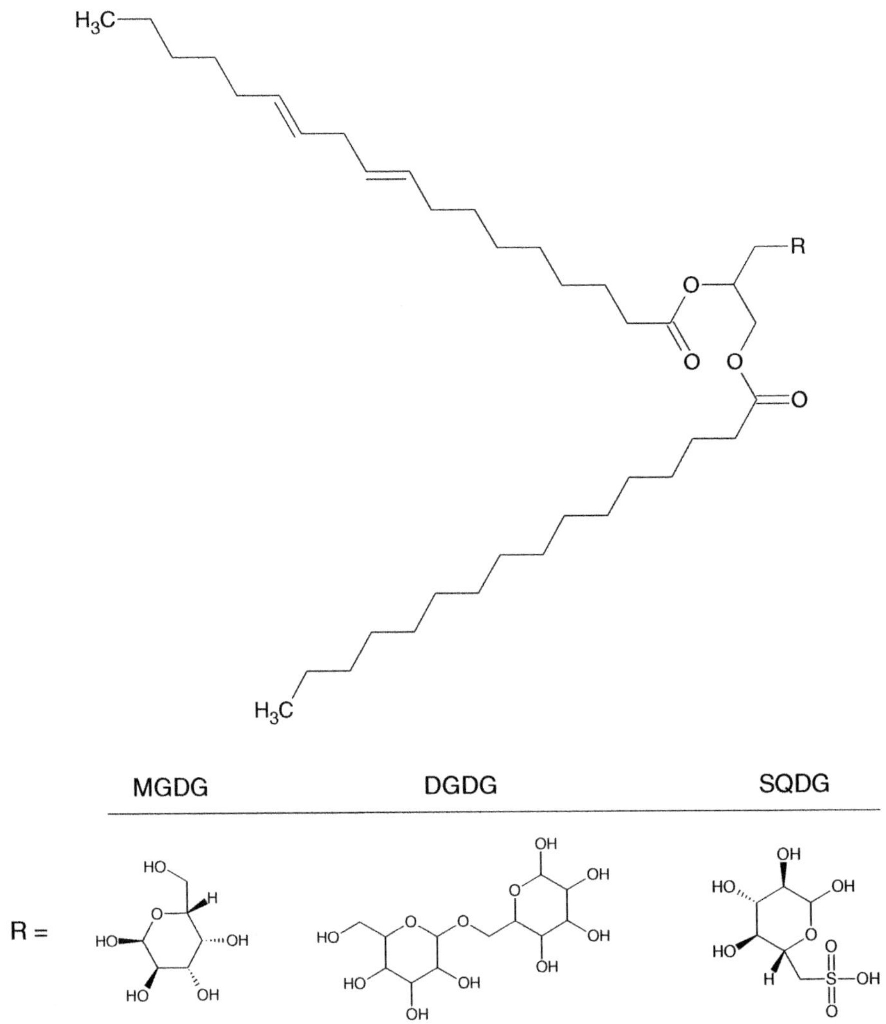

Figure 7.2 Chemical structure of some cereal glycolipids; monogalactosyldiacyl glycerolipids (MGDGs), digalactosyl diaglycerolipids (DGDGs), and sulfoquinovosyl diacylglycerolipids (SQDG).

Masisi et al. 2016). Moreover, these metabolites have different biosynthetic pathways and exhibits various beneficial biological activities such as resistance and defense, antimicrobial, antioxidant properties, and reduction of oxidative stresses (Walter and Marchesan 2011) and therefore perform a protective role against risk of many chronic diseases such as cardiovascular diseases and type 2 diabetes (Gani et al. 2012; Masisi et al. 2016). The chemical structures of some secondary metabolites discussed in this chapter are given in Figure 7.3.

7.2.4.1 Polyphenols Polyphenols are the most abundant low molecular weight phytochemicals in cereals (Khakimov et al. 2014). Polyphenols are divided into several classes; phenolic acids, flavonoids, stillbenes, and lignans (El Gharras 2009).

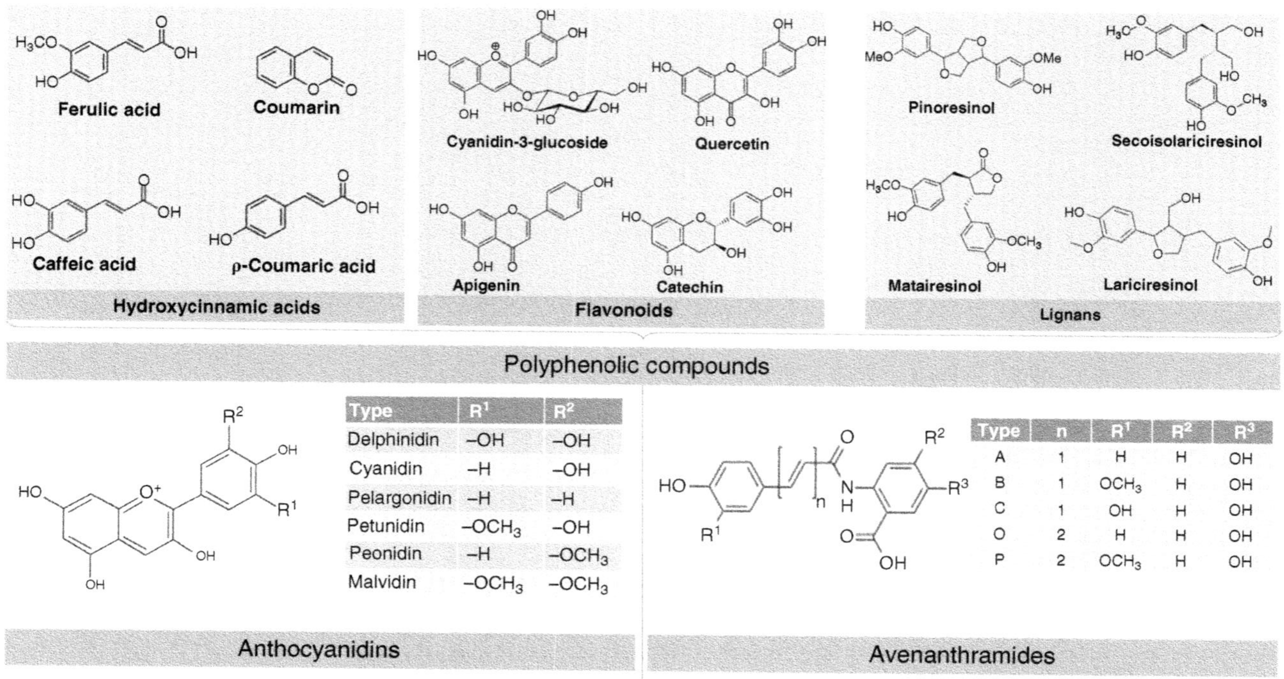

Type	R^1	R^2
Delphinidin	–OH	–OH
Cyanidin	–H	–OH
Pelargonidin	–H	–H
Petunidin	$-OCH_3$	–OH
Peonidin	–H	$-OCH_3$
Malvidin	$-OCH_3$	$-OCH_3$

Type	n	R^1	R^2	R^3
A	1	H	H	OH
B	1	OCH_3	H	OH
C	1	OH	H	OH
O	2	H	H	OH
P	2	OCH_3	H	OH

Figure 7.3 Types and chemical structures of some secondary metabolites from cereals.

Of these, phenolic acids are the most abundant in diets. Generally, polyphenol metabolites contain at least one aromatic ring with one or more hydroxyl group linked to them (Manach et al. 2004). Moreover, polyphenols are present in free and conjugated forms with sugars and other components (Khakimov et al. 2014). The efficacy of the polyphenols depends on the structural factors. For instance, the position and number of the hydroxyl group bonded to the aromatic ring, the character of substituents, and the position of substituents in relation to the hydroxyl group (Pietta 2000).

Polyphenols such as flavonoids, lignans, and hydroxycinnamic acids are synthesized from phenylalamine in the Shikimate pathway (Du Fall and Solomon 2011). They are synthesized during plant development and in response to stress conditions. Phenolic compounds are essential in the reproduction and growth of plants by providing a defense mechanism against ultraviolet radiation, pathogens, and pests (Beckman 2000). These compounds play an important role in color formation and sensory attributes in plants. Flavonoids have a C6–C3–C6 general backbone in which two C6 units (Ring A and B) are of a phenolic nature (Tsao 2010). Flavonoids can be further divided into subclasses which include anthocyanidin, flavanols, flavanones, flavones, chalcones, and isoflavones. Phenolic acids are characterized by single carboxylic acid functionality (Ajila et al. 2010), hence consist of hydroxycinnamic and hydroxybenzoic structures (Manach et al. 2004; Walter and Marchesan 2011). Hydrolysable tannins are derived from gallic acid precursors (Mikulajová et al. 2007).

Polyphenols exhibit beneficial multiple biological effects such as antioxidant, anticarcinogenic, antibacterial, anti-inflammatory, antiallergic, antithrombotic, and antiproliferative properties. Flavonoids can scavenge various free radicals and other reactive species, chelating of transition metals (Tsao 2010), and regulating oxidative stress-mediated enzyme activity (Meyer et al. 2016). As such, their role in prevention of degenerative diseases such as cardiovascular and cancer is increasing (Tsao 2010; Meyer et al. 2016). Polyphenols combined with drugs and other therapies have been reported to modulate gut microbiota (Meyer et al. 2016; Gong et al. 2018). Moreover, their anticancerogenic properties are attributed to the regulation of signaling pathways involved in carcinogenesis, interacting with proteins that control cell cycle progression, and effective modulate the wingless-related integration site signaling pathway in which most conventional therapeutics are ineffective (Meyer et al. 2016).

7.2.4.1.1 Anthocyanidins Anthocyanidins represent the most distinguished subclass of the bioflavonoid group and occur naturally as glycosides that are named anthocyanins. These metabolites are characterized by a C6–C3–C6 basic skeleton that consists of two aromatic rings connected by three carbon links (Walter and Marchesan 2011; Gani et al. 2012). The representatives of anthocyanidins are delphinidin, pelargonidin, malvidin, peonidin, cyanidin, and petunidin (Tsao 2010). They are water soluble pigments found in the pericarp of varieties of barley, maize, rice, rye, and wheat (Khoo et al. 2017; Zykin et al. 2018). Anthocyanidins possess numerous biological effects such as antioxidant, anticancer, and gastro-protective properties (Gani et al. 2012). These bioactive properties strongly depend on their chemical structure (position, number, and types of substituents) and intracellular localization. Anthocyanidins have been shown to have the ability to induce prevention and treatment of several tumors, reduction in cancer cell proliferation, and inhibition of tumor formation and aflatoxin biosynthesis and aid in the prevention of ophthalmological diseases, obesity, and diabetes (Meyer et al. 2016).

7.2.4.2 Avenanthramides Avenanthramides (AVAs) are specific polyphenols unique to oats (Martínez-Villaluenga and Peñas 2017) and consist of 25 distinct entities (Naczk and Shahidi 2006). AVAs are characterized by an anthranilic acid derivative linked to a hydroxycinnamic acid derivative (Gani et al. 2012) and thus are *N*-containing functional substituents. The three major AVAs found in oats are AVAs 1, 3, and 4, also known as avenanthramides B, C, and D, respectively (Gani et al. 2012). These AVAs exhibit antiinflammatory, antiatherogenic, and high antioxidant properties, and hence show potency in modulation of the oxidative defense system in cells, and inhibition of low-density lipoprotein (LDL) oxidation (Tsao 2010; Martínez-Villaluenga and Peñas 2017). Moreover, the antioxidant activity of AVAs is mediated by the induction of heme oxygenase-I (HO-I) expression through the activation of translocation of nuclear factor–E2–related factor 2 (Nrf2) in human kidney cells (Fu et al. 2015).

7.2.5 Other Minor Components

Vitamins and minerals are present in cereals and are distributed in minor quantities. However, cereals contain comparatively high quantities of these contents, which adds to the value of cereals as the best nutritional food crop for humans and animals.

7.2.5.1 Vitamins Cereals possess less quantities of vitamins in the range between 1 and ca. 50 mg/kg. The B-group of vitamins is highly present in cereals (Table 7.2), and these vitamins contribute to about 50–60% of the vitamin source required for humans, especially children. Tocopherols, which are critical and unique fat-soluble vitamins, are present in cereals in the concentration of 20 mg/kg. It should be noted that the cereals contain vitamins in the outer grain layers, specifically in the germ and aleurone layers (Kirchhoff 2002; Mir et al. 2018; Rajni 2018). It has also been reported that the bioavailability of vitamin E in fortified breakfast cereal is greater than that of the encapsulated vitamin supplements (Leonard et al. 2004). These studies showed that cereals can also be considered as an essential dietary vitamin source and consumed as a highly nutritious breakfast.

7.2.5.2 Minerals Cereals contain minerals in the range of ca. 1–2.5%, being equivalent to other raw materials such as vegetables, milk, and meat, but lower than that of pulses. Cereals have been proved to be a key mineral source in the human diet, as they are consumed as a staple food in large quantities (McClure and Muller Jr. 1958). As shown in Figure 7.4, minerals such as calcium, magnesium, phosphorus, and potassium are highly abundant in cereals. Iron, zinc, copper, manganese, and selenium are also reported to be present in cereals (Kumari and Platel 2017a, b). Similar to vitamins, 90% of these minerals were also present in the aleurone layer, germ layer, and outer grain layer known as bran. Thus, it is highly recommended that whole grain products should be included in human nutrition to gain complete cereal mineral content (Borah et al. 2016). In addition to vitamins and minerals, cereals contain metabolic proteins (Kumari and Platel 2017a), carbohydrate-degrading enzymes (Singh et al. 2015a),

Table 7.2 Vitamin contents of some commonly consumed cereals and pseudocereals.

Vitamin		Triticale	Quinoa	Rye	Sorghum	Millet	Oats	Barley	Buckwheat	Brown rice	Wheat	Yellow corn
Vitamin C, total ascorbic acid	mg	0.0	0.0	0.0	0.0	0.0	0.0	0.0	0.0	0.0	0.0	0.0
Thiamin	mg	0.0	0.4	0.3	0.3	0.4	0.5	0.4	0.1	0.4	0.3	0.4
Riboflavin	mg	0.4	0.3	0.3	0.1	0.3	0.2	0.1	0.4	0.1	0.2	0.2
Niacin	mg	0.1	1.5	4.3	3.7	4.7	1.1	6.3	7.0	6.3	5.3	3.6
Pantothenic acid	mg	1.4	0.8	1.5	0.4	0.8	1.1	0.1	1.2	1.6	1.0	0.4
Vitamin B_6	mg	1.3	0.5	0.3	0.4	0.4	0.1	0.4	0.2	0.7	0.2	0.6
Folate, total	µg	0.1	184.0	38.0	20.0	85.0	32.0	8.0	30.0	16.0	28.0	19.0
Choline, total	mg	0.1	70.2	30.4	0.0	0.0	40.4	37.8	0.0		31.2	0.0
Betaine	mg	0.0	630.4	146.1	0.0	0.0		65.5	0.0	0.0	0.0	0.0
Vitamin B_{12}	µg	0.0	0.0	0.0	0.0	0.0	0.0	0.0	0.0	0.0	0.0	0.0
Vitamin A, total	µg	0.0	15.0	12.0	0.0	0.0	0.0	0.0	0.0	0.0	9.0	225.0
Carotene, total	µg	0.0	8.0	7.0	0.0	0.0	0.0	0.0	0.0		5.0	160.0
Cryptoxanthin, beta	µg	0.0	1.0	0.0	0.0	0.0	0.0	0.0	0.0	0.0	0.0	0.0
Lycopene	µg	0.0	14.0	11.0	0.0	0.0	0.0	0.0	0.0	0.0	9.0	214.0
Lutein + zeaxanthin	µg	0.0	163.0	210.0	0.0	0.0	180.0	160.0	0.0	0.0	220.0	1355.0
Tocopherol, total	mg	0.0	5.0	0.0	0.0	0.0	0.2	0.0	0.0	0.0	0.0	0.0
Tocotrienol, total	mg	0.0	0.0	0.0	0.0	0.0	0.0	0.0	0.0	0.0	0.0	0.0
Vitamin D, total	µg	0.0	0.0	0.0	0.0	0.0	0.0	0.0	0.0	0.0	0.0	0.0
Vitamin K (phylloquinone)	µg	0.0	0.0	5.9	0.0	0.9	2.0	2.2	0.0	0.6	1.9	0.3

(*Source*: data taken from usda 2018)

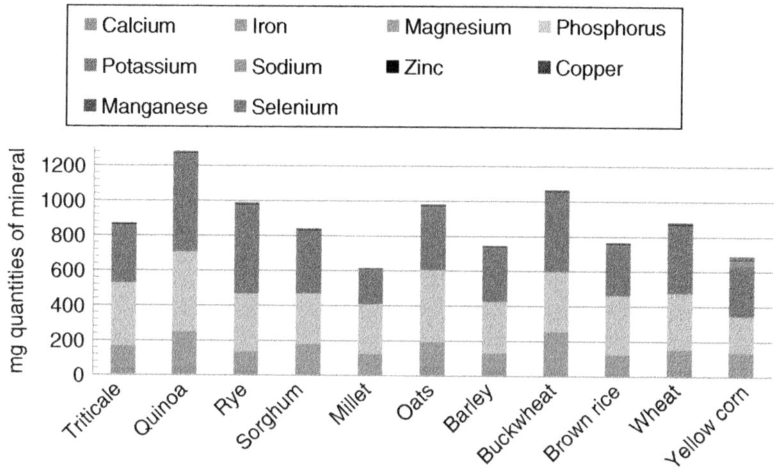

Figure 7.4 Mineral composition of commonly consumed cereals and pseudocereals. *Source*: data taken from USDA (2018).

proteolytic enzymes (Shamloo et al. 2015), hydrolyzing enzymes such as lipase (Matlashewski et al. 1982), phytase (Sandberg and Svanberg 1991), and oxidizing enzymes (Muñoz et al. 2015), which are present in minor quantities.

7.3 Production Strategies for Cereal Ingredients

7.3.1 Production Strategies for Cereal Carbohydrates

The traditional cereal grain processing strategies mainly aim at enhancement of starch accessibility and propensity toward assimilation, while making the products more consumer friendly, in terms of digestibility and prolonged shelf life (Corona et al. 2005). Mechanical and chemical methods are the most widely used techniques in the processing of conjugated carbohydrates and polysaccharides (Singh et al. 2015a). Briefly, mechanical grain processing includes processing right from the start of crop hrvesting, followed by dehusking or dehulling (except for certain crops like wheat, millet, etc.), drying to eliminate moisture, and extensive cleaning mainly for fractionation purposes (Singh et al. 2015a; Zhuang et al. 2018). The processing techniques for obtaining carbohydrates can be broadly grouped under the milling procedure. However, although the basic ideology is the same, the specific steps incorporated during the milling process depend on the type of crop (Yadav and Jindal 2008; Gómez et al. 2009). Chemical processing of cereal grains to isolate carbohydrates is an old practice using soaking with either alkalis or acids, i.e. 1% acetic acid, hydrochloric acid and/or lactic acid, which have distinct advantages such as increased gluiness and clearness overall, gelatinization of rice, and kernel softening of corns and sorghum in the context of wet milling techniques (Ohishi et al. 2007). The palate, coarseness, and shelf life of carbohydrate enriched final products have seen a remarkable improvement in recent years, by harboring enzymatic refining during cereal processing and using enzymes such as cellulases, esterases, xylanases, and β-glucanases

(Singh et al. 2015a). Although not on an industrial scale, the relative assessments have been reported for carbohydrate abstraction and evaluation from cereal grains using capillary electrophoresis (Jager et al. 2007). Ultrasound treatment was reported to be a suitable technique for isolating carbohydrate from cereals (Ebringerová and Hromádková 2010).

7.3.2 Production Strategies for Cereal Proteins and Peptides

Cereal seeds contain predominantly four classes of proteins, based on solubility: prolamines, albumins, glutelins, and globulins (Branlard and Bancel 2007). In their work, Branlard and Bancel (2007) discussed a wide variety of factors that affect the efficacy of protein isolation from cereal seeds, including extent of moisture or water content, endosperm robustness, crop age, and uneven seed disintegration. Salt buffer separation was highlighted to be compatible with the albumin-globulin classes of proteins, whereas a phase segregating procedure was showed to be particularly suitable for amphiphilic proteins. The current focus, however, is a move toward the isolation and purification of biologically active peptides (consisting of ~2–20 amino acids) present in cereal proteins, for their profound applications as therapeutic and nutraceutical agents (Agyei et al. 2015, 2017). An up-to-date discussion on current drifts and critical arguments in cereal bioactive peptide extraction and detection has been accounted for in a very recent review by Piovesana et al. (2018). Bioactive peptides with radical scavenging and antihypertensive actions have been reported in cereals and legumes among others, and past studies have focused on their synthesis, refinement, and scrutiny by using fermentation, enzyme mediated hydrolysis, ultrafiltration, and high throughput chromatography techniques (reversed-phase, hydrophilic interaction, size-inclusion, and ion-exchange) (Piovesana et al. 2018). However, several studies reported the production of bioactive peptides that have been carried out in a laboratory and/or at pilot scale, but still there is no evidence for the demands and supply for these peptides. Optimized and industrial-scale production of bioactive peptides is still a looming challenge (Bazinet and Firdaous 2013), even though a few propositions have been made (Gnasegaran et al. 2017).

7.3.3 Production Strategies for Cereal Lipids

The major composition of lipids present inside the starch granules of cereal endosperms are LPLs, in crops like barley, rye, wheat, and triticale; along with free fatty acids in additional cereals. These lipids are present as inclusion multiplexes with other biomolecules such as starch and/or proteins (Morrison and Coventry 1985). As such, cereal lipids often require specialized protocols, not only for their extraction, but also for their detachment from starch/protein complexes (Morrison and Coventry 1985). The literature corresponding to isolation protocols for lipids from cereal plants are of earlier dates. Fishwick and Wright (1977) compared several protocols for lipid production using a range of solvent systems by measuring phospholipids, sterol lipids, GLs, and the full acyl lipid constituency. The authors highlighted the significance and efficiency of the chloroform-methanol isolation strategy and reported the maximum efficacy of water-logged n-butanol to extract the LPLs. Christie and Morrison (1988) have emphasized the competence of a high-performance liquid chromatography (HPLC) mediated stepwise extraction protocol for discrete GLs and phospholipids in cereal. Morrison and Coventry (1985) documented the significance of lipid isolation strategies from cereal starches using heated aqueous alcohols. According to the authors, two 120-minute and one 60-minute isolation steps offer excellent lipid recovery if carried out at a temperature of 100 °C with a fixed proportion of 75% n-propanol (≥16 ml/g of starch). A detailed discussion of several strategies for separation of starch and lipids were reviewed in the late 1980s by Morrison (1988). The author emphasized that solvent infiltration and lipid discharge are dependent on the extent of starch inflammation and level of alcohol used.

The benefits of a solid-state fermentation strategy on lipid synthesis have been demonstrated by Conti et al. (2001), using fungal strains growing on cereal grains. The authors partitioned extracts of water-soaked barley supporting the growth of fungi belonging to the genus *Mucorales*, and obtained γ-linoleic acid. Gas chromatography analysis of the lipids obtained showed that the operation temperature is the deciding factors for lipid production.

7.3.4 Production Strategies for Cereal-Based Secondary Metabolites

Depending upon their evolution, plant-derived (including cereals) secondary metabolites may be classified into five broad categories: alkaloids, flavonoids, polyketides, phenylpropanoids, and isoprenoids (Gil-Chávez et al. 2013). Isolation of secondary metabolites and natural compounds of significant human physiological benefits, from plants in general and cereals in particular, has been a long haul, with the majority of the earlier studies focusing on minimizing the use of organic solvents. A relevant discussion on earlier strategies, including solvent and/or steam mediated separation and supercritical extraction, their detailed processing techniques, and optimum factors including evolutionary advancements, may be traced to the review by Starmans and Nijhuis (1996). Later, synthesis of culmorin mycotoxin complexes and additional secondary metabolites by *Fusarium* species isolated from Norwegian cereals and subsequent analysis by gas chromatography mass spectrometry was reported (Langseth et al. 2001). Recent works have targeted reviewing the available technologies for isolation and manufacturing of secondary metabolites and their analyses from cereals (Du Fall and Solomon 2011; Gil-Chávez et al. 2013). Gil-Chávez et al. (2013) have added comprehensive discussion on predictable techniques such as solid–liquid and liquid–liquid separation and pressurized-liquid-, ultrasound and microwave-, and supercritical- and subcritical isolation methodologies. Particularly, a detailed account on a pressurized fluid separation technique of phytochemicals from cereals can be found in a more recent work (Kelly et al. 2019). The authors discussed several parameters affecting the pressurized fluid removal process for obtaining phytochemicals from cereals, such as plant area, solvent employed, temperature, pH, pressure, isolation duration, and flow rate, among others. (Du Fall and Solomon 2011) have reviewed available high throughput techniques for analysis of cereal secondary metabolites (cyanogenic glycosides, saponins, terpenoids, benzoxazanoids, and flavonoids), including mass spectrometry, gas and liquid chromatography, nuclear magnetic resonance, and Fourier Transform Infrared Spectroscopy, either in isolation or in combinatorial modes.

7.3.5 Production Strategies for Vitamins and Minerals from Cereal

Early reports of isolation and quantification of vitamin A or retinol, with more than 90% retrieval, exist for cereal products using a high-speed liquid chromatographic method, employing a uniform mobile phase, and subsequent analysis by UV spectroscopy (Dennison and Kirk 1977). More recently, Lebiedzińska and Szefer (2006) reported extraction and analysis of a whole range of vitamins B from cereals, i.e. niacin, riboflavin, pyridoxine, and thiamine. According to the authors, the vitamin proportions were detected using microbiological analytical methods, employing various bacterial strains from the genus *Lactobacillus* and the yeast *Saccharomyces*, applications being tuned to specific vitamins. Recently, Kamankesh et al. (2017) have reported arguably the first investigative scheme to extract and quantify vitamin D_3 or cholecalciferol from cereals using a two-fold technique: an initial isolation using ultrasonic waves, succeeded by a liquid–liquid micro-separation protocol using HPLC. The effects of primary processing conditions and protocols on mineral isolation from cereals have been systematically reviewed recently (Oghbaei and Prakash 2016). A wide variety of multivalent ions and minerals such as iron, calcium, zinc, magnesium, and phosphorus (stored mainly as phytic acid) may be derived from cereals and legumes following sequential extraction steps that include initial water soaking, sprouting, fermentation, and enzymatic treatment of grains. It should be noted that several factors such as contour and coarseness of the grain, and the separation speed may affect the isolation of minerals from cereals. The mechanical procedure of milling is thought to be diminutive toward mineral recovery. However, this trend is somewhat compensated for by a decline in the recuperation of the anti-nutrients such as tannins, trypsin inhibitors, etc. (Oghbaei and Prakash 2013). All these studies were conducted on laboratory- and small-scale generation of both micro- and macro-nutrients.

7.4 Food Applications of Cereal Ingredients

Cereals are essential food crops owing to their energy content, nutrition, and health benefits that can be processed into different types of food products. The nutritional profile of cereals is influenced by genetic and environmental factors. Hence, cereals differ from one another in terms of their nutritional composition and functionality. According to the American Association of Cereal Chemists (AACC 1999), wholegrain cereals consist of true cereals and pseudocereals. There are about 17 different categories of cereals as defined by the Food and Agriculture Organization of the United Nations (FAO 1994) with an estimated production of about 2.85 billion metric tons in 2016 (WorldBank 2018). Among the long list of cereals, maize (corn), wheat, rice, and barley are the most extensively consumed types of cereals worldwide. Most food manufacturers have a high preference for white and clean grains prior to initial processing of cereals due to general consumer preference (Baik and Ullrich 2008). Cereals are indispensable food crops that are often consumed as staple foods for breakfast, with an increasing global market demand. It is estimated that the current revenue generated from the global breakfast cereal market is US$ 7763 million and this is projected to grow annually by 1.6% from 2018 to 2021 (Statista 2018).

7.4.1 Nutritional Applications

Consumer interest in eating cereals and cereal-based foods is high due to their nutritional composition and health benefits. Some of the nutritional benefits of consuming cereals include their total energy content, protein, carbohydrates, fiber, sugar, fat, and mineral constituents, as shown in Figure 7.5. Cereals are used in the preparation of a wide variety of food products including breakfast cereals, porridges, baby foods, geriatric foods, soups, stews, snacks, and bakery flour blends. Formulation of food products from cereals with high nutritional quality can be achieved through arrangement of the hierarchical structures of the nutritional constituents such as protein, starch, and fat in the food matrix (Zúñiga and Troncoso 2012). Mir et al. (2018) highlighted that the usefulness and functionality of a cereal or any grain mainly hinges on the quantity and quality of proteins. Nutritionally, proteins contain three categories of amino acids: non-essential, essential, and conditional essential amino acids. Essential and conditional essential amino acids are crucial for the growth and maintenance of metabolic activities in humans (Mir et al. 2018). The release and availability of these amino acids for nutritional purposes depends on several factors including the complexity of the food matrix, digestion process, presence of anti-nutritional factors, and health status of the consumer. Physical and bioprocessing techniques, such as fermentation, germination, enzymatic treatment, and extrusion, can be used to alter the matrices in order to enhance the bioaccessibility and bioavailability of the essential nutrients in cereals. Other nutritional compositions can further be enhanced through cryogenic pre-treatment, degerming, dehulling, peeling, pearling, cooking, and electrostatic separation (Kołodziejczyk et al. 2018).

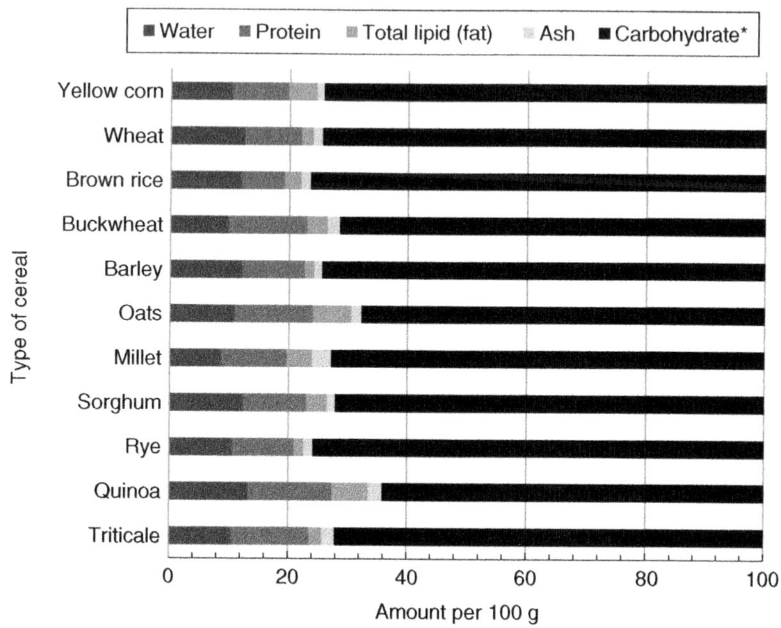

Figure 7.5 Nutritional composition of selected cereals and pseudocereals (per 100 g). *Carbohydrate values are by difference and included dietary fiber. *Source*: data taken from USDA (2018).

Despite the high nutritional qualities, some factors may limit the application of cereals, including consumer taste preferences, perceived poor sensory properties, high prices, insufficient awareness of health benefits, and challenges with enrichment of existing products and meals with cereals (Arvola et al. 2007; Kuznesof et al. 2012; Rosa-Sibakov et al. 2015).

Supplementation of nutritional composition by the addition of cereals is expected to enhance the health benefits of foods. Pathera et al. (2017) developed a ready-to-eat chicken nugget supplemented with wheat bran, leading to an acceptable cooking yield of 97% and a recommended daily allowance of 12–17% dietary fiber when 100 g of the nugget is consumed per day. Notably, different cooking methods affected the nutritional content, pH, and water-holding capacity of the dietary fiber-supplemented meat products due to their differences in heat and mass transfer processes (Singh et al. 2015b; Yadav et al. 2016; Pathera et al. 2017). Foschia et al. (2015) studied the use of dietary fibers (short-chain inulin, long-chain inulin, psyllium) and oat bran in improving the nutritional quality of pasta made with durum wheat semolina. The new pasta products showed acceptable textural properties, optimal cooking time, swelling, and water absorption capacity. In contrast, an adverse effect on the integrity of the protein–starch network was observed when dietary fibers were mixed with psyllium fiber to substitute for durum wheat semolina (Foschia et al. 2015). This observation is similar to the findings of Juszczak et al. (2012) that the quality of pasta is dependent on the manufacturing conditions, type, and amount of dietary fibers included (Juszczak et al. 2012). In a bid to increase the nutritional value of traditional wheat bread and noodles, Chauhan et al. (2018) found that the optimum composition suitable for introducing oat as a substitute for wheat during the preparation of bread and noodles was 20% and 30%, respectively. This supplementation also led to improvement in the compositions of nutrients, including crude protein, fiber, and crude fat in the final products. Moreover, Talukder and Sharma (2010) reported that the optimal inclusion levels of oat and wheat bran were 10 and 15%, respectively, in developing dietary fiber-rich chicken meat pâté. Apart from enhancement of nutritional quality, supplementation of the meat-based food products with cereals can impart additional beneficial properties. For instance, Aravind et al. (2012) demonstrated that substitution of durum semolina with up to 30% of durum bran, or pollard dried at high temperatures, increased the antioxidant capacity and fiber content with minimal influences on the color, sensory properties, and digestibility of the products. The latter is particularly important in ensuring that digestive release and bioaccessibility of nutrients in the cereal-supplemented products is not compromised.

7.4.2 Health Applications

Cereals contain dietary fibers and bioactive phytochemicals that are beneficial to health and are proposed to minimize the risk of developing cardiovascular disease, type 2 diabetes, cancer, and high blood pressure (Rosa-Sibakov et al. 2015). There are two forms of dietary fibers: soluble and insoluble fibers. Soluble fiber is made of β-glucans, pectins, mucilages, gums, and some hemicelluloses, whereas the insoluble fiber includes celluloses, lignin, and some hemicelluloses (Biskup et al. 2017). Cereals have complex cellular and molecular matrices that contain these nutrients and determine their bioaccessibility and bioavailability (Rosa-Sibakov et al. 2015). The health benefits and effectiveness of cereals, upon consumption, depend on the complex and synergistic effects involving the cereal matrix, molecular composition, type, degree of food processing, and enzymatic activity in the gastrointestinal tract (Biskup et al. 2017). For instance, wholegrain barley contains β-glucan and phytochemicals such as phenolic acids, flavonoids, lignans, tocols, phytosterols, and folate (Idehen et al. 2017). These constituents are reported to be essential in minimizing many risk factors associated with chronic diseases. However, they are likely to chemically interact with other ingredients within the cereal matrix, due to their structural diversity, which can alter the rate of their release during digestion.

The heightened prevalence of coronary heart disease worldwide is partly related to dietary factors, abnormal metabolism, and poor lifestyle management. Dietary fibers in cereals, such as β-glucan, are known to alter serum lipids and act as antiatherogenic agents. For instance, Queenan et al. (2007) demonstrated the efficacy of concentrated β-glucan from oats in reducing low-density lipoprotein cholesterol when consumed by hypercholesterolemic men and women, indicating its relevance in managing the risk of cardiovascular disease. Dietary fibers decrease the absorption of dietary cholesterol due to their thickening effect on the content of the small intestine. In addition, dietary fibers can bind and prevent the reabsorption of bile acids, leading to increased hepatic metabolism and lowering of endogenous cholesterol. Another mechanism by which dietary fiber contributes to the reduction of cardiovascular risk factors is through the production of short-chain fatty acids (SCFA) propionate, acetate, and butyrate by colonic microflora (Lin et al. 1995). However, there are conflicting reports on the significant role of propionate, if any at all, in the reduction of cholesterol after colonic fermentation of dietary fibers (Queenan et al. 2007). Furthermore, improvements in postprandial glucose, insulin response, lowering of blood pressure, and body weight as a result of consumption of functional fibers have been demonstrated, and these effects also contribute to the reduction of CVD risk (Wu et al. 2003; Queenan et al. 2007). Colonic fermentation of dietary fiber also results in the production of SCFA butyrate, which is a strong inhibitor of histone deacetylase (Wang et al. 2011a; Triff et al. 2018). Butyrate has several intestinal and extra-intestinal beneficial effects by impacting on cell kinetics, lumen pH, and epigenetics, which subsequently reduces the risk of colon carcinogenesis (Triff et al. 2018). Human studies have confirmed the beneficial effects of cereal fibers compared to other dietary fibers. Du et al. (2009) demonstrated in a prospective cohort study that the consumption of 10 g/day higher cereal fibers contributed significantly to a 77 g/year decrease in weight and a waist circumference change of −0.10 cm/year in European men and women after 6.5 years. However, the study reported that fruit and vegetable fiber intake did not significantly contribute to the reduction of weight change (Du et al. 2009). Cereal fibers used in the study were sourced from rice, pasta, bread, biscuits, crackers, breakfast cereals, and other products made of flour. The difference in effects may be as a result of the type and amounts of the cereal fibers, or other bioactive components present in cereals.

Examples of bioactive phytochemicals in cereals include phenolic compounds, lignans, vitamin E (tocols), phytosterols, and folates (Fardet 2010; Idehen et al. 2017). Phenolic compounds, including those found in cereals, have been reported to have antioxidant promoting effects including vasodilatory, anti-inflammatory, antiapoptotic, antithrombotic, hypolipemic, or antiatherogenic effects (Quiñones et al. 2013). Phenolic acids represent one of the most common types of phenolic compounds in cereals. Biskup et al. (2017) described the potential mechanisms through which bioactive phytochemicals from spelt and common wheat synergistically control blood glucose levels by increasing digestion time. For instance, interactions between phytic acids, soluble fibers, propionates, and arabinoxylans after microflora fermentation delay gastric emptying. Phytic acids, alkylresorcinols, and lectins are able to bind to α-amylase, thereby inhibiting its enzymatic activity and prolonging digestion time. Similarly, phenolic acids and alkylresorcinols can inhibit the enzymatic activity of α-glucosidase (Biskup et al. 2017). Although these effects are crucial for controlling glucose release during digestion, they may also be considered anti-nutritional under certain conditions. Taken together, dietary fibers and phytochemicals found in cereals are important bioactive components for the formulation of cereal-based functional foods, or for enrichment of meat products, with cardiovascular health benefits.

7.5 Conclusion and Future Outlook

Cereals play an important role in global food systems, due to their high contents of nutrients and functional biomolecules such as vitamins, minerals, dietary fiber, proteins, lipids, and carbohydrates. Utilization of cereals as food is impacted significantly by the growing consumer interest in plant-based foods and veganism. This trend is not only recognized in the scientific literature (Kevany et al. 2018) but also in mainstream media, as shown in the Google trends report (GoogleTrends 2018). Cereals will continue to play a critical role in meeting the nutritional and health requirements of consumers as subsequent utilization of cereal bioactive properties gain more interest. Another growing trend among consumers is a preference for food with bioactive properties (Yeung et al. 2018). As such, cereal ingredients that not only supply nutrients but also perform physiologically measurable bioactive effects, particularly in areas such as improved digestive (gut) health, and control of the metabolic syndrome (via antioxidant, anti-inflammatory, immune enhancing properties) will be highly sought after.

That cereals (particularly rice, wheat, and maize) play a significant role in modern diets is well established (Awika 2011). But to maximize the dietary and functional applications of cereals, further research will have to look into the following areas:

a. Exploration of other less utilized but nutritious and gluten-free cereals (e.g. teff) and pseudo cereals (e.g. buckwheat, amaranth, and quinoa) (Mir et al. 2018);
b. Re-utilization or valorization of cereal by-products such as bran and germ for other food or non-food applications (Duţă et al. 2018; Papageorgiou and Skendi 2018);
c. Removing or reducing the levels of the proteinogenic allergenic epitopes via breeding or processing;
d. Development of sustainable processing techniques which are applicable in resource-poor settings and can simultaneously increase bioaccessibility/bioavailability of nutrients while preserving the quality and quantity of nutrients.

References

AACC (1999). AACC members agree on definition of whole grain. http://www.aaccnet.org/news/pdfs/wgPR.pdf (accessed 5 October 2018).

Agyei, D., Danquah, M.K., Sarethy, I.P., and Pan, S. (2015). Antioxidative peptides derived from food proteins. In: *Free Radicals in Human Health and Disease* (eds. V. Rani and U. Yadav), 417–430. New Delhi: Springer.

Agyei, D., Pan, S., Acquah, C. et al. (2017). Structure-informed detection and quantification of peptides in food and biological fluids. *Journal of Food Biochemistry* 43 (1): 12482.

Ajila, C.M., Brar, S.K., Verma, M. et al. (2010). Extraction and analysis of polyphenols: recent trends. *Critical Reviews in Biotechnology* 31: 227–249.

Alvarez, V.A., Guarás, M.P., and Gutiérrez, T.J. (2017). Reactive extrusion for the production of starch-based biopackaging. In: *Biopackaging* (ed. M.A. Masuelli), 308–336. CRC Press.

Amadou, I., Gounga, M.E., and Le, G.-W. (2013). Millets: nutritional composition, some health benefits and processing – a review. *Emirates Journal of Food and Agriculture* 25: 501.

Anandharaj, M., Sivasankari, B., and Parveen Rani, R. (2014). Effects of probiotics, prebiotics, and synbiotics on hypercholesterolemia: a review. *Chinese Journal of Biology* 2014: 572754.

Aravind, N., Sissons, M., and Fellows, C.M. (2012). Effect of soluble fibre (guar gum and carboxymethylcellulose). Addition on technological, sensory and structural properties of durum wheat spaghetti. *Food Chemistry* 131: 893–900.

Arvola, A., Lähteenmäki, L., Dean, M. et al. (2007). Consumers' beliefs about whole and refined grain products in the UK, Italy and Finland. *Journal of Cereal Science* 46: 197–206.

Awika, J.M. (2011). Major cereal grains production and use around the world. In: *Advances in Cereal Science: Implications to Food Processing and Health Promotion* (eds. J.M. Awika, V. Piironen and S. Bean), 1–13. American Chemical Society.

Bae, S.H. (2014). Diets for constipation. *Pediatric Gastroenterology, Hepatology & Nutrition* 17: 203–208.

Baik, B.-K. and Ullrich, S.E. (2008). Barley for food: characteristics, improvement, and renewed interest. *Journal of Cereal Science* 48: 233–242.

Balindong, J.L., Liu, L., Ward, R.M. et al. (2016). Optimisation and standardisation of extraction and HPLC analysis of rice grain protein. *Journal of Cereal Science* 72: 124–130.

Bates, C.J., Mansoor, M.A., Pentieva, K.D. et al. (2010). Biochemical risk indices, including plasma homocysteine, that prospectively predict mortality in older British people: the national diet and nutrition survey of people aged 65 years and over. *British Journal of Nutrition* 104: 893–899.

Bazinet, L. and Firdaous, L. (2013). Separation of bioactive peptides by membrane processes: technologies and devices. *Recent Patents on Biotechnology* 7: 9–27.

Beckman, C.H. (2000). Phenolic-storing cells: keys to programmed cell death and periderm formation in wilt disease resistance and in general defence responses in plants? *Physiological and Molecular Plant Pathology* 57 (3): 101–110.

Belitz, H.-D., Grosch, W., and Schieberle, P. (2009). Cereals and cereal products. In: *Food Chemistry* (eds. H.-D. Belitz, W. Grosch and P. Schieberle), 670–745. Berlin, Heidelberg: Springer.

Bhanja Dey, T., Chakraborty, S., Jain, K.K. et al. (2016). Antioxidant phenolics and their microbial production by submerged and solid state fermentation process: a review. *Trends in Food Science & Technology* 53: 60–74.

Biskup, I., Gajcy, M., and Fecka, I. (2017). The potential role of selected bioactive compounds from spelt and common wheat in glycemic control. *Advances in Clinical and Experimental Medicine* 26: 1015–1021.

Bo-Linn, G.W., Ana, C.A.S., Morawski, S.G., and Fordtran, J.S. (1982). Starch blockers – their effect on calorie absorption from a high-starch meal. *New England Journal of Medicine* 307: 1413–1416.

Borah, A., Mahanta, C.L., and Kalita, D. (2016). Optimization of process parameters for extrusion cooking of low amylose rice flour blended with seeded banana and carambola pomace for development of minerals and fiber rich breakfast cereal. *Journal of Food Science and Technology* 53: 221–232.

Borchani, C., Fonteyn, F., Jamin, G. et al. (2016). Structural characterization, technological functionality, and physiological aspects of fungal β-D-glucans: a review. *Critical Reviews in Food Science and Nutrition* 56: 1746–1752.

Bourtsala, A. and Galanopoulou, D. (2018). *Phospholipases. Reference Module in Food Science*. Elsevier.

Branlard, G. and Bancel, E. (2007). Protein extraction from cereal seeds. In: *Plant Proteomics* (eds. H. Thiellement, M. Zivy, C. Damerval, et al.), 15–25. Humana Press.

Brewer, J.W. (2013). Phospholipids: "greasing the wheels" of humoral immunity. *Biochimica et Biophysica Acta (BBA). – Molecular and Cell Biology of Lipids* 1831: 642–651.

Bruce, S.J., Guy, P.A., Rezzi, S., and Ross, A.B. (2010). Quantitative measurement of betaine and free choline in plasma, cereals and cereal products by isotope dilution LC-MS/MS. *Journal of Agricultural and Food Chemistry* 58: 2055–2061.

Buksa, K., Ziobro, R., Nowotna, A. et al. (2014). Water binding capacity of rye flours with the addition of native and modified arabinoxylan preparations. *Journal of Agricultural Science and Technology* 16 (5): 1083–1095.

Caiafa, K. and Wrabel, M. (2019). Food aid. In: *Encyclopedia of Food Security and Sustainability* (eds. P. Ferranti, E.M. Berry and J.R. Anderson), 122–127. Elsevier.

Carciofi, M., Blennow, A., Nielsen, M.M. et al. (2012). Barley callus: a model system for bioengineering of starch in cereals. *Plant Methods* 8: 36.

Cavazos, A. and Gonzalez De Mejia, E. (2013). Identification of bioactive peptides from cereal storage proteins and their potential role in prevention of chronic diseases. *Comprehensive Reviews in Food Science and Food Safety* 12: 364–380.

Chamlagain, B., Sugito, T.A., Deptula, P. et al. (2018). In situ production of active vitamin B12 in cereal matrices using *Propionibacterium freudenreichii*. *Food Science & Nutrition* 6: 67–76.

Chauhan, D., Kumar, K., Kumar, S., and Kumar, H. (2018). Effect of incorporation of oat flour on nutritional and organoleptic characteristics of bread and noodles. *Current Research in Nutrition and Food Science Journal* 6: 148–156.

Christie, W. and Morrison, W. (1988). Separation of complex lipids of cereals by high-performance liquid chromatography with mass detection. *Journal of Chromatography A* 436: 510–513.

Chuan, T., Lei, L., Waters, D.L.E., and Jin-Song, B. (2016). Association mapping and marker development of genes for starch lysophospholipid synthesis in rice. *Rice Science* 23: 287–296.

Coda, R., Rizzello, C.G., Pinto, D., and Gobbetti, M. (2012). Selected lactic acid bacteria synthesize antioxidant peptides during sourdough fermentation of cereal flours. *Applied and Environmental Microbiology* 78: 1087–1096.

Coda, R., Montemurro, M., and Rizzello, C.G. (2017). Yogurt-like beverages made with cereals. In: *Yogurt in Health and Disease Prevention* (ed. N.P. Shah), 183–201. Academic Press.

Conti, E., Stredansky, M., Stredanska, S., and Zanetti, F. (2001). γ-Linolenic acid production by solid-state fermentation of Mucorales strains on cereals. *Bioresource Technology* 76: 283–286.

Corona, L., Rodriguez, S., Ware, R., and Zinn, R. (2005). Comparative effects of whole, ground, dry-rolled, and steam-flaked corn on digestion and growth performance in feedlot cattle. *The Professional Animal Scientist* 21: 200–206.

Costa, S.M., Mazzola, P.G., Silva, J.C. et al. (2013). Use of sugar cane straw as a source of cellulose for textile fiber production. *Industrial Crops and Products* 42: 189–194.

Craig, S.A. (2004). Betaine in human nutrition. *The American Journal of Clinical Nutrition* 80: 539–549.

Cui, W. and Wood, P. (2000). Relationships between structural features, molecular weight and rheological properties of cereal β-D-glucans. In: *Hydrocolloids* (ed. K. Nishinari), 159–168. Elsevier.

Da Costa, E., Silva, J., Mendonça, S. et al. (2016). Lipidomic approaches towards deciphering glycolipids from microalgae as a reservoir of bioactive lipids. *Marine Drugs* 14: 101.

D'Arrigo, P. and Servi, S. (2010). Synthesis of lysophospholipids. *Molecules* 15: 1354–1377.

De Carvalho, A.J.F. and Trovatti, E. (2016). Biomedical applications for thermoplastic starch. In: *Biodegradable and Biobased Polymers for Environmental and Biomedical Applications* (eds. S. Kalia and L. Avérous), 1–23. Wiley-Scrivener.

Dennison, D. and Kirk, J. (1977). Quantitative analysis of vitamin A in cereal products by high speed liquid chromatography. *Journal of Food Science* 42: 1376–1379.

Doubet, S., Bock, K., Smith, D. et al. (1989). The complex carbohydrate structure database. *Trends in Biochemical Sciences* 14: 475–477.

Du Fall, L.A. and Solomon, P.S. (2011). Role of cereal secondary metabolites involved in mediating the outcome of plant-pathogen interactions. *Metabolites* 1: 64–78.

Du, H., Van Der, A.D.L., Boshuizen, H.C. et al. (2009). Dietary fiber and subsequent changes in body weight and waist circumference in European men and women. *The American Journal of Clinical Nutrition* 91: 329–336.

Duţă, D.E., Culeţu, A., and Mohan, G. (2018). Reutilization of cereal processing by-products in bread making. In: *Sustainable Recovery and Reutilization of Cereal Processing by-Products* (ed. C.M. Galanakis), 279–317. Woodhead Publishing.

Ebringerová, A. (2005). Structural diversity and application potential of hemicelluloses. *Macromolecular Symposia* 232 (1): 1–12.

Ebringerová, A. and Hromádková, Z. (2010). An overview on the application of ultrasound in extraction, separation and purification of plant polysaccharides. *Central European Journal of Chemistry* 8: 243–257.

El Gharras, H. (2009). Polyphenols: food sources, properties and applications – a review. *International Journal of Food Science & Technology* 44: 2512–2518.

Englyst, H.N., Kingman, S., and Cummings, J. (1992). Classification and measurement of nutritionally important starch fractions. *European Journal of Clinical Nutrition* 46: S33–S50.

FAO (1994). Cereals and cereal products. http://www.fao.org/es/faodef/fdef01e.htm (accessed 5 October 2018).

Fardet, A. (2010). New hypotheses for the health-protective mechanisms of whole-grain cereals: what is beyond fibre? *Nutrition Research Reviews* 23: 65–134.

Fardet, A., Rock, E., and Rémésy, C. (2008). Is the in vitro antioxidant potential of whole-grain cereals and cereal products well reflected in vivo? *Journal of Cereal Science* 48: 258–276.

Farrokhi, N., Whitelegge, J.P., and Brusslan, J.A. (2008). Plant peptides and peptidomics. *Plant Biotechnology Journal* 6: 105–134.

Fausch, H., Kündig, W., and Neukom, H. (1963). Ferulic acid as a component of a glycoprotein from wheat flour. *Nature* 199: 287.

Fausto, F., Kacchi, A., and Mehta, D. (1997). Starch products in confectionery. *Beverage & Food World* 24: 4–16.

Fishwick, M.J. and Wright, A.J. (1977). Comparison of methods for the extraction of plant lipids. *Phytochemistry* 16: 1507–1510.

Fontaine, J., Schirmer, B., and Hörr, J. (2002). Near-infrared reflectance spectroscopy (NIRS). Enables the fast and accurate prediction of essential amino acid contents. 2. Results for wheat, barley, corn, triticale, wheat bran/middlings, rice bran, and sorghum. *Journal of Agricultural and Food Chemistry* 50: 3902–3911.

Foschia, M., Peressini, D., Sensidoni, A. et al. (2015). How combinations of dietary fibres can affect physicochemical characteristics of pasta. *LWT- Food Science and Technology* 61: 41–46.

Franco, C.M., Ciacco, C.F., and Tavares, D.Q. (1998). The structure of waxy corn starch: effect of granule size. *Starch-Stärke* 50: 193–198.

Fu, J., Zhu, Y., Yerke, A. et al. (2015). Oat avenanthramides induce heme oxygenase-1 expression via Nrf2-mediated signaling in HK-2 cells. *Molecular Nutrition & Food Research* 59: 2471–2479.

Fuentes-Zaragoza, E., Riquelme-Navarrete, M., Sánchez-Zapata, E., and Pérez-Álvarez, J. (2010). Resistant starch as functional ingredient: a review. *Food Research International* 43: 931–942.

Gani, A., Sm, W., and Fa, M. (2012). Whole-grain cereal bioactive compounds and their health benefits: a review. *Journal of Food Processing & Technology* 3 (3): 146–156.

Garr, J. and Bangudu, A. (1991). Evaluation of sorghum starch as a tablet excipient. *Drug Development and Industrial Pharmacy* 17: 1–6.

Gérard, C., Barron, C., Colonna, P., and Planchot, V. (2001). Amylose determination in genetically modified starches. *Carbohydrate Polymers* 44: 19–27.

Gil-Chávez, G., Villa, J.A., Ayala-Zavala, F.J. et al. (2013). Technologies for extraction and production of bioactive compounds to be used as nutraceuticals and food ingredients: an overview. *Comprehensive Reviews in Food Science and Food Safety* 12: 5–23.

Gnasegaran, G.K., Agyei, D., Pan, S. et al. (2017). Process development for bioactive peptide production. In: *Food Bioactives* (ed. M. Puri), 91–110. Springer.

Gómez, M., Pardo, J., Oliete, B., and Caballero, P.A. (2009). Effect of the milling process on quality characteristics of rye flour. *Journal of the Science of Food and Agriculture* 89: 470–476.

Gong, L., Cao, W., Chi, H. et al. (2018). Whole cereal grains and potential health effects: involvement of the gut microbiota. *Food Research International* 103: 84–102.

GOOGLETRENDS (2018). Veganism. https://trends.google.com/trends/explore?date=all&q=Veganism (accessed 3 October 2018).

Granfeldt, Y., Nyberg, L., and Björck, I. (2008). Muesli with 4 g oat β-glucans lowers glucose and insulin responses after a bread meal in healthy subjects. *European Journal of Clinical Nutrition* 62: 600.

Guerrieri, N. and Cavaletto, M. (2018). Cereals proteins. In: *Proteins in Food Processing*, 2e (ed. R.Y. Yada), 223–244. Woodhead Publishing.

Gul, K., Yousuf, B., Singh, A.K. et al. (2015). Rice bran: nutritional values and its emerging potential for development of functional food – a review. *Bioactive Carbohydrates and Dietary Fibre* 6: 24–30.

Ha, E. and Zemel, M.B. (2003). Functional properties of whey, whey components, and essential amino acids: mechanisms underlying health benefits for active people (review). *The Journal of Nutritional Biochemistry* 14: 251–258.

Hefni, M.E., Schaller, F., and Witthöft, C.M. (2018). Betaine, choline and folate content in different cereal genotypes. *Journal of Cereal Science* 80: 72–79.

Huang, L., Zhang, X., Zhang, H., and Wang, J. (2018). Interactions between dietary fiber and ferulic acid changed the aggregation of gluten in a whole wheat model system. *LWT- Food Science and Technology* 91: 55–62.

Idehen, E., Tang, Y., and Sang, S. (2017). Bioactive phytochemicals in barley. *Journal of Food and Drug Analysis* 25: 148–161.

Inglett, G.E. (1974). *Wheat: Production and Utilization*. The AVI Publishing Company, Inc.

Izydorczyk, M.S. (2017). Functional properties of cereal cell wall polysaccharides. In: *Carbohydrates in Food*, 3e (ed. A.-C. Eliasson), 215–278. CRC Press.

Izydorczyk, M.S. and Biliaderis, C.G. (1995). Cereal arabinoxylans: advances in structure and physicochemical properties. *Carbohydrate Polymers* 28: 33–48.

Jager, A.V., Tonin, F.G., and Tavares, M.F. (2007). Comparative evaluation of extraction procedures and method validation for determination of carbohydrates in cereals and dairy products by capillary electrophoresis. *Journal of Separation Science* 30: 586–594.

Jan, Z., Ellis, P., and Schofield, J. (1995). Mini review: gas cell stabilization and gas retention in wheat bread. *Journal of Cereal Science* 21: 215–230.

Janssen, A., Pischinger, S., and Muether, M. (2010). Potential of cellulose-derived biofuels for soot free diesel combustion. *SAE International Journal of Fuels and Lubricants* 3: 70–84.

Jeong, H.J., Jeong, J.B., Hsieh, C.C. et al. (2010). Lunasin is prevalent in barley and is bioavailable and bioactive in *in vivo* and *in vitro* studies. *Nutrition and Cancer* 62: 1113–1119.

Jiamjariyatam, R., Kongpensook, V., and Pradipasena, P. (2015). Effects of amylose content, cooling rate and aging time on properties and characteristics of rice starch gels and puffed products. *Journal of Cereal Science* 61: 16–25.

Juszczak, L., Witczak, T., Ziobro, R. et al. (2012). Effect of inulin on rheological and thermal properties of gluten-free dough. *Carbohydrate Polymers* 90: 353–360.

Kamankesh, M., Mohammadi, A., Mollahosseini, A. et al. (2017). Vitamin D3: Preconcentration and determination in cereal samples using ultrasonic-assisted extraction and microextraction method. *Cereal Chemistry* 94: 532–538.

Karliner, J. (2002). Lysophospholipids and the cardiovascular system. *Biochimica et Biophysica Acta (BBA) – Molecular and Cell Biology of Lipids* 1582: 216–221.

Kawaguchi, T., Ueno, T., Nogata, Y. et al. (2017). Wheat-bran autolytic peptides containing a branched-chain amino acid attenuate non-alcoholic steatohepatitis via the suppression of oxidative stress and the upregulation of AMPK/ACC in high-fat diet-fed mice. *International Journal of Molecular Medicine* 39: 407–414.

Keenan, M.J., Zhou, J., Mccutcheon, K.L. et al. (2006). Effects of resistant starch, a non-digestible fermentable fiber, on reducing body fat. *Obesity* 14: 1523–1534.

Kelly, N.P., Kelly, A.L., and O'Mahony, J.A. (2019). Strategies for enrichment and purification of polyphenols from fruit-based materials. *Trends in Food Science & Technology* 83: 248–258.

Kevany, K.M., Baur, G., and Wang, G.C. (2018). Shifting food systems: increasing well-being through plant-based approaches. *Explorer* 14 (6): 435–442.

Khakimov, B., Bak, S., and Engelsen, S.B. (2014). High-throughput cereal metabolomics: current analytical technologies, challenges and perspectives. *Journal of Cereal Science* 59: 393–418.

Khoo, H.E., Azlan, A., Tang, S.T., and Lim, S.M. (2017). Anthocyanidins and anthocyanins: colored pigments as food, pharmaceutical ingredients, and the potential health benefits. *Food & Nutrition Research* 61: 1361779.

Kirchhoff, E. (2002). Online-publication of the German food composition table "Souci–Fachmann–Kraut" on the internet. *Journal of Food Composition and Analysis* 15: 465–472.

Kittisuban, P., Lee, B.-H., Suphantharika, M., and Hamaker, B.R. (2014). Slow glucose release property of enzyme-synthesized highly branched maltodextrins differs among starch sources. *Carbohydrate Polymers* 107: 182–191.

Klaochanpong, N., Puttanlek, C., Rungsardthong, V. et al. (2015). Physicochemical and structural properties of debranched waxy rice, waxy corn and waxy potato starches. *Food Hydrocolloids* 45: 218–226.

Klopfenstein, C.F. (1988). The role of cereal beta-glucans in nutrition and health. *Cereal Foods World* 33 (10): 865–869.

Koehler, P. and Wieser, H. (2013). Chemistry of cereal grains. In: *Handbook on Sourdough Biotechnology* (eds. M. Gobbetti and M. Gänzle), 11–45. Boston: Springer.

Kołodziejczyk, P., Makowska, A., Pospieszna, B. et al. (2018). Chemical and nutritional characteristics of high-fibre rye milling fractions. *Acta Scientiarum Polonorum. Technologia Alimentaria* 17: 149–157.

Kong, X., Kasapis, S., and Bao, J. (2015). Viscoelastic properties of starches and flours from two novel rice mutants induced by gamma irradiation. *LWT- Food Science and Technology* 60: 578–582.

Kumari, M. and Platel, K. (2017a). Bioaccessibility of trace elements and chromium speciation in commonly consumed cereals and pulses. *International Journal of Food Properties* 20: 1612–1620.

Kumari, M. and Platel, K. (2017b). Effect of sulfur-containing spices on the bioaccessibility of trace minerals from selected cereals and pulses. *Journal of the Science of Food and Agriculture* 97: 2842–2848.

Kuznesof, S., Brownlee, I.A., Moore, C. et al. (2012). WHOLEheart study participant acceptance of wholegrain foods. *Appetite* 59: 187–193.

Langseth, W., Ghebremeskel, M., Kosiak, B. et al. (2001). Production of culmorin compounds and other secondary metabolites by *Fusarium culmorum* and *F. graminearum* strains isolated from Norwegian cereals. *Mycopathologia* 152: 23–34.

Larkins, B.A., Wu, Y., Song, R., and Messing, J. (2017). Maize seed storage proteins. In: *Maize Kernel Development* (ed. B.A. Larkins), 175–189. Boston: CABI.

Lebiedzińska, A. and Szefer, P. (2006). Vitamins B in grain and cereal–grain food, soy-products and seeds. *Food Chemistry* 95: 116–122.

Lee, N.Y., Kim, Y.-K., Choi, I. et al. (2010). Biological activity of barley (*Hordeum vulgare* L.) and barley by-product extracts. *Food Science and Biotechnology* 19: 785–791.

Lee, C.J., Kim, Y., Choi, S.J., and Moon, T.W. (2012). Slowly digestible starch from heat-moisture treated waxy potato starch: preparation, structural characteristics, and glucose response in mice. *Food Chemistry* 133: 1222–1229.

Leonard, S.W., Good, C.K., Gugger, E.T., and Traber, M.G. (2004). Vitamin E bioavailability from fortified breakfast cereal is greater than that from encapsulated supplements. *The American Journal of Clinical Nutrition* 79: 86–92.

Lin, N. and Dufresne, A. (2014). Nanocellulose in biomedicine: current status and future prospect. *European Polymer Journal* 59: 302–325.

Lin, Y., Vonk, R.J., Slooff, M.J.H. et al. (1995). Differences in propionate-induced inhibition of cholesterol and triacylglycerol synthesis between human and rat hepatocytes in primary culture. *British Journal of Nutrition* 74: 197.

Lu, J., Xie, G., Jia, W., and Jia, W. (2013). Insulin resistance and the metabolism of branched-chain amino acids. *Frontiers of Medicine* 7: 53–59.

Lütteke, T., Frank, M., and Von Der Lieth, C.-W. (2005). Carbohydrate Structure Suite (CSS): analysis of carbohydrate 3D structures derived from the PDB. *Nucleic Acids Research* 33: D242–D246.

Lynch, C.J. and Adams, S.H. (2014). Branched-chain amino acids in metabolic signalling and insulin resistance. *Nature Reviews Endocrinology* 10: 723.

Malaguti, M., Dinelli, G., Leoncini, E. et al. (2014). Bioactive peptides in cereals and legumes: agronomical, biochemical and clinical aspects. *International Journal of Molecular Sciences* 15: 21120–21135.

Manach, C., Scalbert, A., Morand, C. et al. (2004). Polyphenols: food sources and bioavailability. *The American Journal of Clinical Nutrition* 79: 727–747.

Martínez-Villaluenga, C. and Peñas, E. (2017). Health benefits of oat: current evidence and molecular mechanisms. *Current Opinion in Food Science* 14: 26–31.

Masisi, K., Beta, T., and Moghadasian, M.H. (2016). Antioxidant properties of diverse cereal grains: a review on in vitro and in vivo studies. *Food Chemistry* 196: 90–97.

Matlashewski, G., Urquhart, A., Sahasrabudhe, M., and Altosaar, I. (1982). Lipase activity in oat flour suspensions and soluble extracts. *Cereal Chemistry* 59: 418–422.

Mcclure, F. and Muller, A. Jr. (1958). Wheat cereal diets, rat caries, lysine and minerals. *The Journal of Nutrition* 65: 619–631.

Meyer, J., Murray, S.L., and Berger, D.K. (2016). Signals that stop the rot: regulation of secondary metabolite defences in cereals. *Physiological and Molecular Plant Pathology* 94: 156–166.

Mikkelsen, D., Flanagan, B.M., Wilson, S.M. et al. (2015). Interactions of arabinoxylan and (1, 3) (1, 4)-β-glucan with cellulose networks. *Biomacromolecules* 16: 1232–1239.

Mikulajová, A., Takácsová, M., Rapta, P. et al. (2007). Total phenolic contents and antioxidant capacities of cereal and pseudocereal genotypes. *Journal of Food and Nutrition Research* 46: 150–157.

Mir, N.A., Riar, C.S., and Singh, S. (2018). Nutritional constituents of pseudo cereals and their potential use in food systems: a review. *Trends in Food Science & Technology* 75: 170–180.

Morrison, W.R. (1988). Lipids in cereal starches: a review. *Journal of Cereal Science* 8: 1–15.

Morrison, W. and Coventry, A. (1985). Extraction of lipids from cereal starches with hot aqueous alcohols. *Starch-Stärke* 37: 83–87.

Muñoz, R., Rodriguez, J.A., Páez-Hernández, M.E. et al. (2015). Determination of lipoxygenase activity in cereals grains using a silver nanoparticles assay. *Journal of Chemistry* 2015: 786430.

Naczk, M. and Shahidi, F. (2006). Phenolics in cereals, fruits and vegetables: occurrence, extraction and analysis. *Journal of Pharmaceutical and Biomedical Analysis* 41: 1523–1542.

Oghbaei, M. and Prakash, J. (2013). Effect of fractional milling of wheat on nutritional quality of milled fractions. *Trends in Carbohydrate Research* 5 (1): 52–58.

Oghbaei, M. and Prakash, J. (2016). Effect of primary processing of cereals and legumes on its nutritional quality: a comprehensive review. *Cogent Food & Agriculture* 2: 1136015.

Ohishi, K., Kasai, M., Shimada, A., and Hatae, K. (2007). Effects of acetic acid on the rice gelatinization and pasting properties of rice starch during cooking. *Food Research International* 40: 224–231.

Osborne, T.B. (1907). *The Proteins of the Wheat Kernel*. Washington: Carnegie Institute of Washington.

O'Sullivan, A.C. (1997). Cellulose: the structure slowly unravels. *Cellulose* 4: 173–207.

Papageorgiou, M. and Skendi, A. (2018). Introduction to cereal processing and by-products. In: *Sustainable Recovery and Reutilization of Cereal Processing by-Products* (ed. C.M. Galanakis), 1–25. Woodhead Publishing.

Pathera, A.K., Riar, C.S., Yadav, S., and Sharma, D.P. (2017). Effect of dietary fiber enrichment and different cooking methods on quality of chicken nuggets. *Korean Journal for Food Science of Animal Resources* 37: 410–417.

Pereira, C., Cunha, A., Reis, R. et al. (1998). New starch-based thermoplastic hydrogels for use as bone cements or drug-delivery carriers. *Journal of Materials Science: Materials in Medicine* 9: 825–833.

Pietrzyk, S., Fortuna, T., Łabanowska, M. et al. (2018). The effect of amylose content and level of oxidation on the structural changes of acetylated corn starch and generation of free radicals. *Food Chemistry* 240: 259–267.

Pietta, P.G. (2000). Flavonoids as antioxidants. *Journal of Natural Products* 63: 1035–1042.

Piovesana, S., Capriotti, A.L., Cavaliere, C. et al. (2018). Recent trends and analytical challenges in plant bioactive peptide separation, identification and validation. *Analytical and Bioanalytical Chemistry* 410 (15): 1–20.

Poletto, M., Ornaghi, H.L., and Zattera, A.J. (2014). Native cellulose: structure, characterization and thermal properties. *Materials* 7: 6105–6119.

Queenan, K.M., Stewart, M.L., Smith, K.N. et al. (2007). Concentrated oat β-glucan, a fermentable fiber, lowers serum cholesterol in hypercholesterolemic adults in a randomized controlled trial. *Nutrition Journal* 6: 6.

Quiñones, M., Miguel, M., and Aleixandre, A. (2013). Beneficial effects of polyphenols on cardiovascular disease. *Pharmacological Research* 68: 125–131.

Rajni, B. (2018). *Unit-4 the Micronutrients-I: Vitamins*. IGNOU.

Reddy, B.R., Ghosh, A., and Fitzgerald, R. (2013). Method of using polysaccharide based cement additives. United States Patent, US 8513167B2, filed 30 April 2012 and issued 20 August 2013.

Repo-Carrasco-Valencia, R. and Arana, J.V. (2017). Carbohydrates of kernels. In: *Pseudocereals: Chemistry and Technology* (eds. C.M. Haros and R. Schoenlechner), 49–70. Ltd: Wiley.

Richardson, P.H., Jeffcoat, R., and Shi, Y.-C. (2000). High-amylose starches: from biosynthesis to their use as food ingredients. *MRS Bulletin* 25: 20–24.

Rideout, T.C., Harding, S.V., Raslawsky, A., and Rempel, C.B. (2017). Dietary resistant starch supplementation increases high-density lipoprotein particle number in pigs fed a western diet. *Journal of Dietary Supplements* 14: 334–345.

Rohini, A., Agrawal, N., Kumar, H., and Kumar, V. (2018). Emerging role of branched chain amino acids in metabolic disorders: a mechanistic review. *PharmaNutrition* 6 (2): 47–54.

Rosa-Sibakov, N., Poutanen, K., and Micard, V. (2015). How does wheat grain, bran and aleurone structure impact their nutritional and technological properties? *Trends in Food Science & Technology* 41: 118–134.

Rosell, C.M., Barro, F., Sousa, C., and Mena, M.C. (2014). Cereals for developing gluten-free products and analytical tools for gluten detection. *Journal of Cereal Science* 59: 354–364.

Rosentrater, K.A. and Evers, A.D. (2017). Chemical components and nutrition. In: *Kent's Technology of Cereals: An Introduction for Students of Food Science and Agriculture*, 5e (eds. K.A. Rosentrater and A.D. Evers), 267–368. Woodhead Publishing.

Ross, A.B., Zangger, A., and Guiraud, S.P. (2014). Cereal foods are the major source of betaine in the Western diet–analysis of betaine and free choline in cereal foods and updated assessments of betaine intake. *Food Chemistry* 145: 859–865.

Roux, R. and Voigt, A. (2014). Starch composition for use in paper manufacture. United States Patent, US 8801900B2, filed 7 December 2011 and issued 12 August 2014.

Sajilata, M. and Singhal, R.S. (2005). Specialty starches for snack foods. *Carbohydrate Polymers* 59: 131–151.

Sajilata, M.G., Singhal, R.S., and Kulkarni, P.R. (2006). Resistant starch – a review. *Comprehensive Reviews in Food Science and Food Safety* 5: 1–17.

Sandberg, A.S. and Svanberg, U. (1991). Phytate hydrolysis by phytase in cereals; effects on in vitro estimation of iron availability. *Journal of Food Science* 56: 1330–1333.

Šárka, E. and Dvořáček, V. (2017). New processing and applications of waxy starch (a review). *Journal of Food Engineering* 206: 77–87.

Schwartz, D. and Whistler, R.L. (2009). History and future of starch. In: *Starch: Chemistry and Technology* (eds. R. Whistler, J. BeMiller and E. Paschall), 1–10. Academic Press.

Selvendran, R. and Du Pont, M. (1980). An alternative method for the isolation and analysis of cell wall material from cereals. *Cereal Chemistry* 57: 278–283.

Shamloo, M., Eck, P., and Beta, T. (2015). Angiotensin converting enzyme inhibitory peptides derived from cereals. *Journal of Human Nutrition & Food Science* 3: 1057–1067.

Shewry, P.R. (2000). Seed proteins. In: *Seed Technology and Its Biological Basis* (eds. M. Black and J.D. Bewley), 42–84. Sheffield: Sheffield Academic Press.

Shewry, P.R. (2007). Improving the protein content and composition of cereal grain. *Journal of Cereal Science* 46: 239–250.

Shewry, P.R. (2017). Cereal seed storage proteins. In: *Seed Development and Germination* (ed. J. Kigel), 45–72. Abingdon, UK: Routledge.

Shewry, P.R. and Casey, R. (1999). *Seed proteins*. In: *Seed Proteins* (eds. P.R. Shewry and R. Casey), 1–10. Kluwer Academic Publishers.

Shewry, P.R. and Halford, N.G. (2002). Cereal seed storage proteins: structures, properties and role in grain utilization. *Journal of Experimental Botany* 53: 947–958.

Shewry, P.R. and Tatham, A.S. (1990). The prolamin storage proteins of cereal seeds: structure and evolution. *Biochemical Journal* 267: 1–12.

Sima, P., Vannucci, L., and Vetvicka, V. (2018). β-Glucans and cholesterol. *International Journal of Molecular Medicine* 41: 1799–1808.

Singh, J. and Skerritt, J.H. (2001). Chromosomal control of albumins and globulins in wheat grain assessed using different fractionation procedures. *Journal of Cereal Science* 33: 163–181.

Singh, A., Karmakar, S., Jacob, B.S. et al. (2015a). Enzymatic polishing of cereal grains for improved nutrient retainment. *Journal of Food Science and Technology* 52: 3147–3157.

Singh, T., Chatli, M.K., Kumar, P. et al. (2015b). Effect of different cooking methods on the quality attributes of chicken meat cutlets. *Journal of Animal Research* 5: 547–554.

Sinha, A.K., Kumar, V., Makkar, H.P. et al. (2011). Non-starch polysaccharides and their role in fish nutrition – a review. *Food Chemistry* 127: 1409–1426.

Smrcková, P., Saglamtas, M., Hofmanová, T. et al. (2013). Changes of resistant starch during extrusion process. Poster presented at the 9th International Conference on Polysaccharides-Glycoscience, Prague, Czech Republic (6–8 November 2013). Prague: Czech Chemical Society.

Starmans, D.A. and Nijhuis, H.H. (1996). Extraction of secondary metabolites from plant material: a review. *Trends in Food Science & Technology* 7: 191–197.

Statista (2018). Breakfast cereals. https://www.statista.com/outlook/40090100/102/breakfast-cereals/europe (accessed 5 September 2018).

Sun, H., Ma, X., Zhang, S. et al. (2017). Resistant starch produces antidiabetic effects by enhancing glucose metabolism and ameliorating pancreatic dysfunction in type 2 diabetic rats. *International Journal of Biological Macromolecules* 110: 276–284.

Talukder, S. and Sharma, D.P. (2010). Development of dietary fiber rich chicken meat patties using wheat and oat bran. *Journal of Food Science and Technology* 47: 224–229.

Teli, M., Rohera, P., Sheikh, J., and Singhal, R. (2009a). Application of germinated maize starch in textile printing. *Carbohydrate Polymers* 75: 599–603.

Teli, M., Rohera, P., Sheikh, J., and Singhal, R. (2009b). Use of Amaranthus (Rajgeera) starch vis-à-vis wheat starch in printing of vat dyes. *Carbohydrate Polymers* 76: 460–463.

Tigyi, G.J. (2013). Lysophospholipid receptors. In: *Encyclopedia of Biological Chemistry*, 2e (eds. W.J. Lennarz and M.D. Lane), 762–764. Academic Press.

Tong, C., Liu, L., Waters, D.L.E. et al. (2014). Genotypic variation in lysophospholipids of milled rice. *Journal of Agricultural and Food Chemistry* 62: 9353–9361.

Topping, D.L. and Clifton, P.M. (2001). Short-chain fatty acids and human colonic function: roles of resistant starch and nonstarch polysaccharides. *Physiological Reviews* 81: 1031–1064.

Triff, K., Mclean, M.W., Callaway, E. et al. (2018). Dietary fat and fiber interact to uniquely modify global histone post-translational epigenetic programming in a rat colon cancer progression model. *International Journal of Cancer* 143: 1402–1415.
Tsao, R. (2010). Chemistry and biochemistry of dietary polyphenols. *Nutrients* 2: 1231–1246.
USDA (U.S. Department of Agriculture) (2018). Agricultural Research Service, National Nutrient Database for Standard Reference https://ndb.nal.usda.gov/ndb/search/list (accessed 5 October 2018).
Vary, T.C. and Lynch, C.J. (2007). Nutrient signaling components controlling protein synthesis in striated muscle. *The Journal of Nutrition* 137: 1835–1843.
Ververis, C., Georghiou, K., Christodoulakis, N. et al. (2004). Fiber dimensions, lignin and cellulose content of various plant materials and their suitability for paper production. *Industrial Crops and Products* 19: 245–254.
Vomhof, D. and Tucker, T. (1965). The separation of simple sugars by cellulose thin-layer chromatography. *Journal of Chromatography A* 17: 300–306.
Walter, M. and Marchesan, E. (2011). Phenolic compounds and antioxidant activity of rice. *Brazilian Archives of Biology and Technology* 54: 371–377.
Wang, T., Cai, G., Qiu, Y. et al. (2011a). Structural segregation of gut microbiota between colorectal cancer patients and healthy volunteers. *The ISME Journal* 6: 320–329.
Wang, T.J., Larson, M.G., Vasan, R.S. et al. (2011b). Metabolite profiles and the risk of developing diabetes. *Nature Medicine* 17: 448–453.
Wang, S., Wang, J., Yu, J., and Wang, S. (2014). A comparative study of annealing of waxy, normal and high-amylose maize starches: the role of amylose molecules. *Food Chemistry* 164: 332–338.
Wang, W., Xu, M., Wang, G., and Galili, G. (2018). New insights into the metabolism of aspartate-family amino acids in plant seeds. *Plant Reproduction* 31: 203–211.
Waterschoot, J., Gomand, S.V., Fierens, E., and Delcour, J.A. (2015). Production, structure, physicochemical and functional properties of maize, cassava, wheat, potato and rice starches. *Starch-Stärke* 67: 14–29.
Whistler, R.L. and Doane, W.M. (1961). Characterization of intermediary fractions of high-amylose corn starches. *Cereal Chemistry* 38: 251–255.
Wiemer, K. (2018). Breakfast cereals. In: *Food Fortification in a Globalized World* (eds. M.G.V. Mannar and R.F. Hurrell), 183–191. Academic Press.
Włodarczyk-Stasiak, M., Mazurek, A., Kowalski, R. et al. (2017). Physicochemical properties of waxy corn starch after three-stage modification. *Food Hydrocolloids* 62: 182–190.
Woggum, T., Sirivongpaisal, P., and Wittaya, T. (2015). Characteristics and properties of hydroxypropylated rice starch based biodegradable films. *Food Hydrocolloids* 50: 54–64.
Wood, P.J. (2010). Oat and rye β-glucan: properties and function. *Cereal Chemistry* 87: 315–330.
WORLDBANK. (2018). Cereal production (metric tons). https://data.worldbank.org/indicator/AG.PRD.CREL.MT (accessed 5 October 2018).
Wrigley, C.W. and Bekes, F. (2001). *Cereal-Grain Proteins*. Lancaster: Technomic Publishing Co.
Wu, H., Dwyer, K.M., Fan, Z. et al. (2003). Dietary fiber and progression of atherosclerosis: the Los Angeles Atherosclerosis Study. *The American Journal of Clinical Nutrition* 78: 1085–1091.
Yadav, B. and Jindal, V. (2008). Changes in head rice yield and whiteness during milling of rough rice (*Oryza sativa* L.). *Journal of Food Engineering* 86: 113–121.

Yadav, S., Malik, A., Pathera, A. et al. (2016). Development of dietary fibre enriched chicken sausages by incorporating corn bran, dried apple pomace and dried tomato pomace. *Nutrition & Food Science* 46: 16–29.

Yeung, A.W.K., Mocan, A., and Atanasov, A.G. (2018). Let food be thy medicine and medicine be thy food: a bibliometric analysis of the most cited papers focusing on nutraceuticals and functional foods. *Food Chemistry* 269: 455–465.

Yoon, M.-S. (2016). The emerging role of branched-chain amino acids in insulin resistance and metabolism. *Nutrients* 8: 405.

Yu, S. and Tian, L. (2018). Breeding major cereal grains through the lens of nutrition sensitivity. *Molecular Plant* 11: 23–30.

Zaman, S.A. and Sarbini, S.R. (2016). The potential of resistant starch as a prebiotic. *Critical Reviews in Biotechnology* 36: 578–584.

Zambrowicz, A., Timmer, M., Polanowski, A. et al. (2013). Manufacturing of peptides exhibiting biological activity. *Amino Acids* 44: 315–320.

Zeng, H., Huang, C., Lin, S. et al. (2017). Lotus seed resistant starch regulates gut microbiota and increases short-chain fatty acids production and mineral absorption in mice. *Journal of Agricultural and Food Chemistry* 65: 9217–9225.

Zhao, X., Andersson, M., and Andersson, R. (2018). Resistant starch and other dietary fiber components in tubers from a high-amylose potato. *Food Chemistry* 251: 58–63.

Zhuang, X., Zhao, C., Liu, K. et al. (2018). Cereal grain fractions as potential sources of prebiotics: current status, opportunities, and potential applications. In: *Food and Feed Safety Systems and Analysis* (eds. S.C. Ricke, G.G. Atungulu, C.E. Rainwater, et al.), 173–191. Academic Press.

Zielke, C., Stradner, A., and Nilsson, L. (2018). Characterization of cereal β-glucan extracts: conformation and structural aspects. *Food Hydrocolloids* 79: 218–227.

Zúñiga, R.N. and Troncoso, E. (2012). Improving nutrition through the design of food matrices. In: *Scientific, Health and Social Aspects of the Food Industry* (ed. B. Valdez), 264–320. Rijeka: InTech.

Zykin, P., Andreeva, E., Lykholay, A. et al. (2018). Anthocyanin composition and content in rye plants with different grain color. *Molecules* 23: 948.

8
Innovative Gluten-Free Products

Cristina M. Rosell[1], Mehran Aalami[2], and Sahar Akhavan Mahdavi[2]

[1] *Institute of Agrochemistry and Food Technology (IATA-CSIC), Paterna, Spain*
[2] *Department of Food Science & Technology, Gorgan University of Agricultural Sciences and Natural Resources, Gorgan, I.R. Iran*

8.1 Introduction

In recent years, the gluten-free (GF) products market has become one of the most prosperous markets in the field of food and beverages. This is foreseen to continue into the immediate future, as it does not only cater to individuals with medical needs, but also to other consumers who prefer a GF diet. The growing demand for GF food is caused by the increasing number of patients diagnosed with certain pathologies. It is estimated that approximately 0.5–1% of people worldwide should adhere to a GF diet (Capriles and Arêas 2014).

Celiac disease, or coeliac disease, is an immune-mediated enteropathy triggered by the ingestion of gluten in genetically susceptible individuals. More than one genetic factor is necessary for the disease to manifest in a person. The major cause of this disease is the storage proteins (gluten) found in the genus *Triticum* (different species and subspecies of wheat such as *T. aestivum, T. dicoccum, T. durum, T. monococcum*, and *T. spelta*), wheat hybrids (such as *triticale*), other cereals belonging to the *Triticeae* species (barley and rye), and possibly oats. The major protein fractions, which trigger the immune response, are the prolamin fractions, which are part of the storage proteins rich in proline (prol-) and glutamine (-amin). These particular storage proteins can be extracted by 40–70% ethanol and are resistant to proteases and peptidases in the gut. Prolamins are found in different cereal grains with different but related prolamins: wheat (gliadins), barley (hordeins), rye (secalins), and oats (avenins) (*Codex Alimentarius* 2008).

The major symptoms involved in celiac disease are gastrointestinal problems and malabsorption syndrome, which include chronic diarrhea, weight loss, and abdominal distension. It also involves a number of symptoms which influences many major organs of the body (i.e. the immune and nervous systems). Celiac disease is one of the most common lifelong disorders on a worldwide basis. Globally, coeliac disease affects 1 in 100,

Innovative Processing Technologies for Healthy Grains, First Edition.
Edited by Milica Pojić and Uma Tiwari.

although the incidence varies among different regions of the world. Since celiac disease is often atypical or even silent on clinical grounds, it is believed that about 85% of people affected are undiagnosed, leading to the risk of long-term complications, such as osteoporosis, infertility, or cancer. At present, the treatment of coeliac disease requires a strict lifelong GF diet, which leads to the recovery of the intestinal mucosa, improves the symptoms, and reduces the risk of developing complications in most people (American Gastroenterological Association 2001). Furthermore, gluten sensitivity has been coined to describe those individuals who cannot tolerate gluten and experience symptoms similar to those with celiac disease, yet lack the same antibodies and intestinal damage as seen in celiac disease. Gluten sensitivity has been clinically recognized as less severe than celiac disease. It is not accompanied by "the enteropathy, elevations in tissue-transglutaminase, endomysium or deamidated gliadin antibodies, and increased mucosal permeability that are characteristic of celiac disease" (Ludvigsson et al. 2012). In other words, people with gluten sensitivity would not test positive for celiac disease based on blood testing, nor have the same type of intestinal damage found in individuals with celiac disease. As with celiac disease treatment, a non-celiac gluten intolerance diet should also be based on a gluten-free diet (Vinall 2014).

The term "gluten-free" is generally used to indicate a supposed harmless level of gluten rather than its complete absence. The current international *Codex Alimentarius* standard allows 20 ppm of gluten in so-called "gluten-free" foods. Several organizations, such as the Gluten-Free Certification Organization (GFCO), the Celiac Sprue Association (CSA), and the National Foundation for Celiac Awareness (NFCA), certify products and companies as "gluten-free."

The formulation of GF bakery products presents a challenge to both the cereal technologists and the bakers. There is a worldwide trend toward the use of diverse GF products, formulation, and technologies, as alternatives to gluten to improve the structure, mouthfeel, acceptability, and shelf life of GF bakery products. Hence, the following sections will review the recent advances in the preparation of GF products, including innovative processing technologies. Assuming that innovations are related to the latest discoveries, only the most recent reports have been included.

8.2 Gluten-Free Foods

The Association of European Coeliac Societies (AOECS) has defined gluten-free foods and drinks based on the worldwide *Codex Alimentarius* standard, so these types of foods should consist of, or made from, only one or more ingredients which do not contain wheat, rye, barley, oats, or any of their crossbred varieties, and the gluten levels should not exceed 20 mg/kg in total, based on the food as sold or distributed to the consumer. However, some GF foods may also be prepared from gluten-containing grains, which have been specially processed to reduce/remove their gluten content, and their gluten level should not exceed 20 mg/kg in total. It should be noted that oats can be tolerated by most, but not all people, who are intolerant to gluten. Therefore, the allowance of oats, not contaminated with wheat, rye, or barley, covered by this standard may be determined at the national level (AOECS 2016). According to AOECS and Celiac Sprue Association (USA), the approved cereal grains are rice (*Oryza sativa*), maize (*Zea mays*), sorghum (*Sorghum bicolor*), and millet (e.g. *Panicum miliaceum*, *Setaria italica*, *Pennisetum typhoideum*, and *Eleusine coracana*). In addition, carbohydrate-rich pseudocereals, such as buckwheat (*Fagopyrum esculentum*), quinoa

(*Chenopodium quinoa*), and amaranth (*Amaranthus*). are commonly used for the preparation of GF products (AOECS 2005). According to the Gluten Intolerance Group (GIG) (Seattle, Washington) and the Celiac Disease Foundation (CDF) (Studio City, California), these pseudocereals are included in GF diets, but according to the Celiac Sprue Association (CSA) (Omaha, Nebraska), they are non-approved.

8.2.1 Bakery Products

The production of gluten-free bakery products is a technological challenge because it is difficult to achieve the viscoelastic properties of the dough by using GF flour. The utilization of different GF formulation, technologies, and products, which provide alternatives to gluten, has become a worldwide trend to improve the structure, mouthfeel, acceptability, and shelf life of GF bakery products. Water is required in larger quantities for the processing of GF products as compared to wheat flour, so that GF dough is generally homogenized in mixing machines to a batter consistency. Gluten-free products have some deficiencies such as lower volume, poor gas (CO_2) retention ability, a denser crumb, lack of cell structure, slightly crumbly, dry and grainy texture, cracked crust, unpleasant flavor, and appearance of the dry structure, according to literature reviews (Arendt et al. 2011; Capriles and Arêas 2014; Dar 2013; Matos and Rosell 2015). To solve the above-mentioned deficiencies, along with new technologies, many ingredients have been added such as various hydrocolloids (Mir et al. 2016; Morreale et al. 2018), processing aids or enzymes (Renzetti and Rosell 2016), and proteins (Storck et al. 2013), to mimic the viscoelastic properties of gluten and contribute to improving the structure, mouthfeel, acceptability, and shelf life of GF breads.

In the case of GF products, new product ideas are obviously reformulations, where the standard products are reproduced using a formulation excluding gluten-containing ingredients. Nevertheless, the recent concerns about the nutritional quality of GF products have prompted the reformulation of GF breads considering their nutritional pattern (Matos and Rosell 2012). The nutritional value of GF bakery products has been improved by fortification with isolated compounds or addition of raw materials naturally rich in nutritionally valuable constituents (Capriles et al. 2016; O'Shea et al. 2014; Witczak et al. 2016). Innovative raw materials recently used by GF bakeries include rice (Nakano et al. 2018; Matos and Rosell 2013; Matos et al. 2014), sorghum (Marston et al. 2016), pseudocereals like buckwheat (Molinari et al. 2017), or amaranth (Bastos et al. 2016), legumes (Melini et al. 2017), flaxseed (Hargreaves and Zandonadi 2018), psyllium (Fratelli et al. 2018), chestnut (Rinaldi et al. 2017), carob germ flour (Turfani et al. 2017), lupine (Levent and Bilgicli 2011), chia seed (Sandri et al. 2017), teff (do Nascimento et al. 2018), cassava (Garcia et al. 2018), pine nuts (Polet et al. 2019), and yam (Seguchi et al. 2012), among others.

There are a number of products, like wafers and waffles, crepes, as well as cakes and cookies, where the gluten network is not developed, thus GF counterparts are easier to obtain. In order to form such products, the characteristics of the flour should be inspected carefully, i.e. particle size, color, and starch properties of the flour, so that standard quality products can be achieved. Sensory attributes like taste, flavor, color, and size also affect the overall quality of the final product (Gómez and Martínez 2016). Hydration of the flour particles is also affected by the quality of the flour used. Finer particles produce better clear texture of the finished cookies, as it affects the batter/dough emulsion properties (de la Hera et al. 2013). According to the literature, the choice of raw material depends on many factors, but flours from rice, buckwheat, amaranth, teff, and chickpeas along with corn, potato flours and starches, are recommended as the best raw materials for this purpose (Heller 2009; Rai et al. 2014).

8.2.2 Pasta and Extruded Products

Basically, in GF pasta, gluten can be successfully replaced by suitable formulations and recipes using heat-treated flours as the key ingredients, or by adopting non-conventional pasta-making processes to induce new rearrangements of the starch macromolecules. The common ingredients in GF pasta are flours and/or starches with the addition of protein, gums, and emulsifiers, which may partially act as substitutes for gluten (Marti et al. 2013). Many researchers have investigated different formulations and also process specifications for obtaining GF pasta (Benhur et al. 2015; Ferreira et al. 2016; Pagani et al. 2010). Pseudocereals (Mastromatteo et al. 2011), vegetable flours (Padalino et al. 2013), legume flours (Bouasla et al. 2017; Heo et al. 2014), chickpea, unripe plantain (Flores-Silva et al. 2014), and fermented cassava flour (Purwandari et al. 2014) have been reported as suitable raw materials for pasta production. In general, pseudocereals that have shown their potential for the production of good-quality spaghetti and some appropriate combinations include quinoa and oat (Chillo et al. 2009), quinoa, corn, and soy flours (Mastromatteo et al. 2011), amaranth, quinoa, and buckwheat (Schoenlechner et al. 2010), amaranth flour (Borneo and Aguirre 2008; Cárdenas-Hernández et al. 2016; Islas-Rubio et al. 2014) or teff, buckwheat, quinoa, and amaranth (Kahlon and Chiu 2015). Schoenlechner et al. (2010) investigated the use of amaranth, quinoa, and buckwheat for the production of GF pasta with good textural quality, in particular, low cooking loss, optimal cooked weight, and texture firmness. In addition to the flour source, it has been found that protein interactions have a considerable influence on pasta quality. Lorenzo et al. (2018) observed that the cohesiveness of cooked pasta is mainly dependent on the competition between starch and protein molecules to form a continuous network. Cohesiveness of the traditional wheat pasta was significantly higher than that found in pasta made with corn or quinoa flours. This effect was probably related to the cohesive characteristic of the gluten network and its interaction with the gelatinized starch granules. When quinoa flour and egg proteins were replaced with corn flour and zein, the viscosity of the GF pasta dough was similar to wheat pasta dough, but there was an increase in the elastic modulus that resulted in a more fragile structure. Formulating GF pasta requires a thorough knowledge of the component properties of GF flours and starches, but also a proper selection of additives to promote a cohesive mass in the product (Fuad and Prabhasankar 2010). Additives used in GF pasta and extruded products comprise hydrocolloids (Morreale et al. 2018), emulsifiers (Charutigon et al. 2008; Kaur et al. 2005), and proteins (Doxastakis et al. 2007; Krupa-Kozak et al. 2013; Onwulata and Konstance 2006; Savita et al. 2013).

8.2.3 Other Gluten-Free Products

To date, GF beverages, functional drinks, and infant formula, which may be included in the diets of patients with celiac disease, are commercially available. However, they are often based on pure starches, resulting in a dry, sandy mouthfeel and poor overall eating quality. In the last decade, the production of GF cereal based beverages has attracted much attention (Rubio-Flores and Serna-Saldivar 2016). The production of malt and beer from GF cereals has focused on rice (Mayer et al. 2016), sorghum (Ndubisi et al. 2016), teff (Di Ghionno et al. 2017), and pseudocereals such as buckwheat (Agu et al. 2012). Currently, only sorghum, millet, and buckwheat appear to be successful GF beer ingredients, while others have only shown adjunct possibilities. The results collected so far indicate that buckwheat beer might be a promising GF alternative to sorghum beer (Giménez-Bastida et al. 2015). Nevertheless, some marketing efforts are still needed to increase the knowledge of these cereals and pseudocereals (Hager et al. 2014). Also, some functional components from cereals, pseudocereals, or even herbs and herbal extracts, and also antioxidants, vitamins, or minerals, can be directly added to the beverage; although additional research is needed to fully understand their impact (Kreisz et al. 2008).

8.3 Processing Techniques for Improving Gluten-Free Products

Traditionally, the food industry's approach has been to use the GF flours previously mentioned, without modification, whose properties depend on grain characteristics and composition as well as the milling system used. However, flours obtained from traditional processes can be subjected to different physical treatments, ranging from simple sieving to complex hydrothermal treatments, which can modify flour functionality and their adequacy to the different GF elaborations (O'Shea et al. 2014; Tsatsaragkou et al. 2017).

8.3.1 Conventional Physical Treatments

Flour characteristics have a strong influence on the product features. Apart from the intrinsic variations due to genetic or cropping conditions, flour performance can be modified as a function of the production methods, such as type of milling or grinding, flour fractionation, and grain germination (Gómez and Martínez 2016). Flour particle size influences particle hydration and, in consequence, dough rheology (Tsatsaragkou et al. 2017). Moreover, particle size might impact physiological response, particularly the estimated glycemic index (de la Hera et al. 2014). When preparing wheat cakes, it is known that wheat flour quality directly influences the final quality of the product (Kahraman et al. 2008). For example, starch and protein content are known to significantly affect the batter and the final product (Wilderjans et al. 2010), and the effect of particle size of flours on cake quality has been intensely studied (de la Hera et al. 2013; Kim and Shin 2014). Kim and Shin (2014) claimed that rice flour passed through 120–160 mesh sieves (with particle sizes of $< 125\,\mu m$) is appropriate for making rice cakes. According to Dhen et al. (2016), the coarsest soy flours increased the viscosity of the GF layer and sponge batters, whereas the finest soy flours produced sponge cakes with higher specific volume (Dhen et al. 2016).

Related to the particle size of the flours, one of the most interesting physical treatments is micronization (fine grinding) and the subsequent air classification.

This treatment includes a forced reduction of the particle size, which could modify flour functionality and make them more suitable for different processes (Létang et al. 2002). The particle size threshold depends on each cereal, and the size of the starch granules, and the functionality of those flours will clearly be different (Protonotariou et al. 2015). Drastic effects can be obtained by jet milling that allows the micronization of the flour to the level of starch granule size (particle sizes < 40 μm by using high air pressure), which had a noticeable effect on the characteristics of the wheat flour (Protonotariou et al. 2015). Presumably, a holistic approach combining nutritive flours and physical treatments might enhance the benefits from the technological and nutritional points of view (Tsatsaragkou et al. 2017). Although a number of recent studies have been conducted for improving bakery products by jet milling (Angelidis et al. 2016; Protonotariou et al. 2015), there is limited information about its use in GF products. Tsatsaragkou et al. (2017) incorporated carob flour fractions of varying particle size on rice GF breads prepared with carob/rice (15 : 85) flour blends, leading to breads with a significantly lower glycemic index when carob flour had an average particle size of 80 μm. In general, according to the studies, a simple classification of GF flours based on their particle size, and the subsequent selection of the proper fraction, could improve their suitability for the manufacturing of GF products such as bread, cakes, or cookies.

Germination of grains allows the activation of enzyme activities and subsequent hydrolysis of the main constituents and release of bioactive compounds, such as GABA (γ-aminobutyric acid), phenolic compounds, γ-oryzanol (in rice), and an increase in the antioxidant activity, which is related to the mobilization of a variety of enzymes (Cornejo et al. 2015). Nevertheless, it is crucial to define the right germination time for optimizing flour functional properties and bread features. According to Cornejo and Rosell (2015), the germination of rice grains at 28 °C and 100% relative humidity (RH) for 12, 24, and 48 hours led to brown rice flours with significant changes in hydration and pasting properties due to the starch hydrolysis. Regarding the resulting breads, no significant effect in specific volume, humidity, and water activity of the GF bread was found as germination time increased, only a significant softening. After 48 hours of germination, the intense action of α-amylase produced excessive liquefaction and dextrinization, causing inferior bread quality (Cornejo and Rosell 2015). Nevertheless, breads were of superior nutritional quality, on the basis of the higher content of protein, lipids, and bioactive compounds, higher antioxidant activity, and lower phytic acid content and glycemic index (Cornejo et al. 2015). In fact, while the consumption of rice is associated with *diabetes mellitus* due to its high glycemic index, germinated brown rice takes a leading role against diabetes. At the same time, the germination of brown rice reduces the phytic acid content and enhances mineral bioavailability (Kim et al. 2012). The same treatment has been applied to pseudocereals and legumes (Ouazib et al. 2016). Other alternative would be sorghum germination. In fact, Phattanakulkaewmorie et al. (2011) investigated the effect of germination on chemical compositions, physio-chemical properties of malted (germinated) red sorghum flours, and the characteristics of the resulting breads. The higher amylase activity of the germinated sorghum flour led to lower viscosity and set back viscosity and the resulting breads had lower crumb hardness with higher cohesiveness (Phattanakulkaewmorie et al. 2011).

8.3.2 Emerging Technologies

New innovation trends in the production of GF foods employ so-called emerging technologies, focused on the shortening of processing and residence times, the improvement of product quality, and the enhancement of its functionality (Table 8.1). The selection of a processing technology depends upon specific applications and parameters such as flour type, required particle size, physicochemical properties of the final product, processing costs, etc.

8.3.2.1 Non-thermal Processing Technologies High pressure (HP) processing is a "non-thermal" technology that has been developed to obtain microbiologically safe food products while avoiding undesirable changes in the sensory, physico-chemical, and nutritional properties of foods. Most commercial HP treatments are in the

Table 8.1 Overview of common processing technologies and their impact on gluten-free products and/or their properties.

Technique	Advantages	Disadvantages	Source
High pressure	Significant or total inactivation of microorganisms Better functional and nutritional retention of ingredients, improved food quality parameters Food preservation	Expensive processing costs Loss of certain nutrients	Cappa et al. (2016) Vallons et al. (2011) Huttner et al. (2010)
Extrusion	Versatility of product shapes, textures, colors and appearances Energy efficient process Low cost process Destruction of certain naturally occurring toxins Reduction of microorganisms in the final product Denaturation of proteins	Loss of some nutrients Loss of color and flavor Occurrence of Maillard browning and caramelization	Stojceska et al. (2010) Merayo et al. (2011) Giménez et al. (2013)
Hidro-thermal	Low cost Increase in thermal stability Decrease in amylose leaching Increase in gelatinization temperature	Viscosity reduction Lower swelling capacity Long processing time	Purwani et al. (2016) Marston et al. (2016) Onyango et al. (2013)
Dry heating	Low processing cost Possibility of large-scale production in continuous mode	Long processing time Not suitable for thermo-sensitive materials	Marston et al. (2016) D'Amico et al. (2015)
Microwave	Larger pieces can be heated in a shorter time Energy savings	Non-uniform heating Lack of color and flavor development High moisture loss High initial investment	Demirkesen et al. (2013) Turabi et al. (2010)

(continued overleaf)

Table 8.1 (*continued*)

Technique	Advantages	Disadvantages	Source
Ohmic heating	Energy conversion efficiencies are very high Suitable for continuous processing Low capital investment Better product quality Shorter cooking time Causes less nutrient loss	Narrow frequency band Difficult to monitor and control Complex coupling between temperature and electrical field distribution Limited to direct current	An and King (2007) Gaytán-Martínez et al. (2011)
IR heating	Alternate source of energy Fast heating rate Shorter response time Uniform drying temperature High degree of process control High thermal efficiency	Low penetration power Prolonged exposure may cause fracturing	Turabi et al. (2008)
Radio frequency heating	Rapid heating The surface of food does not overheat Minimal heat damage and no surface browning Equipment is small, compact, clean in operation	Difficult to monitor and control	Guo et al. (2008)
Cold plasma	Microbial inactivation Inactivation of enzymes Higher shelf life Increased seed germination rate Reduced cooking time Alteration of hydrophilic/ hydrophobic properties	High initial investments Challenging in scaling up Optimal process parameters must be established for each process and equipment	Sarangapani et al. (2016) Nisoa and Matan (2013)

400–700 MPa range and are applied at refrigerated to moderate temperatures (under 50 °C). Under these conditions, HP is considered a non-thermal method and has become one of the innovative food processing technologies most accepted by the consumers (Torres and Velazquez 2005). Some studies pointed out that application of HP can successfully be applied to cereals, flours, and GF flours, because this treatment alters the structure of biopolymers such as proteins and starch, providing the possibility to create structures and to obtain novel textures (Vallons et al. 2011). Vallons and Arendt (2009) reported a positive effect of high-pressure treatment on buckwheat starch characteristics. In addition, the effect of pressure on the digestibility of common buckwheat starch was studied by Liu et al. (2016). They reported that high hydrostatic pressure significantly increased amylose content, and decreased rapidly digested starch, starch hydrolysis, and viscosity. Comparing the effect of HP on GF products, the main efforts of scientists have concentrated on rice, maize, and oat starch, whereas the investigations of alternative grains as well as different bakery products have been scarce.

On the other hand, recent studies have demonstrated that HP may inactivate the anti-nutritional factors of grains while preserving food quality and constituents, but the optimization of pressure, time, and temperature is required (Cappa et al. 2016). During HP treatment, allergenic proteins from rice grains are solubilized, particularly the 7S globulins; while no apparent alteration in color, shape, or size of treated seeds occurs at moderate pressure (Ahmed and Al-Attar 2017). Other constituents of grains, such as vitamin A, are not significantly affected, while water soluble vitamins (B_1, B_6, and C) are well retained (85%) (Estrada-Girón et al. 2005). Other potential applications of HP include the improvement of the functional properties of foods and food ingredients such as texture, emulsifying, whipping, and dough-forming properties (Rumpold 2005). Nevertheless, HP is not a cheap technology, so the cost-effectiveness of this technology must be determined when developing GF products.

Cold plasma (CP) is an emerging technology, which has attracted the attention of scientists globally. In the last decade, its applications were extended to the food industry as a powerful tool for non-thermal processing, with diverse utilization forms (Ekezie et al. 2017). Plasma is known as the fourth state of matter, consisting of electrons, ions, free radicals, atoms in free and excited state, and large numbers of unionized neutral molecules. CP treatment of foods can yield a wide range of beneficial effects, including longer shelf life, increased seed germination rate, reduction of cooking time of rice, starch modification, microbial inactivation, functionalization, inactivation of enzymes, etc. (Thirumdas et al. 2015). Laboratory plasma can be distinguished into two main groups, i.e. high temperature (or fusion plasma) and low temperature or CP (or gas discharges) (Bogaerts et al. 2002). Regarding GF flours, when CP has been applied to rice starch, an increase in amylose leaching as well as in pasting and final viscosities after the treatment has been observed, but with a decrease in the gelatinization temperature (Thirumdas et al. 2017). This technology has also been applied to brown rice grain, where the changes in the pH, color, hardness, α-amylase activity, and water uptake were obtained (Lee et al. 2016). Sarangapani et al. (2016) studied the efficacy of low-pressure plasma treatment in modification of parboiled rice flour by varying power intensity (30, 40, and 50 W during 5, 10, and 15 minutes), which resulted in improving the gel and flour hydration properties as well as an increase in gelatinization temperature, depending to the starch granule structure.

8.3.2.2 Thermal Processing Techniques The modification of the functionality of starchy ingredients by thermal treatments is attracting great interest, as it allows achieving the properties of chemically modified starches while keeping the "clean label" by avoiding artificial ingredients and chemicals. A simple heating process produces flour dehydration, which could be necessary to preserve flours for longer periods, especially in the case of flours with a moisture content higher than 15%. Moreover, flour functionality can be modified because the treatment can change starch granules, denature proteins, inactivate enzymes, reduce microbial load, and even modify taste and aroma. All these changes can affect flour suitability in the manufacture of the GF products. Thermal treatments that have been used for GF products include dry heat treatment, annealing, heat moisture treatment, extrusion cooking, microwave treatment, and infrared (IR)-based micronization (Houben et al. 2012).

Dry heating treatment (DHT) is a physical modification that changes the physicochemical properties of starch, without destroying its granule structure. This treatment has been successfully applied to selected cereals like rice or millet, after steaming for

20 minutes at ambient pressure. Doughs obtained from those cereals were more cohesive, reducing the breaking or cracking during sheeting/flattening/rolling (Vidya et al. 2013).

When heating is carried out with sufficient water (previous hydration of flour or starch) within the so-called hydrothermal treatments, morphological changes within starch granules take place. Hydrothermal treatments are mainly classified in two groups: those performed below gelatinization temperature, in which the integrity of starch granules is preserved; and those carried out above gelatinization temperature, in which the molecular order of the starch granule is irreversibly destroyed. A similar hydrothermal process, but with severe impact on granular starches, is the heat-moisture-treatment (HMT). HMT is applied at around 25% moisture and 100 °C for several hours, without fostering a complete starch gelatinization but inducing major structural alterations in granular starches. However, the magnitude of these changes is dependent upon the moisture content and starch type (Gómez and Martínez 2016). In addition, HMT flours and starches also have a high emulsifying ability, which can enhance air incorporation in doughs and increase the bread quality. Hydrothermal treatments are being increasingly applied to improve the functionality of starch-based ingredients, and some recent studies pointed out some potential for altering GF flour functionality (Xing et al. 2017). For instance, pre-gelatinized starches obtained by heating in the presence of water are used widely for their technological properties such as solubility in hot or cold water, high viscosity, and smooth texture and can be used in food processing whenever thickening is required, therefore pregelatinized flours have been proposed as bakery improvers for GF breads (Bourekoua et al. 2016).

Extrusion cooking is another powerful food processing operation, considered high-temperature/short-time cooking during which flours are subjected to high temperatures and mechanical shearing at relatively low levels of moisture content. This treatment allows starch pregelatinization, denaturation of protein, enzyme (in)activation, and Maillard reactions, which are dependent on the severity of the extrusion (Masure et al. 2016). Starch granules are mechanically disrupted by high shear forces and drastic pressure changes, resulting in the disappearance of native starch crystallinity, plasticization, expansion of the food structure, reduced paste viscosity, loss of water holding, softer product texture, and changes in color (Onyango et al. 2004). The extrusion also promotes important nutritional changes in the flours, such as an increase in the soluble fiber content and a reduction in the lipid oxidation tendency, the content of anti-nutritional factors, and the microbial population. Besides, the extrusion could increase the content of resistant starch, which is dependent on the treatment intensity (Alsaffar 2011; Hagenimana et al. 2006). The process parameters such as barrel temperature, screw speed, feed composition, moisture content, throughput along with screw configuration, and die geometry result in system variables, such as mechanical and thermal energy inputs, and residence times. These parameters affect nutritional value, texture, flavor, color, and microbial quality and should be optimized for specific purposes.

Traditionally, heat-moisture treatment has been carried out using conventional air ovens or other treatment methods. However, there is an increasing trend toward the use of microwave applications in food processing due to the fact that microwave energy is more efficient than the traditional heating process, since it ensures homogenous operation in the whole volume of substance, greater penetrating depth, and selective absorption. However, there are a limited number of papers which apply microwave energy applications in the study of heat-moisture treatments of GF flours.

Román et al. (2015) found that microwave treatments of maize flour samples tempered with enough water are likely to modify the crystallinity of their starch granules, promoting a swelling and amylose-lipid complexes formation, without causing starch gelatinization. They concluded that microwave treatment could be used to obtain modified maize flours more suitable for some food applications. Yang et al. (2017) evaluated the effect of microwave irradiation on internal molecular structure and physical properties of waxy corn starch, observing an increase in gelatinization temperatures and a decrease in the molecular weight, the relative crystallinity, gelatinization enthalpy (ΔH), viscosities, and syneresis after microwave treatment.

Infrared-based micronization is a process of thermal treatment of grains using near-infrared (NIR) radiation for a relatively short time. This process refers to micron size wavelength of IR (1.8–3.4 μm). During micronization, the high-frequency NIR rays penetrate the material and induce molecular vibration, which consequently brings about intermolecular friction followed by rapid internal heating. Improved starch digestibility by increasing the content of rapidly digestible starch and reducing resistant starch are results of micronization of cereals due to starch gelatinization applied by the rapid heating of grains (~ for 90–180 seconds) in the presence of sufficient amounts of moisture (>25%) (McAllister and Sultana 2011). Micronization also inactivates enzymes and anti-nutritional factors to a significant level, which could prolong the shelf life and enhance nutritional value, respectively (Deepa and Hebbar 2014). Deepa and Hebbar (2017) assessed the potential utilization of micronized wholegrain corn flour in the development of GF pasta. They observed that the weight of cooked pasta prepared from micronized flour resembles that of the control wheat pasta, owing to its more intact structure and inner continuous network that leads to better firmness and color index and improved overall acceptability.

8.3.3 Biotechnological Approaches

Various methodologies have been adopted to meet the challenge of producing GF products, while minimizing any resulting issues. Enzymes are perceived as natural, non-toxic food components and are preferred by consumers over chemical food processing aids (Renzetti and Rosell 2016). Figure 8.1 shows the effect of different enzymes on the cross-section of GF breads made with corn flour. In general, the protein connecting enzymes, by enhancing cross-linking reactions (transglutaminases, oxidases) to create networks, and proteases that decrease the protein hydrophobicity, are desirable for improving the quality of GF bakery products (Renzetti and Rosell 2016).

Conversely, no reports have focused on the lipase action on these types of breads, but results from the author's studies with corn flour revealed a detrimental effect on the bread volume (Figure 7.1). There are many reports available on enzymes in GF products, which describe in detail the mechanism of action of enzymes commonly used in the baking industry (Dłużewska et al. 2015; Gujral et al. 2003b; Kawamura-Konishi et al. 2013; Renzetti and Rosell 2016), the implication at molecular level on the main flour constituents, and their influence on baking properties, textural, and sensorial quality, and nutritional aspects (Dura and Rosell 2016) (Table 8.2). Nevertheless, regarding the use of transglutaminase in GF products, it should be noted that gut tissue transglutaminase plays an important role in celiac disease pathogenesis on exposure to wheat protein, especially gliadins. Accordingly, Gerrard and Sutton (2005) reported that microbial transglutaminase in baked products may act upon

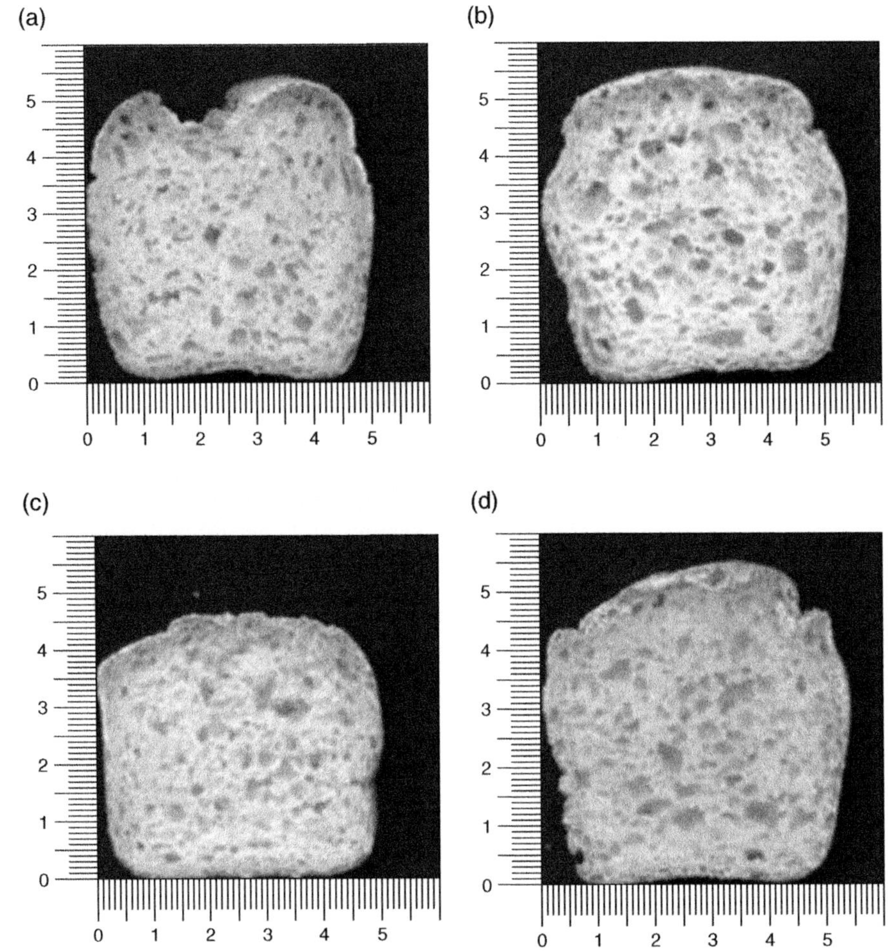

Figure 8.1 Impact of different enzymes on cross-sections of corn bread: (a) Control, (b) Protease, (c) Lipase, (d) Glucose oxidase. *Source*: courtesy of Cristina M. Rosell.

gliadin proteins in dough to generate the epitope associated with the coeliac response. Therefore, the use of transglutaminase in baked products made from wheat, barley, rye, or oats is not recommended, but this concern is still a topic of debate.

Innovation in GF products includes the use of sourdoughs to improve the texture, flavor, taste, nutritional value, and shelf life of bread (Moroni et al. 2009). These effects derive from the complex metabolic activities of the sourdough's lactic acid bacteria and yeasts, such as acidification, production of exopolysaccharides (EPS), proteolytic-amylolytic and phytase activity, and production of antimicrobial substances (Gobbetti and Gänzle 2012). Besides the effects mentioned above, sourdough has been seen as the way to reduce the risk of gluten contamination in GF flours (Di Cagno et al. 2008). To reduce this risk of contamination, Di Cagno et al. (2008) screened 46 strains of lactic acid bacteria, and selected *Lactobacillus sanfranciscensis* LS40 and LS41 and *Lactobacillus plantarum* CF1 for making GF breads. Those breads showed better technological properties (higher specific volume and lower crumb firmness), and also

Table 8.2 Enzyme technology used for modification of gluten-free products.

Enzyme	Gluten-free product	Outcomes	Source
Transglutaminase	Rice flour	Improvement of the dynamic rheological properties of rice flour dough	Gujral and Rosell (2004a)
	Bread (made from brown rice, buckwheat, corn, oat, sorghum, and teff)	Improved structure of buckwheat and brown rice flour breads Detrimental effects on the elastic properties of the corn batter, but improved macroscopic appearance of corn breads	Renze et al. (2008)
	Rice noodle	Decreased cooking loss Enhancement of elastic and loss moduli of rice dough Improved network structure due to cross-linking between proteins	(Kim et al. 2014)
	Breads (corn flour, corn starch and potato starch)	Improvement of physicochemical and sensory features of gluten-free breads	Dłużewska et al. (2015)
	Rice flour, potato starch, corn flour	The extent of protein network formation was determined by protein source and enzyme dosage	Moore et al. (2006)
	Rice bread	Increased volume, decreased hardness	Storck et al. (2013)
		Improvement of crumb texture	Mohammadi et al. (2015)
Glucose oxidase	Rice bread	Increased loaf volume, decreased crumb hardness	Gujral and Rosell (2004b)
Protease	Rice bread	Increased dough consistency, gas retention, bread volume and appearance	Hamada et al. (2013)
Protease		Improvement of the quality of gluten-free rice bread Increased loaf volume, decreased crumb hardness Retardation of bread staling rate	Kawamura-Konishi et al. (2013)

(*continued overleaf*)

Table 8.2 (*continued*)

Enzyme	Gluten-free product	Outcomes	Source
Glucose oxidase and protease	Bread (buckwheat, sorghum, and corn flour)	Enzymes effect depended on the flour used	Renzetti and Arendt (2009)
Laccase, glucose oxidase and protease	Oat bread	Increased bread volume, improvement of crumb structure	Renzetti et al. (2010)
Cyclodextrinase	Rice flour	Increased specific volume, decreased crumb hardness and staling rate	Gujral et al. (2003a)
Protease, lipase and amylases	Rice bread	Lipase and extruded flour increased bread volume and reduced the initial firmness and hardening	Martínez et al. (2013)

improved nutritional values, particularly higher levels of protein and dietary fiber and lower total carbohydrates that the marketed GF breads. Equally interesting results were obtained by the application of sourdoughs to replace the hydrocolloids used in the production of GF baked goods, specifically, EPS formed by lactobacilli from sucrose during sourdough fermentation (Tieking et al. 2003). In fact, the performance of dextran forming *Weissella cibaria* MG1, reuteran producing *Lactobacillus reuteri* VIP, and fructan forming *L. reuteri* Y2 were compared when added as sourdoughs in GF batters (Galle and Arendt 2014). EPS produced during the fermentation led to sorghum breads with softer crumbs and longer shelf life, confirming their potential for use in GF bread production.

8.4 Conclusion and Further Remarks

A requirement to produce a wide range of high-quality gluten-free products is as important as ever, taking into account the prevalence of gluten intolerant patients. In the last decade, many innovations have been carried out for improving the technological and nutritional characteristics of GF foods. Apart from a careful design of the recipes or formulations with alternative ingredients, physical and biotechnological techniques have been applied to modify the flours' functionalities of the bread-making process, and subsequently its food properties. This chapter presents the recent innovative technologies related to GF products, highlighting the technologies for improving them. Among the discussed techniques, the thermal treatment is the most common technique within GF products production. Novel technologies such as HP, radiation, extrusion, etc. are the techniques that offer successful alternatives. The promising results derived from research studies should stimulate further research on the improvement of formulas and optimization of novel technologies in HGF products. Furthermore, extensive research on interfacing food science, nutrition, and health is needed, thus various GF products with both good technological and nutritional properties can be prepared and made available to those with celiac disease, which will help them adhere to a strict GF diet, increase social inclusion, and improve their quality of life.

References

Agu, R., Chiba, Y., Goodfellow, V. et al. (2012). Effect of germination temperatures on proteolysis of the gluten-free grains rice and buckwheat during malting and mashing. *Journal of Agricultural and Food Chemistry* 60: 10147–10154.

Ahmed, J. and Al-Attar, H. (2017). Structural properties of high-pressure-treated chestnut flour dispersions. *International Journal of Food Properties* 20: S766–S778.

Alsaffar, A.A. (2011). Effect of food processing on the resistant starch content of cereals and cereal products–a review. *International Journal of Food Science & Technology* 46: 455–462.

American Gastroenterological Association (2001). American Gastroenterological Association medical position statement: celiac sprue. *Gastroenterology* 120 (6): 1522–1525.

An, H. and King, J. (2007). Thermal characteristics of ohmically heated rice starch and rice flours. *Journal of Food Science* 72: c84–c88.

Angelidis, G., Protonotariou, S., Mandala, I., and Rosell, C.M. (2016). Jet milling effect on wheat flour characteristics and starch hydrolysis. *Journal of Food Science and Technology-Mysore* 53: 784–791.

AOECS (2005). Codex Committee on Nutrition and Foods for Special Dietary Uses. *Association of European Coeliac Societies Tech. Rep*. http://www.fao.org/tempref/codex/Meetings/CCNFSDU/ccnfsdu29/nf2904ae.pdf

AOECS (2016). *AOECS Standard for Gluten-Free Foods: Technical Requirements for Licensing the Crossed Grain Symbol*. Brussels: Association of European Coeliac Societies.

Arendt, E.K., Morrissey, A., Moore, M.M., and Dal Bello, F. (2011). Gluten-free breads. In: *Gluten-Free Cereal Products and Beverages* (eds. E. Arendt and F. Dal Bello), 289–319. Academic Press.

Bastos, G.M., Júnior, M.S.S., Caliari, M. et al. (2016). Physical and sensory quality of gluten-free spaghetti processed from amaranth flour and potato pulp. *LWT - Food Science and Technology* 65: 128–136.

Benhur, D.R., Bhargavi, G., Kalpana, K. et al. (2015). Development and standardization of sorghum pasta using extrusion technology. *Journal of Food Science and Technology* 52: 6828–6833.

Bogaerts, A., Neyts, E., Gijbels, R., and van der Mullen, J. (2002). Gas discharge plasmas and their applications. *Spectrochimica Acta Part B: Atomic Spectroscopy* 57: 609–658.

Borneo, R. and Aguirre, A. (2008). Chemical composition, cooking quality, and consumer acceptance of pasta made with dried amaranth leaves flour. *LWT - Food Science and Technology* 41: 1748–1751.

Bouasla, A., Wójtowicz, A., and Zidoune, M.N. (2017). Gluten-free precooked rice pasta enriched with legumes flours: physical properties, texture, sensory attributes and microstructure. *LWT - Food Science and Technology* 75: 569–577.

Bourekoua, H., Benatallah, L., Zidoune, M.N., and Rosell, C.M. (2016). Developing gluten free bakery improvers by hydrothermal treatment of rice and corn flours. *LWT - Food Science and Technology* 73: 342–350.

Cappa, C., Lucisano, M., Barbosa-Cánovas, G.V., and Mariotti, M. (2016). Physical and structural changes induced by high pressure on corn starch, rice flour and waxy rice flour. *Food Research International* 85: 95–103.

Capriles, V.D. and Arêas, J.A.G. (2014). Novel approaches in gluten-free breadmaking: Interface between food science, nutrition, and health. *Comprehensive Reviews in Food Science and Food Safety* 13: 871–890.

Capriles, V.D., dos Santos, F.G., and Arêas, J.A.G. (2016). Gluten-free breadmaking: improving nutritional and bioactive compounds. *Journal of Cereal Science* 67: 83–91.

Cárdenas-Hernández, A., Beta, T., Loarca-Piña, G. et al. (2016). Improved functional properties of pasta: enrichment with amaranth seed flour and dried amaranth leaves. *Journal of Cereal Science* 72: 84–90.

Charutigon, C., Jitpupakdree, J., Namsree, P., and Rungsardthong, V. (2008). Effects of processing conditions and the use of modified starch and monoglyceride on some properties of extruded rice vermicelli. *LWT - Food Science and Technology* 41: 642–651.

Chillo, S., Civica, V., Iannetti, M. et al. (2009). Properties of quinoa and oat spaghetti loaded with carboxymethylcellulose sodium salt and pregelatinized starch as structuring agents. *Carbohydrate Polymers* 78: 932–937.

Codex Alimentarus (2008). *Codex Standard for Foods for Special Dietary Use for Persons Intolerant to Gluten,* Codex Stan 118–1979, Codex Alimentarus. Rome: Joint FAO/WHO Food Standards Programme (Food and Agriculture Organization of the United Nations).

Cornejo, F. and Rosell, C.M. (2015). Influence of germination time of brown rice in relation to flour and gluten free bread quality. *Journal of Food Science and Technology* 52: 6591–6598.

Cornejo, F., Caceres, P.J., Martínez-Villaluenga et al. (2015). Effects of germination on the nutritive value and bioactive compounds of brown rice breads. *Food Chemistry* 173: 298–304.

D'Amico, S., Mäschle, J., Jekle, M. et al. (2015). Effect of high temperature drying on gluten-free pasta properties. *LWT - Food Science and Technology* 63: 391–399.

Dar, Y.L. (2013). Advances and ongoing challenges in the development of gluten-free baked goods. *Cereal Foods World* 58: 298–304.

Deepa, C. and Hebbar, H.U. (2014). Micronization of maize flour: process optimization and product quality. *Journal of Cereal Science* 60: 569–575.

Demirkesen, I., Sumnu, G., and Sahin, S. (2013). Quality of gluten-free bread formulations baked in different ovens. *Food and Bioprocess Technology* 6: 746–753.

Dhen, N., Román, L., Rejeb, I.B. et al. (2016). Particle size distribution of soy flour affecting the quality of enriched gluten-free cakes. *LWT - Food Science and Technology* 66: 179–185.

Di Cagno, R., Rizzello, C.G., De Angelis, M. et al. (2008). Use of selected sourdough strains of *Lactobacillus* for removing gluten and enhancing the nutritional properties of gluten-free bread. *Journal of Food Protection* 71: 1491–1495.

Di Ghionno, L., Sileoni, V., Marconi, O. et al. (2017). Comparative study on quality attributes of gluten-free beer from malted and unmalted teff [Eragrostis tef (zucc.) trotter]. *LWT - Food Science and Technology* 84: 746–752.

Dłużewska, E., Marciniak-Lukasiak, K., and Kurek, N. (2015). Effect of transglutaminase additive on the quality of gluten-free bread. *CyTA Journal of Food* 13: 80–86.

do Nascimento, K.D.O., Paes, S.D.N.D., de Oliveira, I.R. et al. (2018). Teff: suitability for different food applications and as a raw material of gluten-free, a literature review. *Journal of Food and Nutrition Research* 6: 74–81.

Doxastakis, G., Papageorgiou, M., Mandalou, D. et al. (2007). Technological properties and non-enzymatic browning of white lupin protein enriched spaghetti. *Food Chemistry* 101: 57–64.

Dura, A. and Rosell, C.M. (2016). Enzymes in baking. In: *Microbial Enzyme Technology and Food Applications* (eds. R.C. Ray and C.M. Rosell), 309–328. CRC Press Taylor & Francis Group.

Ekezie, F.-G.C., Sun, D.-W., and Cheng, J.-H. (2017). A review on recent advances in cold plasma technology for the food industry: current applications and future trends. *Trends in Food Science & Technology* 69: 46–58.

Estrada-Girón, Y., Swanson, B., and Barbosa-Cánovas, G. (2005). Advances in the use of high hydrostatic pressure for processing cereal grains and legumes. *Trends in Food Science & Technology* 16: 194–203.

Ferreira, S.M.R., de Mello, A.P., dos Anjos, M.D.C.R. et al. (2016). Utilization of sorghum, rice, corn flours with potato starch for the preparation of gluten-free pasta. *Food Chemistry* 191: 147–151.

Flores-Silva, P.C., Berrios, J.D.J., Pan, J. et al. (2014). Gluten-free spaghetti made with chickpea, unripe plantain and maize flours: functional and chemical properties and starch digestibility. *International Journal of Food Science & Technology* 49: 1985–1991.

Fratelli, C., Muniz, D.G., Santos, F.G., and Capriles, V.D. (2018). Modelling the effects of psyllium and water in gluten-free bread: An approach to improve the bread quality and glycemic response. *Journal of Functional Foods* 42: 339–345.

Fuad, T. and Prabhasankar, P. (2010). Role of ingredients in pasta product quality: a review on recent developments. *Critical Reviews in Food Science and Nutrition* 50: 787–798.

Galle, S. and Arendt, E.K. (2014). Exopolysaccharides from sourdough lactic acid bacteria. *Critical Reviews in Food Science and Nutrition* 54: 891–901.

Garcia, E.L., Mischan, M.M., Leonel, M., and Sahade Brunatti, A.C. (2018). Gluten-free puffed snacks of rice and cassava. *Australian Journal of Crop Science* 12: 185.

Gaytán-Martínez, M., Figueroa, J., Morales-Sanchez, E. et al. (2011). Physicochemical properties of masa and corn tortilla made by ohmic heating. *African Journal of Biotechnology* 10: 16028–16036.

Gerrard, J.A. and Sutton, K.H. (2005). Addition of transglutaminase to cereal products may generate the epitope responsible for coeliac disease. *Trends in Food Science & Technology* 16 (11): 510–512.

Giménez, M.A., González, R., Wagner, J. et al. (2013). Effect of extrusion conditions on physicochemical and sensorial properties of corn-broad beans (*Vicia faba*) spaghetti type pasta. *Food Chemistry* 136: 538–545.

Giménez-Bastida, J.A., Piskuła, M., and Zieliński, H. (2015). Recent advances in development of gluten-free buckwheat products. *Trends in Food Science & Technology* 44: 58–65.

Gobbetti, M. and Gänzle, M. (2012). Physiology and biochemistry of lactic acid bacteria. In: *Handbook on Sourdough Biotechnology* (eds. M. Gobbetti and M. Gänzle), 183–216. Boston: Springer.

Gómez, M. and Martínez, M.M. (2016). Changing flour functionality through physical treatments for the production of gluten-free baking goods. *Journal of Cereal Science* 67: 68–74.

Gujral, H.S. and Rosell, C.M. (2004a). Functionality of rice flour modified with a microbial transglutaminase. *Journal of Cereal Science* 39: 225–230.

Gujral, H.S. and Rosell, C.M. (2004b). Improvement of the breadmaking quality of rice flour by glucose oxidase. *Food Research International* 37: 75–81.

Gujral, H.S., Guardiola, I., Carbonell, J.V., and Rosell, C.M. (2003a). Effect of cyclodextrinase on dough rheology and bread quality from rice flour. *Journal of Agricultural and Food Chemistry* 51: 3814–3818.

Gujral, H.S., Haros, M., and Rosell, C.M. (2003b). Starch hydrolyzing enzymes for retarding the staling of rice bread. *Cereal Chemistry* 80: 750–754.

Guo, W., Tiwari, G., Tang, J., and Wang, S. (2008). Frequency, moisture and temperature-dependent dielectric properties of chickpea flour. *Biosystems Engineering* 101: 217–224.

Hagenimana, A., Ding, X., and Fang, T. (2006). Evaluation of rice flour modified by extrusion cooking. *Journal of Cereal Science* 43: 38–46.

Hager, A.-S., Taylor, J.P., Waters, D.M., and Arendt, E.K. (2014). Gluten free beer – a review. *Trends in Food Science & Technology* 36: 44–54.

Hamada, S., Suzuki, K., Aoki, N., and Suzuki, Y. (2013). Improvements in the qualities of gluten-free bread after using a protease obtained from *Aspergillus oryzae*. *Journal of Cereal Science* 57: 91–97.

Hargreaves, S.M. and Zandonadi, R.P. (2018). Flaxseed and chia seed gel on characteristics of gluten-free cake. *Journal of Culinary Science & Technology* 16 (4): 378–388.

Heller, L. (2009). Commercial aspects of gluten-free products. In: *Gluten-Free Food Science and Technology* (ed. E. Gallagher), 90–106. Wiley.

Heo, S., Jeon, S., and Lee, S. (2014). Utilization of *Lentinus edodes* mushroom β-glucan to enhance the functional properties of gluten-free rice noodles. *LWT - Food Science and Technology* 55: 627–631.

de la Hera, E., Martinez, M., and Gómez, M. (2013). Influence of flour particle size on quality of gluten-free rice bread. *LWT - Food Science and Technology* 54: 199–206.

de la Hera, E., Rosell, C.M., and Gomez, M. (2014). Effect of water content and flour particle size on gluten-free bread quality and digestibility. *Food Chemistry* 151: 526–531.

Houben, A., Höchstötter, A., and Becker, T. (2012). Possibilities to increase the quality in gluten-free bread production: an overview. *European Food Research and Technology* 235 (2): 195–208.

Huttner, E.K., Dal Bello, F., and Arendt, E.K. (2010). Fundamental study on the effect of hydrostatic pressure treatment on the bread-making performance of oat flour. *European Food Research and Technology* 230: 827–835.

Islas-Rubio, A.R., de la Barca, A.M.C., Cabrera-Chávez, F. et al. (2014). Effect of semolina replacement with a raw: popped amaranth flour blend on cooking quality and texture of pasta. *LWT - Food Science and Technology* 57: 217–222.

Kahlon, T.S. and Chiu, M.-C.M. (2015). Teff, buckwheat, quinoa and amaranth: ancient whole grain gluten-free egg-free pasta. *Food and Nutrition Sciences* 6 (15): 1460.

Kahraman, K., Sakıyan, O., Ozturk, S. et al. (2008). Utilization of Mixolab® to predict the suitability of flours in terms of cake quality. *European Food Research and Technology* 227: 565–570.

Kaur, L., Singh, J., and Singh, N. (2005). Effect of glycerol monostearate on the physico-chemical, thermal, rheological and noodle making properties of corn and potato starches. *Food Hydrocolloids* 19: 839–849.

Kawamura-Konishi, Y., Shoda, K., Koga, H., and Honda, Y. (2013). Improvement in gluten-free rice bread quality by protease treatment. *Journal of Cereal Science* 58: 45–50.

Kim, J.-M. and Shin, M. (2014). Effects of particle size distributions of rice flour on the quality of gluten-free rice cupcakes. *LWT - Food Science and Technology* 59: 526–532.

Kim, S.P., Kang, M.Y., Nam, S.H., and Friedman, M. (2012). Dietary rice bran component γ-oryzanol inhibits tumor growth in tumor-bearing mice. *Molecular Nutrition & Food Research* 56: 935–944.

Kim, Y., Kee, J.I., Lee, S., and Yoo, S.-H. (2014). Quality improvement of rice noodle restructured with rice protein isolate and transglutaminase. *Food Chemistry* 145: 409–416.

Kreisz, S., Arendt, E.K., Hübner, F., and Zarnkov, M. (2008). Cereal-based gluten-free functional drinks. In: *Gluten-Free Cereal Products and Beverages* (eds. E.K. Arendt and F. Dal Bello), 373–392. Academic Press.

Krupa-Kozak, U., Baczek, N., and Rosell, C.M. (2013). Application of dairy proteins as technological and nutritional improvers of calcium-supplemented gluten-free bread. *Nutrients* 5: 4503–4520.

Lee, K.H., Kim, H.-J., Woo, K.S. et al. (2016). Evaluation of cold plasma treatments for improved microbial and physicochemical qualities of brown rice. *LWT - Food Science and Technology* 73: 442–447.

Létang, C., Samson, M.F., Lasserre, T.M. et al. (2002). Production of starch with very low protein content from soft and hard wheat flours by jet milling and air classification. *Cereal Chemistry* 79: 535–543.

Levent, H. and Bilgicli, N. (2011). Enrichment of gluten-free cakes with lupin (*Lupinus albus* L.) or buckwheat (*Fagopyrum esculentum* M.) flours. *International Journal of Food Sciences and Nutrition* 62: 725–728.

Liu, H., Wang, L., Cao, R. et al. (2016). *In vitro* digestibility and changes in physicochemical and structural properties of common buckwheat starch affected by high hydrostatic pressure. *Carbohydrate Polymers* 144: 1–8.

Lorenzo, G., Sosa, M., and Califano, A. (2018). Alternative proteins and pseudocereals in the development of gluten-free pasta. In: *Alternative and Replacement Foods* (eds. A. Grumezescu and A.M. Holban), 433–458. Academic Press.

Ludvigsson, J.F., Leffler, D.A., Bai, J.C. et al. (2012). The Oslo definitions for coeliac disease and related terms. *Gut* 62 (1): 43–52.

Marston, K., Khouryieh, H., and Aramouni, F. (2016). Effect of heat treatment of sorghum flour on the functional properties of gluten-free bread and cake. *LWT - Food Science and Technology* 65: 637–644.

Marti, A., Caramanico, R., Bottega, G., and Pagani, M.A. (2013). Cooking behavior of rice pasta: effect of thermal treatments and extrusion conditions. *LWT - Food Science and Technology* 54: 229–235.

Martínez, M.M., Marcos, P., and Gómez, M. (2013). Texture development in gluten-free breads: effect of different enzymes and extruded flour. *Journal of Texture Studies* 44: 480–489.

Mastromatteo, M., Chillo, S., Iannetti, M. et al. (2011). Formulation optimisation of gluten-free functional spaghetti based on quinoa, maize and soy flours. *International Journal of Food Science & Technology* 46: 1201–1208.

Masure, H.G., Fierens, E., and Delcour, J.A. (2016). Current and forward looking experimental approaches in gluten-free bread making research. *Journal of Cereal Science* 67: 92–111.

Matos, M.E. and Rosell, C.M. (2012). Relationship between instrumental parameters and sensory characteristics in gluten-free breads. *European Food Research and Technology* 235: 107–117.

Matos, M.E. and Rosell, C.M. (2013). Quality indicators of rice-based gluten-free bread-like products: relationships between dough rheology and quality characteristics. *Food and Bioprocess Technology* 6: 2331–2341.

Matos, M.E. and Rosell, C.M. (2015). Understanding gluten-free dough for reaching breads with physical quality and nutritional balance. *Journal of the Science of Food and Agriculture* 95: 653–661.

Matos, M.E., Sanz, T., and Rosell, C.M. (2014). Establishing the function of proteins on the rheological and quality properties of rice based gluten free muffins. *Food Hydrocolloids* 35: 150–158.

Mayer, H., Ceccaroni, D., Marconi, O. et al. (2016). Development of an all rice malt beer: a gluten free alternative. *LWT - Food Science and Technology* 67: 67–73.

McAllister, T. and Sultana, H. (2011). Effects of micronization on the *in situ* and *in vitro* digestion of cereal grains. *Asian-Australasian Journal of Animal Sciences* 24: 929–939.

Melini, F., Melini, V., Luziatelli, F., and Ruzzi, M. (2017). Current and forward-looking approaches to technological and nutritional improvements of gluten-free bread with legume flours: a critical review. *Comprehensive Reviews in Food Science and Food Safety* 16: 1101–1122.

Merayo, Y.A., González, R.J., Drago, S.R. et al. (2011). Extrusion conditions and zea mays endosperm hardness affecting gluten-free spaghetti quality. *International Journal of Food Science & Technology* 46: 2321–2328.

Mir, S.A., Shah, M.A., Naik, H.R., and Zargar, I.A. (2016). Influence of hydrocolloids on dough handling and technological properties of gluten-free breads. *Trends in Food Science & Technology* 51: 49–57.

Mohammadi, M., Azizi, M.-H., Neyestani, T.R. et al. (2015). Development of gluten-free bread using guar gum and transglutaminase. *Journal of Industrial and Engineering Chemistry* 21: 1398–1402.

Molinari, R., Costantini, L., Timperio, A.M. et al. (2017). Tartary buckwheat malt as ingredient of gluten-free cookies. *Journal of Cereal Science* 80: 37–43.

Moore, M.M., Heinbockel, M., Dockery, P. et al. (2006). Network formation in gluten-free bread with application of transglutaminase. *Cereal Chemistry* 83: 28–36.

Moroni, A.V., Dal Bello, F., and Arendt, E.K. (2009). Sourdough in gluten-free bread-making: an ancient technology to solve a novel issue. *Food Microbiology* 26: 676–684.

Morreale, F., Garzon, R., and Rosell, C.M. (2018). Understanding the role of hydrocolloids viscosity and hydration in developing gluten-free bread. A study with hydroxypropylmethylcellulose. *Food Hydrocolloids* 77: 629–635.

Nakano, A., Kokawa, M., and Kitamura, Y. (2018). Development and characterization of gluten-free bread using rice gel and rice flour. *Journal of the Japanese Society for Food Science and Technology* 65: 124–129.

Ndubisi, C., Okafor, E., Amadi, O. et al. (2016). Effect of malting time, mashing temperature and added commercial enzymes on extract recovery from a Nigerian malted yellow sorghum variety. *Journal of the Institute of Brewing* 122: 156–161.

Nisoa, K.S.N.M.M. and Matan, N. (2013). Low pressure radio frequency plasma effects on the mould control, physical quality, nutritional value, mineral content and trace element content of brown rice snack bars. *Journal of Food and Nutrition Research* 52: 87–94.

Onwulata, C. and Konstance, R. (2006). Extruded corn meal and whey protein concentrate: effect of particle size. *Journal of Food Processing and Preservation* 30: 475–487.

Onyango, C., Henle, T., Ziems, A. et al. (2004). Effect of extrusion variables on fermented maize–finger millet blend in the production of uji. *LWT - Food Science and Technology* 37: 409–415.

Onyango, C., Mewa, E.A., Mutahi, A.W., and Okoth, M.W. (2013). Effect of heat-moisture-treated cassava starch and amaranth malt on the quality of sorghum-cassava-amaranth bread. *African Journal of Food Science* 7: 80–86.

O'Shea, N., Arendt, E., and Gallagher, E. (2014). State of the art in gluten-free research. *Journal of Food Science* 79: R1067–R1076.

Ouazib, M., Garzon, R., Zaidi, F., and Rosell, C.M. (2016). Germinated, toasted and cooked chickpea as ingredients for breadmaking. *Journal of Food Science and Technology-Mysore* 53: 2664–2672.

Padalino, L., Mastromatteo, M., Lecce, L. et al. (2013). Manufacture and characterization of gluten-free spaghetti enriched with vegetable flour. *Journal of Cereal Science* 57: 333–342.

Pagani, M., Bonomi, F., Bottega, G. et al. (2010). Characterizing sorghum-based pasta: a multi-disciplinary approach. International Symposium on Gluten-Free Cereal Products & Beverages, Finnish Coeliac Society, Tampere, Finland (8–11 June 2010).

Phattanakulkaewmorie, N., Paseephol, T., and Moongngarm, A. (2011). Chemical compositions and physico-chemical properties of malted sorghum flour and characteristics of gluten free bread. *World Academy of Science, Engineering and Technology* 5: 532–538.

Polet, J.P., de Oliveira, V.R., de Oliveira Rios, A., and de Souza, C.G. (2019). Physico-chemical and sensory characteristics of gluten-free breads made with pine nuts (*Araucaria angustifolia*) associated to other flours. *Journal of Culinary Science & Technology* 17: 136–145.

Protonotariou, S., Mandala, I., and Rosell, C.M. (2015). Jet milling effect on functionality, quality and in vitro digestibility of whole wheat flour and bread. *Food and Bioprocess Technology* 8: 1319–1329.

Purwandari, U., Hidayati, D., Tamam, B., and Arifin, S. (2014). Gluten-free noodle made from gathotan (an Indonesian fungal fermented cassava). Flour: cooking quality, textural, and sensory properties. *International Food Research Journal* 21 (4): 1615–1621.

Purwani, E., Widaningrum, W., Thahir, R., and Muslich, M. (2016). Effect of heat moisture treatment of sago starch on its noodle quality. *Indonesian Journal of Agricultural Science* 7: 8–14.

Rai, S., Kaur, A., and Singh, B. (2014). Quality characteristics of gluten free cookies prepared from different flour combinations. *Journal of Food Science and Technology* 51: 785–789.

Renzetti, S. and Arendt, E.K. (2009). Effects of oxidase and protease treatments on the breadmaking functionality of a range of gluten-free flours. *European Food Research and Technology* 229: 307–317.

Renzetti, S. and Rosell, C.M. (2016). Role of enzymes in improving the functionality of proteins in non-wheat dough systems. *Journal of Cereal Science* 67: 35–45.

Renzetti, S., Dal Bello, F., and Arendt, E.K. (2008). Microstructure, fundamental rheology and baking characteristics of batters and breads from different gluten-free flours treated with a microbial transglutaminase. *Journal of Cereal Science* 48: 33–45.

Renzetti, S., Courtin, C., Delcour, J., and Arendt, E. (2010). Oxidative and proteolytic enzyme preparations as promising improvers for oat bread formulations: rheological, biochemical and microstructural background. *Food Chemistry* 119: 1465–1473.

Rinaldi, M., Paciulli, M., Caligiani, A. et al. (2017). Sourdough fermentation and chestnut flour in gluten-free bread: a shelf-life evaluation. *Food Chemistry* 224: 144–152.

Román, L., Martínez, M.M., Rosell, C.M., and Gómez, M. (2015). Effect of microwave treatment on physicochemical properties of maize flour. *Food and Bioprocess Technology* 8: 1330–1335.

Rubio-Flores, M. and Serna-Saldivar, S.O. (2016). Technological and engineering trends for production of gluten-free beers. *Food Engineering Reviews* 8: 468–482.

Rumpold, B.A. (2005). Impact of high hydrostatic pressure on wheat, tapioca, and potato starches. Ph.D. Thesis. Berlin: Technical University of Berlin.

Sandri, L.T., Santos, F.G., Fratelli, C., and Capriles, V.D. (2017). Development of gluten-free bread formulations containing whole chia flour with acceptable sensory properties. *Food Science & Nutrition* 5: 1021–1028.

Sarangapani, C., Thirumdas, R., Devi, Y. et al. (2016). Effect of low-pressure plasma on physico-chemical and functional properties of parboiled rice flour. *LWT - Food Science and Technology* 69: 482–489.

Savita, S., Arshwinder, K., Gurkirat, K., and Vikas, N. (2013). Influence of different protein sources on cooking and sensory quality of pasta. *International Journal of Engineering* 3: 1757–1763.

Schoenlechner, R., Mandala, I., Kiskini, A. et al. (2010). Effect of water, albumen and fat on the quality of gluten-free bread containing amaranth. *International Journal of Food Science & Technology* 45 (4): 661–669.

Seguchi, M., Ozawa, M., Nakamura, C., and Tabara, A. (2012). Development of gluten-free bread baked with yam flour. *Food Science and Technology Research* 18: 543–548.

Stojceska, V., Ainsworth, P., Plunkett, A., and Ibanoglu, S. (2010). The advantage of using extrusion processing for increasing dietary fibre level in gluten-free products. *Food Chemistry* 121: 156–164.

Storck, C.R., da Rosa Zavareze, E., Gularte, M.A. et al. (2013). Protein enrichment and its effects on gluten-free bread characteristics. *LWT - Food Science and Technology* 53: 346–354.

Thirumdas, R., Sarangapani, C., and Annapure, U.S. (2015). Cold plasma: a novel non-thermal technology for food processing. *Food Biophysics* 10: 1–11.

Thirumdas, R., Trimukhe, A., Deshmukh, R., and Annapure, U. (2017). Functional and rheological properties of cold plasma treated rice starch. *Carbohydrate Polymers* 157: 1723–1731.

Tieking, M., Korakli, M., Ehrmann, M.A. et al. (2003). In situ production of exopolysaccharides during sourdough fermentation by cereal and intestinal isolates of lactic acid bacteria. *Applied and Environmental Microbiology* 69: 945–952.

Torres, J.A. and Velazquez, G. (2005). Commercial opportunities and research challenges in the high pressure processing of foods. *Journal of Food Engineering* 67: 95–112.

Tsatsaragkou, K., Kara, T., Ritzoulis, C. et al. (2017). Improving carob flour performance for making gluten-free breads by particle size fractionation and jet milling. *Food and Bioprocess Technology* 10: 831–841.

Turabi, E., Sumnu, G., and Sahin, S. (2008). Optimization of baking of rice cakes in infrared–microwave combination oven by response surface methodology. *Food and Bioprocess Technology* 1: 64–73.

Turabi, E., Regier, M., Sumnu, G. et al. (2010). Dielectric and thermal properties of rice cake formulations containing different gums types. *International Journal of Food Properties* 13: 1199–1206.

Turfani, V., Narducci, V., Durazzo, A. et al. (2017). Technological, nutritional and functional properties of wheat bread enriched with lentil or carob flours. *LWT - Food Science and Technology* 78: 361–366.

Vallons, K.J. and Arendt, E.K. (2009). Effects of high pressure and temperature on the structural and rheological properties of sorghum starch. *Innovative Food Science & Emerging Technologies* 10: 449–456.

Vallons, K.J., Ryan, L.A., and Arendt, E.K. (2011). Promoting structure formation by high pressure in gluten-free flours. *LWT - Food Science and Technology* 44: 1672–1680.

Vidya, S., Ravi, R., and Bhattacharya, S. (2013). Effect of thermal treatment on selected cereals and millets flour doughs and their baking quality. *Food and Bioprocess Technology* 6: 1218–1227.

Vinall, P. (2014). Celiac disease and non-celiac gluten sensitivity. *MD Conference Express* 13 (21): 17–18.

Wilderjans, E., Luyts, A., Goesaert, H. et al. (2010). A model approach to starch and protein functionality in a pound cake system. *Food Chemistry* 120: 44–51.

Witczak, M., Ziobro, R., Juszczak, L., and Korus, J. (2016). Starch and starch derivatives in gluten-free systems – a review. *Journal of Cereal Science* 67: 46–57.

Xing, J.-j., Liu, Y., Li, D. et al. (2017). Heat-moisture treatment and acid hydrolysis of corn starch in different sequences. *LWT - Food Science and Technology* 79: 11–20.

Yang, Q., Qi, L., Luo, Z. et al. (2017). Effect of microwave irradiation on internal molecular structure and physical properties of waxy maize starch. *Food Hydrocolloids* 69: 473–482.

9

Cereal-Based Animal Feed Products

Abirami R. Ganesan[1] and Gaurav Rajauria[2]

[1] *School of Applied Sciences, College of Engineering, Science and Technology, Fiji National University, Nabua Campus, Suva, Nabua, Fiji Islands*
[2] *School of Agriculture and Food Science, University College Dublin, Lyons Research Farm, Celbridge, Co. Kildare, Ireland*

9.1 Introduction

The major cereals, representing nearly 90% of cereal-grain production, are wheat, maize, and rice, while the economic significance of triticale, rye, barley, oats, sorghum, and the millets are lagging behind due to the lower volume of their production (Wrigley 2017). Around 750 million tons of the total cereals produced are redirected to animal feed, bioethanol, and other nonfood industries (Serna-Saldivar 2016). The wide utilization of cereal grains in animal feeding is due to the amount of energy they provide, associated with high starch content, but also with non-starch polysaccharides (NSP), proteins, and fat. Cereal grains are the predominant feed for pigs, sheep, poultry, dairy cows, and feed-lot steers accounting for 60–80% of the total feed (Black 2016). Moreover, cereal by-products from milling and processing, consisting of the pericarp-seed coat, aleurone layer and germ, are also widely used in animal feed (Papageorgiou and Skendi 2018).

Almost all rice grown is utilized directly for human food, while a comparable portion of wheat is consumed by humans and their half portion is used for livestock diet. However, the dominant feed grain is corn which forms the primary diet for cattle, sheep, and poultry in several countries of the world. Since non-ruminants such as pigs and poultry are incompetent to digest forages, their diet consists mainly of corn or cereal by-products. Brazil, the United States, and Asian countries use corn as a primary source of energy for poultry feed. However, wheat is the chief energy supplier for poultry diets in Europe, Canada, Australia, New Zealand, and the Russian Federation (www.poultryhub.org). The rising demand for livestock is due to increasing meat product consumption as a result of population growth, rise in income, and changes in lifestyles and eating habits. Therefore, animal feed has a foremost role in the global food industry and is a crucial component to increase animal protein.

Innovative Processing Technologies for Healthy Grains, First Edition.
Edited by Milica Pojić and Uma Tiwari.

Cereal grains and/or their by-products have different structural, physical, and chemical natures. Structurally, cereal grains consist of a seed coat or bran, nutritive reserve or endosperm, and a germ. The seed coat is mainly composed of the aleurone layer (i.e. a layer below the seed coat) and pericarp, which is rich in oil, insoluble dietary fibers, minerals, protein, and vitamins. The endosperm represents the main part of the cereal grain, containing mostly starch, small amounts of proteins and fibers and commonly milled as white flour to be used in bread-making. The germ is the reproductive part of the cereal grain that germinates to grow into the plant and is rich in lipid, sterols, vitamins, minerals, and antioxidants (Evers and Millar 2002; Brouns et al. 2012). Compositionally, cereals are low in protein and high in carbohydrate, consisting of 12–14% water, 2–5% ash, 53–77% starch, 1–6% fat, 2–11% crude fiber, and 7–17% protein. Among cereals, oats contain a high amount of protein (17.1%), fats (6.4%), and crude fiber (11.3%), but are shown to have a low amount of starch (52.8%). Comparatively, rice contains a high amount of starch (77.2%), while millet contains a low amount of protein (7.3%) (Saldivar 2003).

The nutritional composition of cereals is not homogenously distributed throughout the grain. In wheat, bran makes up about 13–19% of the total wheat grain weight, but consists of about 53% dietary fiber, vitamins, and a very high concentration of minerals and ash compared to the endosperm (Onipe et al. 2015). Each ingredient in the cereal can differ depending on its origin, while different cultivars of a given type of cereal could exhibit compositional variability as well (Saldivar 2003). For instance, a different cultivar of millet will show varying compositions of protein, fat, crude fiber, ash, and starch content. Although cereals have high nutritive values, the bioavailability of these nutrients is lowered due to the presence of naturally-occurring anti-nutritional factors (ANFs) such as phytates, polyphenols, tannins, and trypsin inhibitory factors. The ANFs reduce the bioavailability of nutrients, particularly when the grains are consumed in unprocessed or raw form. However, applying various processing methods, including heating, dehulling, and milling, can modify or reduce their impact (Nikmaram et al. 2017).

Processing methods can denature the vitamins or amino acids of the feed, therefore the essential processing conditions should be chosen carefully to remove ANFs and to preserve the nutritional value of the food (Kumar et al. 2019). Some of the compounds such as polyphenols or tannins which were considered as ANFs earlier, due to their metal chelating and enzyme inhibition activities, are now considered as nutraceuticals, which are treated as high-value functional ingredients in the food industry (Devi et al. 2014). Various ANFs in different feed ingredients and the processing methods applied to reduce/remove their impact from the feed are listed in Table 9.1.

The internal structure of a cereal grain has a significant role in its energy release (Black 2016). Grains of hard endosperm texture are associated with low degradation capacity, which requires special treatment like steaming, milling, parboiling etc., to provide better and effective fermentation in the rumen. The feed commonly requires a grinding process to obtain specific particle size to simplify the particle absorption and improve feed intake and nutrient digestibility (Lyu et al. 2020). Apart from grinding, the improvement of digestibility, nutritional quality, sensory properties, nutrients availability, microbial safety, and/or other physical and chemical properties of feed is achieved by using various traditional or modern feed processing technologies described later in this chapter (Dalbhagat et al. 2019). Processing also reduces the presence of toxic substances and enhances the shelf life of the products (He et al. 2010). Thermal processing technologies (e.g. extrusion, autoclaving, puffing, infrared/IR, and

Table 9.1 Anti-nutritional factors (ANFs) in feed ingredients and processing conditions to remove them from feed.

Ingredient name	Scientific name	Description	Anti-nutritional factor	Part of plant	Treatment
Wheat flour by-product (red dog), less than 4% fiber	*Triticum aestivum/T. vulgare/T. sativum/T. durum*	Cereal grain product, plant product, low-protein	Phytohaemagglutinins	Whole grain	Aqueous heat treatment (100 °C: 10 minutes)
Wheat flour by-product (red dog), less than 4% fiber	*Triticum aestivum/T. vulgare/T. sativum/T. durum*	Cereal grain product, plant product, low-protein	Flatulence factor	Whole grain	No method listed
Wheat flour by-product (red dog), less than 4% fiber	*Triticum aestivum/T. vulgare/T. sativum/T. durum*	Cereal grain product, plant product, low-protein	Phytic acid	Whole grain	Aqueous extraction (18 hours)
Wheat flour by-product (red dog), less than 4% fiber	*Triticum aestivum/T. vulgare/T. sativum/T. durum*	Cereal grain product, plant product, low-protein	Protease inhibitor	Whole grain	Autoclaving (121 °C: 15–30 minutes)
Wheat flour by-product (red dog), less than 4% fiber	*Triticum aestivum/T. vulgare/T. sativum/T. durum*	Cereal grain product, plant product, low-protein	Phytic acid	Whole grain	Addition of dietary phytase
Wheat flour by-product (red dog), less than 4% fiber	*Triticum aestivum/T. vulgare/T. sativum/T. durum*	Cereal grain product, plant product, low-protein	Phytic acid	Whole grain	Milling to remove the outer layer of the seed
Wheat flour by-product (red dog), less than 4% fiber	*Triticum aestivum/T. vulgare/T. sativum/T. durum*	Cereal grain product, plant product, low-protein	Amylase inhibitor	Whole grain	No method listed
Wheat flour by-product (red dog), less than 4% fiber	*Triticum aestivum/T. vulgare/T. sativum/T. durum*	Cereal grain product, plant product, low-protein	Phytic acid	Whole grain	Incubation in water (60 °C: 10 hours)
Wheat flour by-product (red dog), less than 4% fiber	*Triticum aestivum/T. vulgare/T. sativum/T. durum*	Cereal grain product, plant product, low-protein	Protease inhibitor	Whole grain	Aqueous heat treatment (100 °C: 10 minutes)
Wheat flour by-product (red dog), less than 4% fiber	*Triticum aestivum/T. vulgare/T. sativum/T. durum*	Cereal grain product, plant product, low-protein	Estrogenic factors	Whole grain	No method listed
Wheat flour by-product (red dog), less than 4% fiber	*Triticum aestivum/T. vulgare/T. sativum/T. durum*	Cereal grain product, plant product, low-protein	Phytic acid	Whole grain	Incubation in water (60 °C: 10 hours)
Wheat flour by-product (red dog), less than 4% fiber	*Triticum aestivum/T. vulgare/T. sativum/T. durum*	Cereal grain product, plant product, low-protein	Protease inhibitor	Whole grain	Aqueous heat treatment (100 °C: 10 minutes)
Wheat flour by-product (red dog), less than 4% fiber	*Triticum aestivum/T. vulgare/T. sativum/T. durum*	Cereal grain product, plant product, low-protein	Estrogenic factors	Whole grain	No method listed
Soybean meal, 46% protein, expeller	*Glycine max*	Oilseed protein product, plant protein	Estrogenic factors	Whole grain	No method listed

(*continued overleaf*)

Table 9.1 (continued)

Ingredient name	Scientific name	Description	Anti-nutritional factor	Part of plant	Treatment
Sorghum	*Sorghum bicolor syn. S. vulgare*	Cereal grain product, plant product, low-protein	Tannins	Whole grain	De-hulling the seed
Sorghum	*Sorghum bicolor syn. S. vulgare*	Cereal grain product, plant product, low-protein	Phytic acid	Whole grain	Incubation in water (60 °C: 10 hours)
Sorghum	*Sorghum bicolor syn. S. vulgare*	Cereal grain product, plant product, low-protein	Phytic acid	Whole grain	Milling to remove the outer layer of the seed
Sorghum	*Sorghum bicolor syn. S. vulgare*	Cereal grain product, plant product, low-protein	Tannins	Whole grain	De-hulling the seed
Sorghum	*Sorghum bicolor syn. S. vulgare*	Cereal grain product, plant product, low-protein	Phytic acid	Whole grain	Incubation in water (60 °C: 10 hours)
Sorghum	*Sorghum bicolor syn. S. vulgare*	Cereal grain product, plant product, low-protein	Cyanogens	Whole grain	Heat treatment
Sorghum	*Sorghum bicolor syn. S. vulgare*	Cereal grain product, plant product, low-protein	Tannins	Whole grain	Reconstitution
Sorghum	*Sorghum bicolor syn. S. vulgare*	Cereal grain product, plant product, low-protein	Protease inhibitor	Whole grain	Autoclaving (121 °C: 15–30 minutes)
Sorghum	*Sorghum bicolor syn. S. vulgare*	Cereal grain product, plant product, low-protein	Cyanogens	Whole grain	Steam heating
Sorghum	*Sorghum bicolor syn. S. vulgare*	Cereal grain product, plant product, low-protein	Phytic acid	Whole grain	Fermentation
Sorghum	*Sorghum bicolor syn. S. vulgare*	Cereal grain product, plant product, low-protein	Phytic acid	Whole grain	Steam autoclaving (120 °C: 2 hours)
Sorghum	*Sorghum bicolor syn. S. vulgare*	Cereal grain product, plant product, low-protein	Tannins	Whole grain	Extrusion
Sorghum	*Sorghum bicolor syn. S. vulgare*	Cereal grain product, plant product, low-protein	Cyanogens	Whole grain	Soaking in water
Sorghum	*Sorghum bicolor syn. S. vulgare*	Cereal grain product, plant product, low-protein	Tannins	Whole grain	Fermentation with lactic acid bacteria

Sorghum	*Sorghum bicolor syn. S. vulgare*	Cereal grain product, plant product, low-protein	Protease inhibitor	Whole grain	Aqueous heat treatment (100 °C: 10 minutes)
Sorghum	*Sorghum bicolor syn. S. vulgare*	Cereal grain product, plant product, low-protein	Tannins	Whole grain	Autoclaving (121 °C: 15–30 minutes)
Sorghum	*Sorghum bicolor syn. S. vulgare*	Cereal grain product, plant product, low-protein	Phytic acid	Whole grain	Aqueous extraction (18 hours)
Sorghum	*Sorghum bicolor syn. S. vulgare*	Cereal grain product, plant product, low-protein	Phytic acid	Whole grain	Addition of dietary phytase
Sorghum	*Sorghum bicolor syn. S. vulgare*	Cereal grain product, plant product, low-protein	Tannins	Whole grain	Treatment with alkali
Barley, grain	*Hordeum vulgare/H. distichum*	Cereal grain products, plant products, low-protein	Phytic acid	Whole grain	Steam autoclaving (120 °C: 2 hours)
Barley, grain	*Hordeum vulgare/H. distichum*	Cereal grain products, plant products, low-protein	Phytohaemagglutinins	Whole grain	Aqueous heat treatment (100 °C: 10 minutes)
Barley, grain	*Hordeum vulgare/H. distichum*	Cereal grain products, plant products, low-protein	Phytic acid	Whole grain	Aqueous extraction (18 hours)
Barley, grain	*Hordeum vulgare/H. distichum*	Cereal grain products, plant products, low-protein	Phytic acid	Whole grain	Steam autoclaving (120 °C: 2 hours)
Barley, grain	*Hordeum vulgare/H. distichum*	Cereal grain products, plant products, low-protein	Phytohaemagglutinins	Whole grain	Aqueous heat treatment (100 °C: 10 minutes)
Barley, grain	*Hordeum vulgare/H. distichum*	Cereal grain products, plant products, low-protein	Phytic acid	Whole grain	Aqueous extraction (18 hours)
Barley, grain	*Hordeum vulgare/H. distichum*	Cereal grain products, plant products, low-protein	Phytic acid	Whole grain	Addition of dietary phytase
Barley, grain	*Hordeum vulgare/H. distichum*	Cereal grain products, plant products, low-protein	Protease inhibitor	Whole grain	Autoclaving (121 °C: 15–30 minutes)

(continued overleaf)

Table 9.1 *(continued)*

Ingredient name	Scientific name	Description	Anti-nutritional factor	Part of plant	Treatment
Barley, grain	*Hordeum vulgare/H. distichum*	Cereal grain products, plant products, low-protein	Phytic acid	Whole grain	Milling to remove the outer layer of the seed
Barley, grain	*Hordeum vulgare/H. distichum*	Cereal grain products, plant products, low-protein	Phytic acid	Whole grain	Incubation in water (60 °C: 10 hours)
Corn/maize grain, yellow	*Zea mays*	Cereal grain product, plant product, low-protein	Protease inhibitor	Whole grain	Aqueous heat treatment (100 °C: 10 minutes)
Corn/maize grain, yellow	*Zea mays*	Cereal grain product, plant product, low-protein	Invertase inhibitor	Whole grain	No method listed
Corn/maize grain, yellow	*Zea mays*	Cereal grain product, plant product, low-protein	Phytic acid	Whole grain	Aqueous extraction (18 hours)
Corn/maize grain, yellow	*Zea mays*	Cereal grain product, plant product, low-protein	Phytic acid	Whole grain	Addition of dietary phytase
Corn/maize grain, yellow	*Zea mays*	Cereal grain product, plant product, low-protein	Phytic acid	Whole grain	Milling to remove the outer layer of the seed
Corn/maize grain, yellow	*Zea mays*	Cereal grain product, plant product, low-protein	Phytic acid	Whole grain	Incubation in water (60 °C: 10 hours)
Corn/maize grain, yellow	*Zea mays*	Cereal grain product, plant product, low-protein	Protease inhibitor	Whole grain	Autoclaving (121 °C: 15–30 minutes)
Corn/maize grain, yellow	*Zea mays*	Cereal grain product, plant product, low-protein	Phytic acid	Whole grain	Aqueous extraction (18 hours)
Corn/maize grain, yellow	*Zea mays*	Cereal grain product, plant product, low-protein	Phytic acid	Whole grain	Addition of dietary phytase
Corn/maize grain, yellow	*Zea mays*	Cereal grain product, plant product, low-protein	Phytic acid	Whole grain	Milling to remove the outer layer of the seed
Corn/ maize, dent, yellow, grain	*Zea mays*	Cereal grain product, plant product, low-protein	Phytic acid	Whole grain	Aqueous extraction (18 hours)
Corn/ maize, dent, yellow, grain	*Zea mays*	Cereal grain product, plant product, low-protein	Phytic acid	Whole grain	Addition of dietary phytase

Corn/ maize, dent, yellow, grain	*Zea mays*		Cereal grain product, plant product, low-protein	Phytic acid	Whole grain	Milling to remove the outer layer of the seed
Corn/ maize, dent, yellow, grain	*Zea mays*		Cereal grain product, plant product, low-protein	Phytic acid	Whole grain	Incubation in water (60 °C: 10 hours)
Corn/ maize, dent, yellow, grain	*Zea mays*		Cereal grain product, plant product, low-protein	Protease inhibitor	Whole grain	Autoclaving (121 °C: 15–30 minutes)
Corn/ maize, dent, yellow, grain	*Zea mays*		Cereal grain product, plant product, low-protein	Invertase inhibitor	Whole grain	No method listed
Corn/ maize, dent, yellow, grain	*Zea mays*		Cereal grain product, plant product, low-protein	Phytic acid	Whole grain	Fermentation
Corn/ maize, dent, yellow, grain	*Zea mays*		Cereal grain product, plant product, low-protein	Phytic acid	Whole grain	Steam autoclaving (120 °C: 2 hours)
Corn/ maize, dent, yellow, grain	*ffgbv.*		Cereal grain product, plant product, low-protein	Protease inhibitor	Whole grain	Aqueous heat treatment (100 °C: 10 minutes)
Corn/ maize, dent, yellow, grain	*Zea mays*		Cereal grain product, plant product, low-protein	Trypsin inhibitor	Whole grain	Extrusion at temperatures between 120 and 140 °C

(*Source*: adapted from FAO 2013)

toasting) have found a significant practical application in destroying mycotoxins and *Fusarium* toxins in the cereal grains, thus improving the quality and safety of cereal grains/by-products (Adegbeye et al. 2020). To enable the bioavailability of nutrients for improving the performance of the gut, various processing technologies convert raw material into a more edible form of feed. However, the selection of processing conditions of different thermal treatments, including processing time, moisture content, temperature, and turbulence of the raw material, must be controlled and validated (Levic and Sredanovic 2010). Vukmirović et al. (2017) reported that pelleting had an impact comparable to fine grinding because the pellet pressing process reduces the granule size. An ideal particle size of pig feed could be acquired during grinding, but a blend of processing methods such as hammer and roller milling enabling targeted granule sizes should to be avoided due to diminished feed consumption and reduced accessibility of nutrients from coarse particles in the gastrointestinal tract of pigs. Nutrient composition, feed formulation, particle size, and the geometric mean diameter of grain directly affect the growth of the gastrointestinal tract and productive performance in livestock (Wang et al. 2014).

9.2 Cereal Grains and By-Products as Feedstuff

After the food industry, the animal feed industry is the second largest user of cereal grains, whether used in wholegrain form or milling/processing by-products (Rosentrater and Evers 2018). Although cereal grains provide highly digestible energy to domestic animals, their proteins and unbalanced essential amino acids composition are not adequate for optimal growth, especially in the early stages of development (Serna-Saldivar 2016). Therefore, the amount of energy available to and required by the animals depends on cereal species, cultivars, individual grain samples, and animal type (Black 2016). The chemical composition of cereal by-products commonly present in animal feed is given in Table 9.2.

9.2.1 Nutritional Value of Cereal Grains Used for Animal Feed Products

9.2.1.1 Rice Rice grains, as well as rice milling products (polished rice and broken rice) are primarily food products and are not commonly used for livestock feeding due to their high cost. According to the Food and Agriculture Organization of the United Nations (FAO 2016), the global production of milled rice reached 448.2 million metric tons (Statista 2019), wherein a quadrant portion is used as livestock feed. Unlike other cereal grains, rice is not completely used for animal feed, but nowadays an increasing number of farmers in Japan are replacing corn with domestic rice as pig and cattle feed, not only to meet consumer demands for food safety, but also to brand the livestock products, contributing to local economic recovery (*Japan Times* 2017). Moreover, feed rice varieties have been bred to be suitable for animal diets, among which a rice cultivar Momiroman appears to be suitable for poultry diets due to its large seed (Sittiya et al. 2011). White polished rice of low quality, as well as damaged and broken kernels from rice milling process, can be redirected for use in the livestock industry. Paddy rice, also called "rough rice," consists of the whole rice grain with the

Table 9.2 Chemical composition of cereal by-products on a dry matter basis.

Nutrients (%)	Wheat middlings[1]	Wheat bran[1]	Sorghum DDGS[1]	Corn DDGS[1]	Corn gluten[1] meal	Corn gluten feed[1]	Corn germ meal[1]	Corn bran[1]
Ash	6.1	6.9	5.2	4.8	1.8	6.3	4.3	2.2
Gross energy (kcal/kg)	4446	4312	5249	5185	5667	4708	4432	4589
Crude protein	18.4	17.2	30.5	29.2	66.8	25.8	26.2	9.8
Starch	16.8	16.9	3.1	8.6	13.0	11.1	18.7	22.6
S-NCP	1.2	3.8	3.5	3.4	0.7	5.1	6.6	5.2
I-NCP	22.7	24.3	14.4	15.8	1.9	16.3	19.7	26.8
Cellulose	6.7	6.4	6.7	5.8	0.9	7.1	8.1	10.1
Total NSP	30.7	34.5	24.7	25.0	3.6	28.7	34.5	41.8
Soluble dietary fiber	1.2	3.8	3.5	3.4	0.7	5.1	6.6	5.2
Insoluble dietary fiber	36.9	37.6	29.4	25.5	12.4	26.5	30.1	40.6
Dietary fiber	38.1	41.4	32.9	28.9	13.2	31.6	36.7	45.7
Indispensable amino acids[2], (g/kg)	60.4	54	131.7	114.3	243.3	76.6	25	45.1
Dispensable amino acids[2], (g/kg)	84.9	122.3	168.3	181.9	312.4	98.9	25	54.8

S-NCP = soluble non-cellulosic polysaccharides, I-NCP = insoluble non-cellulosic polysaccharides, NSP = non-starch polysaccharides. Total NSP = S-NCP + I-NCP + cellulose. *- g/kg
(*Source*: [1]Jaworski et al. 2015; Stein et al. 2016; [2]Zhou et al. 2018)

hull accounting for 20% of the grain. Although paddy rice is of lower nutritional value due to its higher fiber content, in the Far East its significance is increasing as a component of livestock feed and as an alternative for other cereal grains (e.g. poultry feed). Researchers have proved that paddy rice enhances the digestibility of crude fiber, gross energy, and metabolizable energy in poultry feed, and possesses high bioaccessibility and bioavailability of nutrients, although their levels are low.

In order to reduce the cost of paddy processing (such as grinding), many Asian countries are using paddy rice as a feed for poultry. As reported, the gizzard is efficient in feed grinding, even more efficient then hammer mills used for grinding the whole grains. Therefore, a new feed processing technique has been developed offering pelleting of high amounts of unmilled cereals. This technique offers a number of benefits, from both the technological or poultry health point of view: improved feed utilization, higher quality of the pelleted feed, reduced feed production costs, and improved poultry gut health (Svihus 2001). The paddy rice is perfect for the physiological functions in poultry as it contains appropriate particle size in feed, improves the gizzard activity, and also increases retention time of the digesta which results in the suppression of *Campylobacter* growth in chicks (Nishii et al. 2016).

The standardized ileal digestibility (SID) of amino acid in white polished rice is almost 94%, being higher when compared to other nutrient-dense feed (Cervantes-Pahm et al. 2014). In particular, various combinations of amino acids in the diet determine the effectiveness of protein, especially in terms of essential amino acids and their proportion in the animal feed. These protein requirements vary during various physiological stages of the animal. Some studies showed that the crude protein (CP) (167.9 g/kg) present in ground hulled rice and rice husk improve the bodyweight of geese, such as slaughter yield (87.73%), breast yield (17.38%), and high yield (17.35%), compared to the control diet with ground maize and soybean meal (Wang et al. 2014). The micronutrients were found to be high in rough rice compared to polished rice, since the essential micronutrients remain in the husk and hull fraction removed during processing. For lactating animals, the replacement of a corn diet with brown rice to produce cattle feed containing a high starch content (25.4%) and digestibility value (96%), was reported in Japan (Miyaji et al. 2012).

9.2.1.2 Wheat Wheat is the second-most important animal feed raw material. Based on the volume of worldwide production and trade prices, wheat represents one of the most attractive feedstocks replacing a portion of other cereals grains. It can be utilized either as whole grain or as wheat by-products, comprised of damaged wheat kernels, bran, germ, and low-grade flour streams segregated from storage and milling processing accounting for 25–30% of the total wheat (Papageorgiou and Skendi 2018).

The utilization of whole wheat grains in poultry feed not only reduces the feed processing costs, but also improves poultry gut health, but with the negative impact on performance, welfare, and carcass quality (Singh et al. 2014). Ahmed et al. (2018) indicated a reduction in the nutritional quality of growing turkey diet due to reduced growth performance and feed efficiency. The effectiveness of feeding the whole wheat grains appeared to be variable, depending on the type of poultry and the method of feeding.

Due to its high starch content (50–80%), wheat is considered as an important energy component for animal feed. Although the protein content is lower than common animal feed protein ingredients such as soybean meal (11–15%), wheat proteins are characterized by high dietary inclusion level (30–70%) and a significant amount of indispensable amino acids. Wheat can supply up to 60% of the animals' total dietary requirements, with up to 70% of indispensable amino acids (Rosenfelder et al. 2013). However, wheat contains a significant amount of NSP (e.g. arabinoxylans), which act as antinutritive factors and limit the nutrient digestibility in the small intestine of monogastric animals. Arabinoxylans contribute to high viscosity of digesta in which the beneficial nutrients are entrapped and not able to be digested, especially because non-ruminant animals do not possess enzymes for their hydrolysis (Zhang et al. 2018). Additionally, bran contains a high measure of phytic acid, and phytate which primarily stores phosphorus but affects the bioavailability of divalent cations, including calcium, magnesium, zinc, and iron.

A biochemical analyses on broiler chickens showed a lower level of creatine kinase (141.11 U/mg protein) and calpain (30.04 ng/mg protein) in breast muscle of those fed with wheat/wheat middling in comparison to alfalfa meal (creatine kinase 142.59 U/mg protein, calpain 40.09 ng/mg protein). Creatine kinase activity is an indicator of muscular damage; therefore, muscle structure was well preserved by the diet containing wheat (Jiang et al. 2018). The geometric mean diameter and hardness of the endosperm are equally important for the performance of poultry.

9.2.1.3 Maize Maize is the cereal grain most widely used for animal feed, followed by wheat and barley. It is estimated that 65–85% of the total world maize production is used for animal feed (Serna-Saldivar 2016; Rosentrater and Evers 2018). Favorability of maize as animal feed is due to its high starch content (565–627 g/kg), providing high energy values (Moss et al. 2020), low content of NSP, and high digestibility as compared to other plant-based cereal feeds (Rosentrater and Evers 2018). However, its protein content is lower than that of other cereals (8–9%). Due to slower fermentation of maize starch, it combines well with rapidly digested grains (wheat, barley, and high-moisture maize). It is utilized either as whole grain or processed affecting the rate of digestibility – high moisture maize is more digestible than dry maize, while ground, cracked, or steam-flaked maize grains are more digestible than unprocessed grains.

Around 60% of starch contributes to apparent metabolizable energy (AME), and starch digestibility of poultry feed. It has been shown that corn possesses a high available energy content and low soluble NSP as compared to other plant-based feeds. The high starch content might cause acidosis in dairy cattle; however, the corn diet has less risk of acidosis when compared to barley and wheat diets. Furthermore, with its moderately high-calorie value and low fiber content, incorporation of maize in routine feeding during hot weather might decrease acidosis, lessen the heat effects on dairy animals, and sustain metabolizable energy for milk production. Feeding steam-flaked corn to dairy cow shows more dietary benefits such as an increase in protein and decrease in fat content of milk in contrast to cows fed with grounded corn counterparts. Similarly, dairy cows fed with finely rolled maize (1.3 mm size) show a rise in milk production (33 l/day) and higher protein content in milk when compared to other forms of maize, such as milled (~30 l/day) or coarse (<30 l/day) (Ahmadi et al. 2020).

9.2.1.4 Barley Barley is an important energy source for animal feed, which covers 60% of all cereal fed, primarily used in cattle feed such as beef, dairy cattle, and pigs (Nikkhah 2012). It is characterized by its protein content in the range of 12.5–17%, and high content of β-glucans, arabinoxylans, and cellulose (~23–41%), which makes it unsuitable for poultry feed. However, there are two types of barley suitable for monogastric animals – naked or hull-less, and high-lysine barley, characterized with a lower fiber content and an improved essential amino acid profile (especially lysine), respectively (Serna-Saldivar 2016). The hull-less barley provides more metabolizable energy and better feed conversion rates, while high-lysine barley contains 20–40% more protein and lysine compared to regular barley.

Barley NSP are more fermentable than that of wheat and maize. The high amount of neutral detergent fiber (NDF) and acid detergent fiber (ADF) reduce nutrient digestibility due to the formation of a viscous solution in the presence of water, which affects the reduction of digestibility and nutrient availability, and forms sticky feces in the animal's intestine, thus affecting the apparent total tract digestibility (ATTD) in pigs (Clarke et al. 2018). Therefore, barley feed requires special pre-treatment such as the addition of enzymes in the feed formulation for its proper utilization in livestock diet. Enzymes, for example β-glucanase and β-xylanase, lower the viscosity of β-glucans and increase the nutrient utilization in piglet diet. It has been noticed that the barley diet in weanling pigs lowered the incidence of diarrhea due to the prebiotic action of β-glucan in the colon (Owusu-Asiedu et al. 2010). Conversely, these two enzymes used in fish feed reduce apparent digestibility and negatively affect growth (Sinha et al. 2011). However, some crucial nutrients like calcium, magnesium, copper, potassium, manganese, and zinc, increased during digestion in the presence of phytase enzyme in the fish feed containing barley (Cheng and Hardy 2002). Diminished digestibility of fiber associated with a barley diet is due to the rapid fermentation and is likely caused by decreased pH level of the rumen. Amino acid composition of protein and fiber proportion of barley is a perfect food for growing bovines. Young bovines fed with barley protein in the starter phase is found to be favorable for increased body weight. Hence, barley protein can be used as an alternative to soybean protein, showing its potential as a competitive substitute for other ingredients (Anderson and Schroeder 2009). According to Ametaj et al. (2010), barley grain consumption increases the metabolites in the rumen of dairy cows, such as glucose, alanine, maltose, propionate, uracil, valerate, xanthine, ethanol, and phenylacetate. However, unhealthy toxic metabolites in rumen were seen in the cattle fed with 30–45% of barley grain diet. A higher percentage of barley in cow feed reduces rumen 3-phenylpropionate, which aids in low methane (CH_4) emission.

9.2.1.5 Sorghum Sorghum, a native of Africa, is cultivated in warm and dryland regions, but also in the United States and Australia as a fodder crop. Four classes of sorghum are known: brown or high tannin, yellow or red, white, and mixed sorghum, Yellow or red have the most importance in animal feed, mainly as a substitute for maize, especially in poultry or swine, although the yellow or red varieties are characterized by their slightly higher protein content, lower fat content, and lower metabolizable energy compared to maize.

The sorghum starch granules are tightly bound to the protein matrix, especially in the subaleurone layer and hard endosperm, which limits the penetration of water, digestive enzymes, and ruminal microorganisms. Kafirin is the predominant protein (prolamine) fraction present in the sorghum protein, accounting for 54.1% of its total protein that contains more disulfide cross-bonds compared to zeins of maize. Other protein fractions such as glutelin, globulin, and albumin account for 33.4, 7.0 and 5.6% of sorghum endosperm protein, respectively. However, waxy sorghums containing more than 95% amylopectin and a weaker protein matrix are considered to have the same nutritional value and similar digestibility as maize. However, suitable processing methods can enhance the amino acid proportion in sorghum. For instance, the total essential amino acid of steamed and flaked sorghum (189.5 and 276.55 mg/g protein) is found to be superior to the raw and boiled sorghum (91.07 and 141.91 mg/100 g protein). Threonine (19.3 and 20.69 mg/g protein) and tryptophan (14.69 and 12.31 mg/g protein) decreased in steaming and flaking process compared to fermented sorghum samples, i.e. 24.28 and 16.88 mg/g protein, respectively (Mohapatra et al. 2019). Brown sorghum contains high amounts of condensed tannins, acting as antinutritional compounds, which decrease the protein digestibility and the overall nutritional value (Serna-Saldivar 2016). Tannins hinder the bioavailability of protein during metabolism, due to their complex interactions with different organic nitrogen compounds and the formation of tannin–protein complexes (Adamczyk et al. 2017). Although in some countries a traditional variety of sorghum with high tannin content is still cultivated, low-tannin and 99% tannin-free diets for livestock feed have been introduced in the US. It is likely that both kafirin and phenolic interrupt energy utilization in broilers fed on sorghum-based diets. Therefore, the cultivation of white sorghum containing "non-tannin" phenolic compounds seems to be a suitable option. When the starter and grower phase of broiler diet consisting of low tannin sorghum (500 g/kg) was replaced with corn feed, the increase of final body weight (1418 g) was achieved as compared to that of corn diet (1224.7 g). Moreover, the best carcass weight and carcass dressing was found in a total sorghum fed diet (1067 g, 75.64%), followed by substituted sorghum (883–1056 g, 73.24–73.89%), and corn diet (749 g, 68.55%) (Tandiang et al. 2014). Coarsely grounded sorghum, with a mean particle size of 3 mm was reported to have a better performance in broilers as compared to finely ground sorghum with a mean particle size of 1 mm (Rodgers et al. 2009). Similar results were reported by Selle et al. (2019), who reported significantly higher starch and protein digestibility coefficients of the feed containing coarser particles (mean particle size of 1405 μm) in comparison to that containing finer sorghum particles (mean particle size of 794 μm). A steamed flake sorghum grain diet given as a replacement for corn grain decreases the ratio of acetate to propionate, due to the improved ruminal fermentation and nitrogen utilization in beef cattle (Wang et al. 2018).

9.2.2 Nutritional Value of Cereal By-Products Used for Animal Feed

Cereal by-products result from various cereal processing like dry milling, wet milling (for starch and glucose production), and brewing. Generally, about 25–35% of cereal mass is removed during processing, whereby its physiochemical and nutritional characteristics are impacted by the type of grain and the type of processing operations (Serna-Saldivar 2016).

9.2.2.1 Rice By-Products The huge production of rice in Asia proportionally contributes to the production of rice by-products – husk (outer layer of rice known as hull) and rice bran (Papademetriou 2000). During the polishing of rice, around 60–72% of the rice grain is converted into an edible form for human food. In comparison, the remaining 28–35% of the grain is separated out as by-products, mainly for animal feed. Therefore, rice by-products include hull, husk, bran, and damaged and broken rice kernels (Singh et al. 2013). Rice by-products are a rich source of nutrients and phytochemicals such as tocopherols, tocotrienols, polyphenols (mainly ferulic acid and α-lipoic acid), phytosterols, γ-oryzanol, and carotenoids (carotene, lycopene, lutein, and zeaxanthin), with strong antioxidant, anti-inflammatory, and chemopreventive properties. In animal feed, rice bran and fat may comprise up to 40% of the daily intake of pigs, cows, and poultry (De Godoy et al. 2013). Rice hull possesses lower nutritional values compared to other cereal by-products. For instance, Kumari et al. (2018) reported lower CP (11%), crude fiber (10.90%), total carbohydrates (49.92%), and total dietary fiber (21.63%) in full fat rice bran compared to defatted rice bran (13.80, 13.10, 61.46, and 24.50%) respectively. The hulled rice and rice husk are effectively used for poultry feed; for instance, the goose gizzard can easily break down the fiber of hulled rice and rice husk; therefore, they can well use this fiber-rich diet. This shows improvement in carcass weight and digestive tract development of poultry (Wang et al. 2014).

9.2.2.2 Wheat By-Products Wheat processing results in different types of by-products that can be used as animal feedstuffs – wheat bran, shorts, middling, distillers' grain, feed-flour, and straw (Huang et al. 2014). Wheat middlings (WM) are commonly used in the formulation of ruminant feed. WM are composed of wheat bran, shorts, germ, flour, and low-grade flour streams (Adedokun et al. 2015). Around 25–30% of WM are obtained as by-products from 70–75% of wheat grains through the wheat flour milling industry. WM are suitable for cattle feed consumption. As per USDA, WM are ranked a close second to soybean meal, and represent one of the leading commercial by-products used in the feed industry (Blasi et al. 1998). The total dietary fiber of WM is 36.45% and starch 20.28%. They are found to be superior compared to soybean meal and their protein content (CP 17.97%) is better than corn (CP 6.80% as fed basis) (Casas et al. 2018). Wheat germ meal (WGM) consists mainly of wheat germ together with some bran and middling or shorts. It contains 277 g/kg dry mass (DM) CP and 115 g/kg DM crude fat. This possesses exclusive nutritional and functional properties related to the presence of crude fiber (31 g/kg DM), comprising of cellulose, lignin, hemicellulose, pectin, gum, and other oligosaccharides present in the bran layer of wheat, thus WGM serve as an essential nutrient for fish (Reis et al. 2019). On the other hand, feeding wheat to cattle requires certain precautions due to the pasty consistency of wheat protein and decreased motility in the rumen. Therefore, special processing techniques are required for the conservation of nutrients. Some of the techniques, such as dry-rolling, coarse grinding, and steamrolling, produce thick flakes that are included in the manufacturing steps to improve the feeding value. In general, fine grinding processes decline the intake of feed ratio, which may cause acidosis and bloat of the rumen. However, the extrusion processes enhanced nutrients in wheat bran such as protein, phosphorus, and soluble dietary fiber (148, 15.8, 39 g/kg) compared to native wheat bran (147.3, 15.4, 32 g/kg) and fermented wheat bran (145.2, 15.5, 34 g/kg). This process also results in a higher coefficient of total tract apparent digestibility of dry mass *in vivo* (Kraler et al. 2014).

9.2.2.3 Maize By-Products Maize by-products are the most predominant feed ingredients commonly used in ruminant and other types of animal feed. Distillers dried grains with solubles (DDGS) from ethanol production are considered a low-cost animal and fish feed ingredient. At the point when starch from maize is fermented to ethanol, the residual fiber, fat, and protein are concentrated in the distiller's grains. Maize DDGS is an alternative nutrient source for fish, which is comprised of protein 26–33%, fat 9–14%, and NDF 33–44% (Brown et al. 2016). DDGS is the potential feedstuff for omnivorous fish, due to their higher tolerance of fiber than that of carnivorous fish. As compared to maize grain, DDGS exhibit a profound impact on the beef cattle, when 30% of wet distiller's grain solubles (WDGS) are given at the finishing phase of beef cattle. In that case, an increase in trans-octadecenoic acid, linoleic acid, total trans fatty acids, polyunsaturated fatty acids, omega-6, and omega-6 to omega-3 ratio, in all muscles was observed. The rise in muscle fat is likely related to a higher percent of maize oil in the distiller's grains in comparison to the parent grains. The high amount of unsaturated fatty acids in the lean muscle tends to oxidize easily and compromise the shelf life of steak (de Mello et al. 2018). When 40% WDGS or DDGS was included in the feedlot steers' diet, an increase in fecal enterohemorrhagic *Escherichia coli* (EHEC O157) was observed. When DDGS were given in the finishing feeding phase in another group of steers, a similar effect was noticed. This experiment also showed that the WDGS diet incorporated with direct-fed microbials (DFM) had a low incidence of EHEC O157 (Wilson et al. 2016; Schneider et al. 2018). Therefore, DDGS/WDGS, with a mixture of enzyme or DFM, would exclude this undesirable impact in beef cattle.

9.2.2.4 Sorghum By-Products The main sorghum by-products originate from the sorghum distilled spirits and biofuel production, and comprise sorghum wet distiller's grains with solubles (sWDGS) and sorghum dry distiller's grains with solubles (sDDGS) (Sotak et al. 2011). sDDGS and sWDGS are commonly added as a substitute in the main cereal diet for beef cattle, dairy cows, and pigs, consisting of crude fat (8.8–13%), CP (32.9–35.9%), and amino acids (0.38–6.92%) (Sotak et al. 2011). While 95% of sDDGS comprise of 0.06–3.18% of amino acids, which include methionine 0.12%, lysine 0.41%, tryptophan 0.06%, threonine 0.45%, and arginine 0.42%, this implies that sDDGS is deficient in essential amino acids for poultry (Hansen 2016). As compared with the parent grain, the DM and non-fiber carbohydrate content of sWDGS were found to be lower, whereas ether extract (EE), NDF, acid detergent fiber (ADF), CP, neutral detergent-insoluble CP, ammonia, acid detergent-insoluble CP, lignin, and tannins (see Table 9.3) were considerably higher. Heifer diet containing 0–15% sWDGS along with steam-flaked corn improved CP and ADF content in the feed, as compared to a diet without sWDGS (da Silva et al. 2019). However, sWDGS did not improve the body weight of the steer carcass with both types of corn-based diet. sDDGS can be included in the diet for all categories of poultry; for instance, 8–16% sDDGS in the diet of geese had higher average feed intake, feed/gain ratio, breast meat, leg meat, subcutaneous fat and skin, while abdominal fat were not affected by the sDDGS diet (Wang et al. 2018). Data about incorporation rates for sorghum DDGS is extremely scarce; however, it is trusted that up to 30% might be incorporated into finisher pig feed and reproducing pigs (Stein et al. 2016). Despite this, a higher percentage of sDDGS (up to 45%) in pigs increased backfat iodine value, while fat became less red. It appeared that 30% of corn DDGS and sorghum DDGS showed similar characteristics, such as average daily gain, gain:feed, and average daily feed intake, with no difference in growth performance of finishing pigs (Sotak et al. 2015).

Table 9.3 Summary of the effect of various processing techniques on the cereal grain.

Treatment process	Disrupts seed layer; Expose endosperm	Reduces particle size	Separates starch granules; Disrupts endosperm matrix	Disrupts starch granules; Cause hydration and gelatinization
Dry rolling	+++	+	–	–
Grinding/milling	+++	+++		
Steam flaking	+++	++	+	+
Extrusion	+++	–	++	+
Pelleting	+++	–	+	+
Reconstitution/ensiling	+	—	—	—
Micronization	+	+	—	—
Popping	++	–	+	+++

Effect of the treatment processes is currently not known; + indicates a minor effect, ++ moderate effect, and +++ major effect on grain and structure or digestion. (*Source*: adapted from Rowe et al. 1999)

9.3 Processing Methods of Cereal Grains for Feed Purposes

Processing of cereal grains for animal feed can be achieved by a number of mechanical, physical, and chemical methods. Each type of cereal requires a specific *primary* and *secondary* processing treatment. *Primary processing* involves cleaning, dehulling, mixing, milling, rolling, pounding, grinding, cracking, tempering, soaking, drying, and sieving, resulting in products that are still not consumable (Singh et al. 2013). The *secondary processing* stage adds values to the cereals and makes products consumable for large categories of animals. This processing step involves pelleting, ensiling, fermentation, baking, micronization, puffing, flaking, frying, and extrusion (Hasting and Higgs 1978). Grain processing can also be categorized in terms of dry and wet processing. Dry processing comprises grinding, dry rolling, popping, extruding, micronizing, and roasting, while wet processing encompasses soaking, steam-rolling, steam processing and flaking, pressure cooking, exploding, pelleting, and reconstitution (Dehghan-Banadaky et al. 2007). However, the primary aim of any of these methods is to break down the seed coat, substantially reduce the particle size, improve palatability, inactivate ANFs, and expose the surface area to the digestive enzymes (Tosta et al. 2020). Each of the operations listed can influence the feed quality, feed efficiency, feed intake, animal performance, and shelf life, as well as the final profitability. Table 9.3 summarizes the major effects of various cereal grains processing techniques. The main processing methods utilized in the animal feed industry are described below.

9.3.1 Primary Processing Methods

9.3.1.1 Cracking, Dry Rolling, and Grinding or Milling Cracking, dry rolling, and grinding or milling, are different processing operations, but their common mode of action is to break down the kernel coat and reduce the particle size and hence increase the surface area for better digestion. Cracking is the minimal processing method required to expose the endosperm by removing the pericarp which ensures efficient digestion by most animals. Rolling is also used to crack the seed coat and expose the grains for digestive enzymes and allow beneficial action of bacteria. Since rolling still retains the larger particle size of grains, it limits the digestion or fermentation. Grinding or milling is the major and most common method for feed processing. It is the simplest process and results in extremely fine particles providing the exposure of more surface area for the action of digestive enzymes. However, the outcomes of the grinding process mainly depend upon the hardness of the grain endosperm. Starch granules are more susceptible to shearing and shattering if the grain endosperm is harder, whereas they tend to remain intact in the case with a softer endosperm. Grinding or milling can be accomplished by different types of mechanical mills, such as hammer, attrition, roller mills, and cutters. With a hammer mill, the kernel is grinded by shattering it into small pieces, while milling with roller mills results from crushing, cutting, and shearing of kernels. In hammer mills, the particle size is controlled by the size of the openings in the screen, whereas in roller mills it is achieved by adjusting the gap between the rolls and subsequent sieving (Womac et al. 2007). Grinding generally improves feed digestibility and homogeneity, increases bulk density and facilitates further secondary processing such as extrusion and pelleting.

9.3.2 Secondary Processing Methods

9.3.2.1 Pelleting Pelleting is a common commercial process wherein fine grounded or milled sample, sometimes dusty, unpalatable, and difficult-to-handle material, is combined into a larger particle by using a combination of pressure, heat, and moisture. The mechanical operations engaged during pelleting include compression, extrusion, and adhesion. The resulting larger particles in the form of pellets are more palatable, thus affecting improved feeding efficiency and intake compared to unpelleted feed (Lv et al. 2015). During pelleting, a fine mixture of feed passes through a conditioning chamber where moisture (usually in the form of steam) is added. This moistened soft feed is forced through the holes of a metal ring-type die at high temperature, which converts it into a finished pellet with tight layers of feed mixture using compression and extrusion mechanism. Moisture provides lubrication for compression and extrusion. The starch present on the surface of grains is partially gelatinized by heat, steam, and by the friction produced as the feed passes through the die, contributing to the better adhesion of material and increase degradability of starch (Tosta et al. 2020). The pellet quality is determined by the particle size. A fine grounded sample exposes more surface area that allows a higher water absorption during the conditioning phase, which results in higher feed temperature, increased starch gelatinization, and better adhesion to feed ingredients. Additional factors that may influence the pellet quality include chemical composition, bulk density, and texture, as well as the relative humidity of soft feed.

9.3.2.2 Steam Flaking Steam-flaking or flaking is based on a gelatinization process occurring when heat, moisture, and pressure are applied to cereal grains to transform them into flakes. It is especially used by the beef cattle feedlot industry in the US (Matsushima 2006; Vasconcelos and Galyean 2007). It is a process of steaming the whole grain for 20–40 minutes at atmospheric pressure, and subsequently rolling it to varying degrees of flake density 437–283 g/l. Different grains require different flake density in order to get optimal performance and feed efficiency (Brown et al. 2000). The process breaks the seed coat and endosperm, causing sufficient disruption of the starch–protein matrix, resulting in higher starch digestibility in the rumen or in the total gastrointestinal tract (Armbruster 2006). Flaking has a greater response to grains with a high starch content as well as grains that have less variation in the starch content (Armbruster 2006). Qiao et al. (2015) argued that steam flaking can increase the energy value and organic matter digestibility of rice. It can improve the nutrition value of wheat and rice if the maize diet in the ruminant is partially replaced with any of these grains.

9.3.2.3 Extrusion Extrusion is a thermomechanical processing which involves a combined impact of moisture, high temperature and pressure on the cereal grain material in a process comprising operations such as mixing, cooking, kneading, shearing, shaping, and forming (Xu et al. 2020). Extruders are widely used to manufacture a variety of cereals and cereal by-product-based food products. During the extrusion process, the cereal grain material is conveyed through a steam jacketed barrel and exposed to the influence of heat and pressure (Strabler et al. 2009; Rokey et al. 2010). By varying processing conditions, such as temperature, moisture, pressure, and time, the extrusion may enhance the nutritional value, chemical composition, and efficiency of feed ingredients (Rahman et al. 2015). Therefore, all processing condition should be optimized prior to attaining maximum results from extrusion processing.

The primary aim of extrusion is to change the microstructure, chemical characteristics, and the macroscopic shape of the starting cereal material to be more usable by the animals. This is achieved by starch gelatinization, and protein denaturation forming complexes among lipids, starch, and proteins. However, other biomolecules such as fibers are also affected during the extrusion processing, whose content is decreased due to extrusion, and insoluble fibers are converted into soluble ones (Dalbhagat et al. 2019).

Complete starch gelatinization is generally achieved at 120 °C or above, with 20–30% or lower moisture content, while proteins also undergo structural unfolding and/or aggregation when subjected to moist heat or high shear. Due to extreme processing conditions, the nutritional value of lipid components and lipid content also undergoes reduction due to oxidation of unsaturated fatty acids to lipid hydroperoxides and lipid–protein complex formation (Levic and Sredanovic 2010; Dalbhagat et al. 2019). The degradation of vitamins during extrusion is the result of increased temperature, screw speed, feed moisture, feed rate, and die diameter (Dalbhagat et al. 2019). Therefore, the extrusion process may have a negative effect or no effect on the bioavailability and digestibility of feed components. For instance, Solanas et al. (2007) reported that extruded cereal (maize and barley, 60 : 40) had lower CP degradability (63.7%) compared to non-extruded cereal feed (71.2%) determined by *in vitro* ruminal nutrient digestion. However, the combination of mechanical and thermal processes involved in the extrusion process destroyed the ANFs such as phytate, polyphenols, oxalates, and trypsin inhibitors, the levels of which were reduced by 54.5, 73.3, 36.8, and 72.3% at high die temperature (115, 140, and 165 °C) and high screw speed (400 rpm) in wheat bran, rice bran, barley bran, and oat bran (Nikmaram et al. 2017). Moreover, a positive impact occurred on the extruded sorghum bran which was enhanced by 14.10% of free phenolic compounds and 15.5% of antioxidant capacity, which was not observed in the unextruded sorghum bran. This report suggested that feed moisture of 30% with a temperature of 160 °C in the fourth zone of the extruder increased the free total phenolic compounds (7428.95 µg GAE/g) and antioxidant activity by 2,2-diphenyl-1-picrylhydrazyl (DPPH) of free total phenolics compounds (14.12 µg TE/g) (Ortiz-Cruz et al. 2020).

On the basis of the method of the operation and the type of extruder construction (i.e. single- and twin-screw extruder), extrusion can be classified as hot and cold extrusion based (Dalbhagat et al. 2019). Single-screw extruders include cold forming (pasta-type), high-pressure forming, low-shear cooking, high-shear cooking, and collet extruders, while twin-screw extruders include non-intermeshed co-rotating, non-intermeshed counter rotating, intermeshed co-rotating, and intermeshed counter rotating extruders, with the co-rotating, intermeshed screw type widely accepted in the food and feed industry (Sobowale 2018). Though twin-screw extruders are generally more expensive than single-screw extruders, they are still attractive to food manufacturers due to their positive conveying mechanism, the degree of quality control, more uniform flow of the product through the barrel, and processing flexibility that they offer.

9.3.2.4 Ensiling Ensiling is a forage preservation method in which the water-soluble carbohydrates are fermented and partially converted to organic acids (principally lactic acid and to a lesser extent to acetic acid) by lactic acid bacteria. Due to the production of organic acids, the pH of the ensiled material reduces, which inhibits the spoilage microbes and help in feed preservation (Saylor et al. 2020). After ensiling, water-soluble carbohydrates are readily available for microbial enzymatic fermentation. Ensiling is a very common practice in silage preparation wherein a crop, forage, or agricultural by-product with high moisture (>50%) content is fermented. Ensiling flows through four phases: (i) aerobic phase (involving respiration and proteolysis); (ii) fermentation phase (involving the action of primarily lactic acid bacteria); (iii) stable phase (involving very little biological activity); and (iv) feedout phase (involving losses of highly digestible nutrients due to unrestricted access of oxygen to the silage) (Bolsen et al. 2007). The high moisture levels in feed materials allow endogenous enzyme activity until the pH drops as a result of fermentation during silage preparation, provided anaerobic conditions are maintained (Pieper et al. 2011). They reported that ensiling of pre-mature cereal grains with lactic acid bacteria could be an alternative to other techniques of preservation. Parmenter et al. (2018) evaluated the effectiveness of ensiling of wet brewer's grains. The results demonstrated that ensiling altered the crude fat content, while CP and lactic acid content were unaffected. It was concluded that wet brewer's grains can be successfully ensiled in a similar way to conventional feeds, but further research is needed to determine the feasibility and cost-effectiveness of the process.

9.3.2.5 Micronization Micronization is thermal processing treatment which significantly improves digestibility and nutritional value of feed products such as cereals, pulses, beans, and oilseeds. The process uses infrared radiation at wavelengths of 3.4–1.8 µm, which are found to be highly efficient in attaining the required temperature over a short time period (Radosavljević et al. 2010). Prior to micronization, cereal grains are conditioned to specific moisture content and exposed to the stream of infrared waves for a certain period of time (Scanlon et al. 2005). During micronization, the emitted radiation rapidly increases the temperature inside the grain, affecting moisture evaporation by 30–40%, depending on the type of treated material. For instance, infrared radiation quickly increases the internal temperature of cereal grains to between 90 and 100 °C in 50 seconds, thus the grain swells and breaks. Thereafter, the swelled grains are rolled up into thin and elastic flakes with improved palatability, flavor, color, appearance, and nutritional values (www.micronizing.com). Micronization not only reduces the moisture content, but also adequately increases the particle area, the solubility of carbohydrate. and starch digestibility. The most significant changes in micronized cereal grains occur in starch, as the structure of starch granules is disrupted with lost double-helical structure due to heat and water. The process is known as gelatinization, which occurs within the temperature range 52–85 °C for wheat starch and 62–80 °C for maize starch (Thomas and Atwell 1999). Every type of starch has a characteristic temperature range within which gelatinization occurs; therefore, micronization differently affects taste, edibility, and nutritive value of different types of grains (Radosavljević et al. 2010). This process is also capable of reducing surface microflora as well as protease inhibitor contents (e.g. trypsin inhibitor) (Levic and Sredanovic 2010). The micronized products have wider applications in animal feed, pet food, infant food, and beverage industries.

9.4 Safety Risk and Hazards

Cereal-based feeds pose a lower risk to food safety than many other foods. However, a number of potential hazards such as natural, chemical, or microbiological can arise at specific points in the cereal grain production and processing which, if managed incorrectly, could pose a significant threat to human as well as livestock health (Alldrick 2014). Food safety management or quality assurance requires a risk-based assessment of potential hazards presented both at grain production and processing stages. The safety risk can be mitigated or minimized by adopting quality assurance principles utilizing both Good Agricultural Practice (GAP) and Hazard Analysis Critical Control Point (HACCP) (*Codex Alimentarius* 2003). To maintain the quality and safety, agronomy plays a key role in the safety of both the cereal crop and eventually the ingredients and finished products prepared from it. These include the natural food toxicants or ANFs, chemical contaminants such as naturally-occurring mycotoxins or pesticides (Alldrick 2014). ANFs do not possess a significant challenge in terms of food safety but mycotoxins are poisonous substances produced by certain filamentous fungi, thus presenting a serious health hazard, and are considered the most significant issue related to the cereals. Mycotoxins are very hard to decontaminate, but their levels can be reduced by removing the contaminated fraction by implementing a series of cleaning methods such as size separation, density separation, or optical sorting (Schaarschmidt and Fauhl-Hassek 2018). Potential mycotoxins associated with cereals including aflatoxins, ochratoxin A (Ryu et al. 2019), fumonisins, deoxynivalenol, and zearalenone, have a set limit under the RASFF (Rapid Alert System for Food and Feed) system operating within the EU (Rapid Alert System for Food and Feed 2009). The use of chemical pesticides, one of the most divisive attributes of modern agriculture, presents an adverse health effect; their residue in foods/feeds is articulated in Regulation (EC) 396/2005 (European Parliament and Council 2005).

Intriguingly, the RASFF data generated for EU Member States reveals that the safety risks associated with cereal-based products are not related to how the raw material is produced, but to how it is processed. These risks include, for instance, chemical contamination, infestation, improper use of food additives or colorants, or failures in allergen control systems, which are a part of good manufacture practice (GMP) (Alldrick 2014). Although microbial issues are not so important in cereals-based food products, due to some forms of thermal treatment applied, a vast amount of knowledge revealed that *Salmonella* may survive industrial processing in low-water activity foods (Margas 2016). *Salmonella* is one of the challenging microorganisms in animal feed, because many strains survive drying and require severe heat treatment for inactivation, therefore end-product testing for verifying microbial product safety is required (Margas 2016). The selection of proper processing not only reduces the risk of chemical and microbial contamination in the cereal-based product, but also makes it palatable and maintains the product quality, productivity, and contribution to sustainability (Margas 2019).

Another risk that may pose a serious hazard in the animal feed sector is a constant increase in the market share of transgenic (genetically modified, GM and genetically edited, GE) crops. The major GM crops, such as soybean, maize, cotton, canola, sugar beet, papaya, squash, eggplant, potatoes, and apples, are commercially available (I.S.A.A.A. 2017). Maize is the only cereal crop that is GM, which shockingly took over 32% of the global area used for maize crops (59.7 million hectares) (Giraldo et al. 2019). Most of these GM crops are mixed with feed and are fed to livestock, thus strong regulations are needed to ensure complete safety of livestock and humans. It requires a detailed feed safety examination as to whether genetic alteration could inadvertently alter the nutritional characteristics and increase the toxicity or allergenicity of the transgenic crop (Pauwels et al. 2015). The legislation for GM and GE crops and their products is subjected to rigorous evaluations as a part of several regulatory requirements and requires GM traceability and labeling system that enables tracking of transgenic feed products all along the supply chain (Giraldo et al. 2019). The method chosen to comply with traceability should be sensitive enough to detect the tolerance threshold level of the transgene(s) corresponding to jurisdiction (Ramessar et al. 2007). The majority of techniques for GM safety assessment for human food have been developed, but a specific methodology for the assessment of GM crops intended for animal feed is needed. This will provide a more accurate assessment to GM feed safety and will facilitate a more efficient use of resources and avoid needless feed safety risk (Giraldo et al. 2019).

9.5 Conclusion and Future Perspectives

Many cereal grains and their by-products are not directly usable as animal feed. In some cases, the nutritional value of certain grains/by-products is lower than required for animal feed, based on their individual nutrient contents while, in other cases, chemical and physical properties reduce the biological value and digestibility of raw materials. To produce the best value from cereal grain by-products, the characteristics of the raw material must be equated with the digestive requirements of the animal. Hence, grain processing is common practice of the feed industry to enable the optimal use of nutrients and enhance digestibility. Processing not only alters the chemical and physical structure of macronutrients such as starch, lipid, and protein, but also reduces the impact of ANFs such as NSP and lignin, which can have a negative effect on fermentation and intestinal digestion. Although the grain processing technologies have been known for centuries, recent applications of micronization have helped feed manufacturers to customize specific feed to manipulate feed digestion. However, a combined approach including selection of cereal grain by-products, understanding animal requirements, recognizing the importance of site of digestion, and by utilizing the combination of enzymatic, chemical methods and specific grain processing technologies, can open up new avenues for cereal by-products for animal feeding. To maintain product safety, many of the hazards presented by both the cereal grain and its products can be reduced by applying GMP, GAP, and HACCP principles, both at grain producers to grain processors level. However, a good knowledge of cereal grains, their supply chain, and any hazards that might be present, as well as resources to measure and enforce compliance, are required. Additionally, increasing cultivation of GM crops and their applications in animal feed may pose a challenge, thus requiring a new framework for their risk assessment. Finally, the use of cereal grain by-products in the food chain entails that these products should comply with quality and safety regulations without compromising food and feed safety. To achieve this, a lot of understanding in terms of regulations to facilitate the use of grain by-products in the food chain, as well as novel innovative methods to recover the cereals by-products, is required.

References

Adamczyk, B., Simon, J., Kitunen, V. et al. (2017). Tannins and their complex interaction with different organic nitrogen compounds and enzymes: old paradigms versus recent advances. *ChemistryOpen* 6 (5): 610–614.

Adedokun, S.A., Jaynes, P., Payne, R.L., and Applegate, T.J. (2015). Standardized ileal amino acid digestibility of corn, corn distillers' dried grains with solubles, wheat middlings, and bakery by-products in broilers and laying hens. *Poultry Science* 94 (10): 2480–2487.

Adegbeye, M.J., Reddy, P.R.K., Chilaka, C.A. et al. (2020). Mycotoxin toxicity and residue in animal products: prevalence, consumer exposure and reduction strategies-a review. *Toxicon* 177: 96–108.

Ahmadi, F., Ghorbani, G.R., Sadeghi-Sefidmazgi, A. et al. (2020). Performance and feeding behavior of dairy cows fed high-concentrate diets containing steam-flaked or ground corn varying in particle size. *Journal of Dairy Science* 103: 3191–3203.

Ahmed, R., Juniper, D., Tonks, A., and Rymer, C. (2018). The effect of incremental inclusion of whole grain wheat in the diet of growing turkeys on growth performance, feed conversion ratio, cecal health, and digesta characteristics. *Livestock Science* 214: 36–41.

Alldrick, A.J. (2014). Chemical safety of cereal-based foods: risk management considerations. *Quality Assurance and Safety of Crops & Foods* 6 (1): 3–14.

Ametaj, B.N., Zebeli, Q., Saleem, F. et al. (2010). Metabolomics reveals unhealthy alterations in rumen metabolism with an increased proportion of cereal grain in the diet of dairy cows. *Metabolomics* 6 (4): 583–594.

Anderson, V. and Schroeder, J.W. eds. (2009). Feeding barley to dairy cattle. Fargo, North Dakota: North Dakota State University Extension Service.

Armbruster, S. (2006). Steam flaking grains for feedlot cattle: a consultant's perspective. Proceedings of Cattle Grain Processing Symposium, Tulsa, Oklahoma, 46–55. Stillwater, OK, USA: Oklahoma Agricultural Experiment Station.

Black, J.L. (2016). Cereal grains as animal feed. In: *Encyclopedia of Food Grains*, 2e (eds. C. Wrigley, H. Corke, K. Seetharaman, et al.), 215–222. Academic Press.

Blasi, D.A., Kuhl, G.L., Drouillard, J.S. et al. (1998). Wheat middlings: Composition, feeding value, and storage guidelines. *Kansas State University Tech. Rep. MF-2353.*

Bolsen, K.K., Brent, B.E., and Pope, R.V. (2007). The ensiling process: Basic principles. https://www.progressiveforage.com/forage-types/silage/0307-fg-the-ensiling-process-basic-principles (accessed 31 October 2018).

Brouns, F., Hemery, Y., Price, R., and Anson, N.M. (2012). Wheat aleurone: separation, composition, health aspects, and potential food use. *Critical Reviews in Food Science and Nutrition* 52: 553–568.

Brown, M.S., Krehbiel, C.R., Duff, G.C. et al. (2000). Effect of degree of corn processing on urinary nitrogen composition, serum metabolite and insulin profiles, and performance by finishing steers. *Journal of Animal Science* 78: 2464–2474.

Brown, M.L., Schaeffer, T.W., Rosentrater, K.A. et al. (2016). Feeding DDGS to finfish. In: *Distillers Grains: Production, Properties, and Utilization* (eds. K. Liu and K.A. Rosentrater), 341–390. CRC Press.

Casas, G.A., Rodriguez, D.A., and Stein, H.H. (2018). Nutrient composition and digestibility of energy and nutrients in wheat middlings and red dog fed to growing pigs. *Journal of Animal Science* 96 (1): 215–224.

Cervantes-Pahm, S.K., Liu, Y., and Stein, H.H. (2014). Digestible indispensable amino acid score and digestible amino acids in eight cereal grains. *British Journal of Nutrition* 111 (9): 1663–1672.

Cheng, Z.J. and Hardy, R.W. (2002). Effect of microbial phytase on apparent nutrient digestibility of barley, canola meal, wheat and wheat middlings, measured in vivo using rainbow trout (*Oncorhynchus mykiss*). *Aquaculture Nutrition* 8 (4): 271–277.

Clarke, L.C., Sweeney, T., Curley, E. et al. (2018). Effect of β-glucanase and β-xylanase enzyme supplemented barley diets on nutrient digestibility, growth performance and expression of intestinal nutrient transporter genes in finisher pigs. *Animal Feed Science and Technology* 238: 98–110.

Codex Alimentarius (2003). Hazard Analysis and Critical Control Point (HACCP) system and guidelines for its application. CAC/RCP 1–1969, Rev 4. http://www.fao.org/input/download/standards/23/CXP_001e.pdf (accessed 20 March 2020).

Dalbhagat, C.G., Mahato, D.K., and Mishra, H.N. (2019). Effect of extrusion processing on physicochemical, functional and nutritional characteristics of rice and rice-based products: a review. *Trends in Food Science & Technology* 85: 226–240.

De Godoy, M.R., Kerr, K.R., and Fahey, G.C. Jr. (2013). Alternative dietary fiber sources in companion animal nutrition. *Nutrients* 5 (8): 3099–3117.

Dehghan-Banadaky, M., Corbett, R., and Oba, M. (2007). Effects of barley grain processing on productivity of cattle. *Animal Feed Science and Technology* 137 (1–2): 1–24.

Devi, P.B., Vijayabharathi, R., Sathyabama, S. et al. (2014). Health benefits of finger millet (*Eleusine coracana* L.) polyphenols and dietary fiber: a review. *Journal of Food Science and Technology* 51 (6): 1021–1040.

European Parliament and Council (2005). Regulation (EC) No 396/2005 of The European Parliament and of The Council of 23 February 2005 on maximum residue levels of pesticides in or on food and feed of plant and animal origin and amending Council Directive 91/414/EEC foodstuffs. *Official Journal of the European Union* L70: 1–16.

Evers, T. and Millar, S. (2002). Cereal grain structure and development: some implications for quality. *Journal of Cereal Science* 36 (3): 261–284.

FAO (2013) Anti-nutritional factors within feed ingredients. Aquaculture Feed and Fertilizer Resources Information System. http://www.fao.org/fishery/affris/feed-resources-database/anti-nutritional-factors-within-feed-ingredients/en (accessed 16 April 2020).

FAO (2016). Rice market monitor XIX(4). http://www.fao.org/fileadmin/templates/est/COMM_MARKETS_MONITORING/Rice/Images/RMM/RMM-Dec16_H.pdf (accessed 7 May 2018).

Giraldo, P., Shinozuka, H., Spangenberg, G. et al. (2019). Safety assessment of genetically modified feed: is there any difference from food? *Frontiers in Plant Science* 10: 1592.

Hansen, C.L. (2016). Using DDGS as a feedstock for bioenergy via anaerobic digestion. In: *Distillers Grains: Production, Properties, and Utilization* (eds. K. Liu and K.A. Rosentrater), 465–474. CRC Press.

Hasting, W.H. and Higgs, D. (1978). Feed milling processes. In: *Fish Feed Technology*. Rome, Italy: Food and Agriculture Organization of the United Nations http://www.fao.org/3/x5738e/x5738e0j.htm (accessed 31 October 2018).

He, J., Zhou, T., Young, J.C. et al. (2010). Chemical and biological transformations for detoxification of trichothecene mycotoxins in human and animal food chains: a review. *Trends in Food Science & Technology* 21 (2): 67–76.

Huang, Q., Shi, C.X., Su, Y.B. et al. (2014). Prediction of the digestible and metabolizable energy content of wheat milling by-products for growing pigs from chemical composition. *Animal Feed Science and Technology* 196: 107–116.

I.S.A.A.A. (International Service for the Acquisition of Agri-biotech Applications) (2017). Global status of commercialized biotech/GM crops in 2017: Biotech crop adoption surges as economic benefits accumulate in 22 years. https://www.isaaa.org/resources/publications/briefs/53/executivesummary/default.asp (accessed 20 April 2020).

Japan Times (2017). More Japanese livestock farmers using feed rice. *Japan Times* (27 November). www.japantimes.co.jp/news/2017/11/27/business/japanese-livestock-farmers-using-feed-rice/#.XpB4FPkzYr8 (accessed 10 March 2020).

Jaworski, N.W., Lærke, H.N., Bach Knudsen, K.E., and Stein, H.H. (2015). Carbohydrate composition and in vitro digestibility of dry matter and nonstarch polysaccharides in corn, sorghum, and wheat and coproducts from these grains. *Journal of Animal Science* 93 (3): 1103–1113.

Jiang, S., Gou, Z., Li, L. et al. (2018). Growth performance, carcass traits and meat quality of yellow-feathered broilers fed graded levels of alfalfa meal with or without wheat. *Animal Science Journal* 89 (3): 561–569.

Kraler, M., Schedle, K., Domig, K.J. et al. (2014). Effects of fermented and extruded wheat bran on total tract apparent digestibility of nutrients, minerals and energy in growing pigs. *Animal Feed Science and Technology* 197: 121–129.

Kumar, K., Kumar, S., Singh, T.P., and Sharma, M. (2019). Anti-nutritional factors in cereals and pulses and their management. Proceedings of the 2nd International Conference on Innovations in Chemical, Biological and Environmental Sciences (Science and Technology for a Sustainable Future), New Delhi, India (27–28 February 2019). Kolkata, West Bengal, India: The Indian Science Congress Association.

Kumari, N., Vinita, N.K., and Rani, P. (2018). Nutrient composition of full fat and defatted rice bran. *Asian Journal of Dairy and Food Research* 37 (1): 77–80.

Levic, J., and Sredanovic, S. (2010). Heat treatments in animal feed processing. Proceedings of the 2nd Workshop Feed-to-Food FP7 REGPOT-3. Extrusion technology in feed and food processing, Novi Sad, Serbia (19–21 October, 2010). Lazara, Novi Sad, Serbia: Institute for Food Technology.

Lv, M., Yan, L., Wang, Z. et al. (2015). Effects of feed form and feed particle size on growth performance, carcass characteristics and digestive tract development of broilers. *Animal Nutrition* 1 (3): 252–256.

Lyu, Z., Wang, L., Wu, Y., and Huang, C. (2020). Effects of particle size and lipid form of corn on energy and nutrient digestibility in diets for growing pigs. *Asian-Australasian Journal of Animal Sciences* 33 (2): 286.

Margas, E. (2016). The survival and control of Salmonella in low-aw foods environments. Doctoral thesis. The University of Nottingham, UK.

Margas, E. (2019). Re-assessment of food safety risks in cereals processing. *Miller Magazine*. https://www.millermagazine.com/english/re-assessment-of-food-safety-risks-in-cereals-processing/.html (accessed 20 April 2020).

Matsushima, J.K. (2006). History of feed processing. Proceedings of Cattle Grain Processing Symposium, Tulsa, Oklahoma, 1–16. Stillwater, OK, USA: Oklahoma Agricultural Experiment Station.

de Mello, A.S., Jenschke, B.E., Senaratne, L.S. et al. (2018). Effects of finishing diets containing wet distillers grains plus solubles on beef quality attributes and fatty acid profile. *Meat Science* 136: 16–22.

Miyaji, M., Matsuyama, H., Hosoda, K., and Nonaka, K. (2012). Effect of replacing corn with brown rice grain in a total mixed ration silage on milk production, ruminal fermentation and nitrogen balance in lactating dairy cows. *Animal Science Journal* 83 (8): 585–593.

Mohapatra, D., Patel, A.S., Kar, A. et al. (2019). Effect of different processing conditions on proximate composition, anti-oxidants, anti-nutrients and amino acid profile of grain sorghum. *Food Chemistry* 271: 129–135.

Moss, A.F., Khoddami, A., Chrystal, P.V. et al. (2020). Starch digestibility and energy utilisation of maize-and wheat-based diets is superior to sorghum-based diets in broiler chickens offered diets supplemented with phytase and xylanase. *Animal Feed Science and Technology* 264: 114475.

Nikkhah, A. (2012). Barley grain for ruminants: a global treasure or tragedy. *Journal of Animal Science and Biotechnology* 3 (1): 22.

Nikmaram, N., Leong, S.Y., Koubaa, M. et al. (2017). Effect of extrusion on the anti-nutritional factors of food products: an overview. *Food Control* 79: 62–73.

Nishii, M., Yasutomi, M., and Sone, Y. (2016). Effects of a whole-grain paddy rice diet on the pH distribution in the gizzard and retention time of digesta in the crop of broiler chicks. *The Journal of Poultry Science* 53 (3): 181–191.

Onipe, O.O., Jideani, A.I., and Beswa, D. (2015). Composition and functionality of wheat bran and its application in some cereal food products. *International Journal of Food Science & Technology* 50 (12): 2509–2518.

Ortiz-Cruz, R.A., Ramírez-Wong, B., Ledesma-Osuna, A.I. et al. (2020). Effect of extrusion processing conditions on the phenolic compound content and antioxidant capacity of sorghum (Sorghum bicolor [L.] Moench) bran. *Plant Foods for Human Nutrition* 75: 252–257.

Owusu-Asiedu, A., Simmins, P.H., Brufau, J. et al. (2010). Effect of xylanase and β-glucanase on growth performance and nutrient digestibility in piglets fed wheat–barley-based diets. *Livestock Science* 134 (1–3): 76–78.

Papademetriou, M.K. (2000). Rice production in the Asia-Pacific region: issues and perspectives. In: Bridging the rice yield gap in the Asia-Pacific region, 220. Bangkok, Thailand: Food and Agriculture Organization of the United Nations.

Papageorgiou, M. and Skendi, A. (2018). Introduction to cereal processing and by-products. In: *Sustainable Recovery and Reutilization of Cereal Processing By-Products* (ed. C. Galanakis), 1–25. Woodhead Publishing.

Parmenter, R.T., James, D.M., Townsend, H.E., and Rickard, J.W. (2018). Effectiveness of ensiling wet brewer's grains. *Journal of Animal Science* 96: 238–239.

Pauwels, K., De Keersmaecker, S.C., De Schrijver, A. et al. (2015). Next-generation sequencing as a tool for the molecular characterisation and risk assessment of genetically modified plants: added value or not? *Trends in Food Science & Technology* 45 (2): 319–326.

Pieper, R., Hackl, W., Korn, U. et al. (2011). Effect of ensiling triticale, barley and wheat grains at different moisture content and addition of lactobacillus plantarum (DSMZ 8866 and 8862) on fermentation characteristics and nutrient digestibility in pigs. *Animal Feed Science and Technology* 164 (1–2): 96–105.

Qiao, F.Q., Wang, F., Ren, L.P. et al. (2015). Effect of steam-flaking on chemical compositions, starch gelatinization, in vitro fermentability, and energetic values of maize, wheat and rice. *Journal of Integrative Agriculture* 14 (5): 949–955.

Radosavljević, M., Jovanović, R., Pajić, Z. et al. (2010). The development of a new assortment of food and feed by the application of micronisation and extrusion. Proceedings of the 2nd Workshop Feed-to-Food FP7 REGPOT-3. Extrusion technology in feed and food processing, Novi Sad, Serbia (19–21 October, 2010), 168–180. Lazara, Novi Sad, Serbia: Institute for Food Technology.

Rahman, M.A.U., Rehman, A., Chuanqi, X. et al. (2015). Extrusion of feed/feed ingredients and its effect on digestibility and performance of poultry: a review. *International Journal of Current Microbiology and Applied Sciences* 4 (4): 48–61.

Ramessar, K., Peremarti, A., Gómez-Galera, S. et al. (2007). Biosafety and risk assessment framework for selectable marker genes in transgenic crop plants: a case of the science not supporting the politics. *Transgenic Research* 16 (3): 261–280.

Rapid Alert System for Food and Feed (2009). http://ec.europa.eu/food/food/rapidalert/index_en.htm (accessed 20 April 2020).

Reis, B., Marques, A., Campos, I. et al. (2019). Wheat germ as an alternative ingredient to a fair average quality fishmeal in diets for European seabass. *Aquaculture Nutrition* 25 (4): 932–945.

Rodgers, N., Iji, P.A., Mikkelsen, L.L. et al. (2009). Effect of grain particle size and milling method on broiler performance and apparent metabolisable energy. Proceedings of the 20th Australian Poultry Science Symposium, 133–136. Sydney: Poultry Research Foundation.

Rokey, G.J., Plattner, B., and Souza, E.M.D. (2010). Feed extrusion process description. *Revista Brasileira de Zootecnia* 39: 510–518.

Rosenfelder, P., Eklund, M., and Mosenthin, R. (2013). Nutritive value of wheat and wheat by-products in pig nutrition: a review. *Animal Feed Science and Technology* 185: 107–125.

Rosentrater, K.A. and Evers, A.D. (2018). *Kent's Technology of Cereals*, 4e. Woodhead Publishing.

Rowe, J.B., Choct, M., and Pethick, D.W. (1999). Processing cereal grains for animal feeding. *Australian Journal of Agricultural Research* 50 (5): 721–736.

Ryu, D., Kowalski, R.J., Ganjyal, G., and Lee, H.J. (2019). Reduction of ochratoxin a in oats and rice by twin-screw extrusion processing with baking soda. *Food Control* 105: 21–28.

Saldivar, S. (2003). Cereals: dietary importance. In: *Encyclopedia of Food Sciences and Nutrition* (eds. B. Caballero, L. Trugo and P. Finglas), 1027–1033. Academic Press.

Saylor, B.A., Casale, F., Sultana, H., and Ferraretto, L.F. (2020). Effect of microbial inoculation and particle size on fermentation profile, aerobic stability, and ruminal *in situ* starch degradation of high-moisture corn ensiled for a short period. *Journal of Dairy Science* 103 (1): 379–395.

Scanlon, M.G., Cenkowski, S., Segall, K.I., and Arntfield, S.D. (2005). The physical properties of micronised lentils as a function of tempering moisture. *Biosystems Engineering* 92 (2): 247–254.

Schaarschmidt, S. and Fauhl-Hassek, C. (2018). The fate of mycotoxins during the processing of wheat for human consumption. *Comprehensive Reviews in Food Science and Food Safety* 17 (3): 556–593.

Schneider, L.G., Klopfenstein, T.J., Stromberg, Z.R. et al. (2018). A randomized controlled trial to evaluate the effects of dietary fibre from distillers grains on enterohemorrhagic *Escherichia coli* detection from the rectoanal mucosa and hides of feedlot steers. *Zoonoses and Public Health* 65 (1): 124–133.

Selle, P.H., Truong, H.H., Khoddami, A. et al. (2019). The impacts of hammer-mill screen size and grain particle size on the performance of broiler chickens offered diets based on two red sorghum varieties. *British Poultry Science* 60 (3): 209–218.

Serna-Saldivar, S.O. (2016). *Cereal Grains: Properties, Processing, and Nutritional Attributes*. CRC press.

da Silva, J.C., Cole, N.A., Ponce, C.H. et al. (2019). Effects of supplemental fat concentration on feeding logistics, animal performance, and nutrient losses of heifers fed finishing diets based on steam-flaked corn and sorghum-based distiller's grains. *Journal of Animal Science* 97 (6): 2583–2597.

Singh, A., Das, M., Bal, S., and Boinejee, R. (2013). Rice processing. In: *Engineering Aspects of Cereals and Cereal Based Products* (eds. R.D.P.F. Guine and P.M. dos Reis Correia), 71–97. Boca Raton: CRC Press.

Singh, Y., Amerah, A.M., and Ravindran, V. (2014). Whole grain feeding: methodologies and effects on performance, digestive tract development and nutrient utilisation of poultry. *Animal Feed Science and Technology* 190: 1–18.

Sinha, A.K., Kumar, V., Makkar, H.P. et al. (2011). Non-starch polysaccharides and their role in fish nutrition–a review. *Food Chemistry* 127 (4): 1409–1426.

Sittiya, J., Yamauchi, K.E., and Morokuma, M. (2011). Chemical composition, digestibility of crude fiber and gross energy, and metabolizable energy of whole paddy rice of Momiroman. *The Journal of Poultry Science* 48 (4): 259–261.

Sobowale, S.S. (2018). Design, construction and performance evaluation of a twin screw extruder. *Agricultural Engineering International: CIGR Journal* 19 (4): 181–186.

Solanas, E., Castrillo, C., and Calsamiglia, S. (2007). Effect of extruding the cereal and/or the legume protein supplement of a compound feed on in vitro ruminal nutrient digestion and nitrogen metabolism. *Journal of Animal Physiology and Animal Nutrition* 91 (5–6): 269–277.

Sotak, K.M., Goodband, R.D., Tokach, M.D. et al. (2011). The effects of sorghum dried distillers grains with solubles on nursery pig performance. Proceedings of Kansas Swine Day 2011 (17 November 2011), 118–128. Manhattan, KS, USA: Kansas State University.

Sotak, K.M., Houser, T.A., Goodband, R.D. et al. (2015). The effects of feeding sorghum dried distillers grains with solubles on finishing pig growth performance, carcass characteristics, and fat quality. *Journal of Animal Science* 93 (6): 2904–2915.

Statista (2019). World production volume of milled rice from 2008/2009 to 2018/2019. https://www.statista.com/statistics/271972/world-husked-rice-production-volume-since-2008 (accessed 23 March 2020).

Stein, H.H., Lagos, L.V., and Casas, G.A. (2016). Nutritional value of feed ingredients of plant origin fed to pigs. *Animal Feed Science and Technology* 218: 33–69.

Strabler, R., Crisler, A.J., Depenbusch, B.E., and Drouillard, J. S. (2009). Extruded complete feed for finishing cattle. Kansas Agricultural Experiment Station Research Reports 1: 54–61.

Svihus, B. (2001). Norwegian poultry industry converts to whole grain pellets. *World Politics* 17: 20–21.

Tandiang, D.M., Diop, M.T., Dieng, A. et al. (2014). Effect of corn substitution by sorghum grain with low tannin content on broilers production: animal performance, nutrient digestibility and carcass characteristics. *International Journal of Poultry Science* 13 (10): 568.

Thomas, D.J. and Atwell, W.A. (1999). Starch structure. In: *Starches: Practical Guide for the Food Industry* (eds. D.J. Thomas and W.A. Atwell), 1–12. St. Paul: Eagan Press.

Tosta, M.R., Prates, L.L., Christensen, D.A., and Yu, P. (2020). Effect of processing methods (rolling, steam-flaking, pelleting) on protein molecular structure profile, rumen degradation, and intestinal digestion of cool-climate adapted oats grain in comparison with barley grain in western Canada. *Livestock Science* 232: 103901.

Vasconcelos, J.T. and Galyean, M.L. (2007). Nutritional recommendations of feedlot consulting nutritionists: the 2007 Texas Tech University survey. *Journal of Animal Science* 85: 2772–2781.

Vukmirović, Đ., Čolović, R., Rakita, S. et al. (2017). Importance of feed structure (particle size) and feed form (mash vs. pellets) in pig nutrition–a review. *Animal Feed Science and Technology* 233: 133–144.

Wang, Z.Y., Yang, H.M., Lu, J. et al. (2014). Influence of whole hulled rice and rice husk feeding on the performance, carcass yield and digestive tract development of geese. *Animal Feed Science and Technology* 194: 99–105.

Wang, C., Liu, Q., Guo, G. et al. (2018). Effects of substituting corn with steam-flaked sorghum on growth, digestion and blood metabolites in young cattle fed feedlot diets. *Animal Production Science* 58 (2): 299–306.

Wilson, B.K., Holland, B.P., Step, D.L. et al. (2016). Feeding wet distillers grains plus solubles with and without a direct-fed microbial to determine performance, carcass characteristics, and fecal shedding of Escherichia coli O157: H7 in feedlot heifers. *Journal of Animal Science* 94 (1): 297–305.

Womac, A.R., Igathinathane, C., Bitra, P. et al. (2007). Biomass pre-processing size reduction with instrumented mills. Proceedings of 2007 ASAE Annual Meeting, Minneapolis, Minnesota (17–20 June 2007). St. Joseph, MI, USA: American Society of Agricultural and Biological Engineers.

Wrigley, C. (2017). The cereal grains: providing our food, feed and fuel needs. In: *Cereal Grains* (eds. C. Wrigley, I. Batey and D. Miskelly), 27–40. Woodhead Publishing.

Xu, E., Campanella, O.H., Ye, X. et al. (2020). Advances in conversion of natural biopolymers: a reactive extrusion (REX)–enzyme-combined strategy for starch/protein-based food processing. *Trends in Food Science & Technology* 99: 167–180.

Zhang, Z., Tun, H.M., Li, R. et al. (2018). Impact of xylanases on gut microbiota of growing pigs fed corn-or wheat-based diets. *Animal Nutrition* 4 (4): 339–350.

Zhou, P., Theil, P.K., Wu, D., and Knudsen, K.E.B. (2018). In vitro digestion methods to characterize the physicochemical properties of diets varying in dietary fibre source and content. *Animal Feed Science and Technology* 235: 87–96.

10
The Consumption of Healthy Grains: Product, Health, and Wellness Trends

Catherine Barry-Ryan[1], Marco Vassallo[2], and Milica Pojić[3]

[1] *School of Food Science and Environmental Health, Technological University Dublin, Dublin, Ireland*
[2] *CREA – Research Centre for Agricultural Policies and Bio-Economy, Rome, Italy*
[3] *Institute of Food Technology, University of Novi Sad, Novi Sad, Serbia*

10.1 Introduction

Recent years have been marked by changing consumer requirements, needs, and expectations from the food industry, requiring not only to satisfy hunger and obtain necessary nutrients from a food product, but also to improve health, prevent nutrition-related diseases, and enhance physical and mental well-being (Foschia et al. 2013). With increased consumers' consciousness of health and well-being, their consciousness of sustainability of food processing also rose, which has created opportunities for the food industry to develop innovative food products. The consumer acceptance of novel and innovative food alternatives is a complex and challenging issue dependent not only on personal sensory preferences, but also on social and cultural backgrounds (de Beukelaar et al. 2019; Tuorila and Hartmann 2020). The general acceptance of a healthier diet that involves plant-based products and the consumption of more plant proteins are mostly favorable for grain and cereal food categories (Costa and Johnson 2019).

The pattern of cereal consumption has been dominated by several trends: a shift from the traditional major staple cereals to the minor, ancient, and pseudocereals, increased requirements for whole or minimally processed grains and cereals, and increased interest in *free-from* foods. The global free-from food market is expected to increase at a compound annual growth rate of 5% from 2018 through 2023, with dominant gluten-, lactose-, and allergen-free food products (Costa and Johnson 2019). The listed trends and consumer interest in healthy cereal products resulted in a growing number of

Innovative Processing Technologies for Healthy Grains, First Edition.
Edited by Milica Pojić and Uma Tiwari.

product launches and increasing number of nutrition and health statements – claims on packaged products to help the marketing of these products (Van der Kamp 2008).

10.2 Benefits of Wholegrain Consumption and Consumers

The overall health benefits of grains, especially whole grains, have been univocally recognized as *healthy* due to their potential to alter the body composition, fecal microbiota, fasting blood glucose, total cholesterol, high density lipoprotein (HDL), low density lipoprotein (LDL), and triglycerides and are therefore recommended for a healthy diet (Cooper et al. 2017; WHO 2018). Consumers have no doubts that grains and grain-based products are recommended for preserving health (Van der Kamp and Lupton 2013), and therefore their attitudes toward buying/consuming these goods are generally positive (Shepherd et al. 2012). Furthermore, this attitude contributes to their intention to buy and consume food (Demartini et al. 2019) and it has been widely modeled in food choice research (Vassallo et al. 2016). Therefore, it is reasonably expected that a positive attitude toward healthy grains would lead the consumers to increase their consumption. On the other hand, the rate of consumption of these foods, especially whole-grain and fiber-enriched products, does not seem to be so enthusiastic worldwide (Cooper et al. 2017; McGill and Devareddy 2015; Schaffer-Lequart et al. 2017; Shepherd et al. 2012). The reason for this contradiction between general positive attitude and actual consumption of healthy grain products needs to be investigated.

10.3 Consumers' Attitudes Toward Behavior

What does an attitude toward *healthy* food, like grains, really reflect and what are product stimuli composed of? To answer these two questions, it is important to introduce the concept of an attitude toward a behavior. An attitude toward a behavior is a psychological construct that essentially reflects "a person's judgment that performing the behavior is good or bad, that he is in favor of or against performing behavior" (Fishbein and Ajzen 1980). This is a direct measure of the attitude, a self-reported explicit measure that constitutes a valuation response, a judgment, based on an instrumental process (e.g. giving a bad or good, harmful or beneficial evaluation) and/or an experiential aspect (e.g. giving an unpleasant or pleasant evaluation) (Ajzen 2019). These direct measures are also known as reflective measures because they reflect the consumers' instrumental and experiential conscious judgment. The common correlations among the reflective measures give rise to attitude variability that should, in turn, be explained by the behavioral beliefs the consumers gain on a possibly direct experience of consuming the products. These further measures are so-called formative measures of the attitude because they form an attitude by explaining variability in the consumers' judgments. This attitude process, adapted from Ajzen's (2019) theories, and formally expressed as a latent variable model[1] about consuming healthy grains foods, is shown in Figure 10.1. Within the boxes, the measures are depicted (i.e. they are usually items scored on the 5- or 7-point Likert

[1]The estimation procedure is developed under the Structural Equation Modeling (SEM) statistical technique with Multiple Indicators Multiple Causes (MIMIC)-type models (Bollen and Davis 2009).

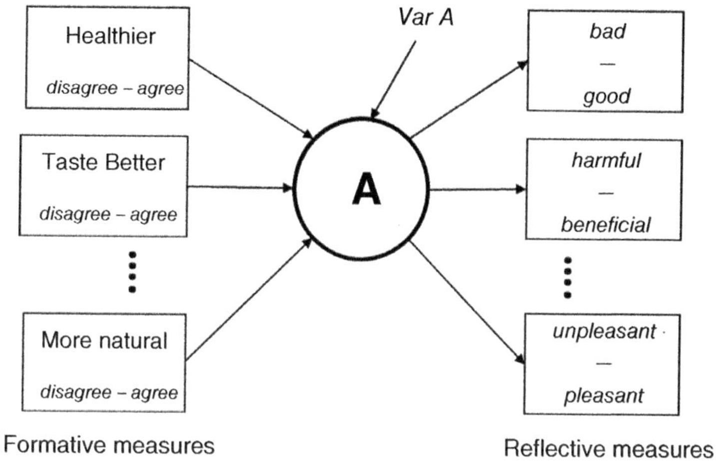

Figure 10.1 Conceptual model of the attitude toward healthy grain food. *Source*: adapted from Ajzen (2019).

scale), while the attitude construct with its variability or variance (i.e. var. A), estimated by the common correlations of the reflective measures and explained by each formative measure, is depicted in the circle. The arrows indicate the direction of the causations. In practical words, these behavioral beliefs explain why an attitude toward consuming healthy food is good, beneficial, and pleasant (or the contrary). This means that the attitude might have a positive/negative overall judgment evoked by stimuli or consumers' beliefs, on direct experience of consumption. The causations between the behavioral beliefs and the reflective judgments provide a validation to the attitude construct.

Understanding this mechanism allows the reader to catch what presumably happens in the consumers' mind while their attitude takes shape. However, all the possible behavioral beliefs should be found positively/negatively oriented to explain a subsequent positive/negative attitudinal view, and thus validating the attitude construct. For example, if consumers positively judge healthy grain products as good, beneficial, and pleasant, they presumably agree that healthy grain products taste better than refined grain products and agree that they are even healthier and more natural. However, the problem is that not all the beliefs gain a positive score, but possibly tenuous scores or indifferent or negative scores are given alternatively. This makes it difficult to explain the common variance A in the reflective judgments, even if they are apparently positively oriented, but can possibly be explained by other unknown causes other than the beliefs. This situation might, in turn, sequentially attenuate the attitude effect on the intention to consume and thus influence the actual consumption. It is also expected that factors other than the attitude can influence the actual consumption of healthy grain products like, for instance, availability in the shops and price barriers, but if a consumer has no doubts and/or any misleading or incomplete information about the health benefits, the taste, the correctness on what these benefits are, etc., the aforementioned barriers will become affordable. So then, having clear and correct information about all the properties of the healthy grain products, and thereby having a satisfied direct experience as well, leads to a validated attitude toward consuming these foods that, in turn, smooths over differences about supply and commercial barriers.

10.4 Consumers' Attitudes Toward Consumption of Healthy Grains

Nowadays, the consumers' attitude toward *healthy* grains is struggling with two obstacles: (i) the consumers' subjective knowledge of the health properties of the grain-based products, which is heterogenous, driven by health interests, eating disorders, and/or even specific diseases often requiring health, food, and nutrition literacy (Dean et al. 2012; Velardo 2015); and (ii) the consumers perception of their eating behavior, being perceived more or less healthy. Consumers often feel the need to adhere to a *healthy* diet, although this is often driven by a hedonistic choice rather than recommended by guidelines. All this generates indeciveness in consumption of *healthy* food like grains, even though the attitudes are essentially optimistic (de Ridder et al. 2013).

Regarding the first obstacle, although the official foods and nutrients recommendation guidelines are very similar from one country to another, the ways these recommendations reach the consumers are inconsistent (FAO 2016). The ways of recommending nutritional food quantities to follow a healthy and balanced diet differ from country to country by different means (e.g. visual food guides, different content and amount health information, targeting to specific groups), causing confusion to consumers and making it difficult for them to implement these recommendations into their daily diets (de Ridder et al. 2013, 2017). Moreover, the current digital era is contributing to the worsening of this communication deficiency, because the consumers are literally overloaded by internet food-oriented programs and social media users' opinions (de Ridder et al. 2017). As a consequence, the consumers are not able to distinguish the food scientific-oriented recommendations from all the others (de Ridder et al. 2017). As a result, the healthy food knowledge itself is corrupted and ambiguous or even over-interpreted or mis-interpreted, when consumers attempt to put into practice healthy food choices. All this encompasses *healthy* grain products, both as staple foods like breads, breakfast cereals, and more hedonistic typology like biscuits.

Concerning the second obstacle, although strongly related to the first, the circumstances are more complex because today it seems that many people are affected by the syndrome of the "self-healthy person." In most cases these people are not following either a healthy diet or a healthy lifestyle. These consumers sporadically read some information about the health properties of food spread by magazines, TV channels, and internet portals and adapt their diet to this mass of information for the purposes of a clear conscience about following a healthy diet, while they are essentially only satisfying their hedonism. In order to address these unreasonable situations, research is focusing on two approaches: the self-referencing (SR) task and food labeling. Both strategies can significantly contribute to stabilize attitudes toward consuming healthy food, like grains, in the consumer's mind and thus strengthen the consensus postulated in the model given in Figure 10.1.

10.4.1 The Role of Self-Referencing Task in Food Choice

The self-referencing (SR) task is a new and interesting concept of food choice, relevant to healthy food behavior, even though specific application of the SR task to healthy grain products is still lacking (Demartini et al. 2019; Mattavelli et al. 2017). The SR consists in an association between the self and the object (product), based on the innate principle of positive self-esteem of the person/consumer. By stimulating the self with mental representations of the behavior toward that product and its evaluation, it is possible to measure the associations that reflect unconscious attitudes, or implicit attitudes of the consumers (Demartini et al. 2019; Gawronski et al. 2006). Demartini et al. (2019) argues that the aforementioned information provided by food labels to consumers, or by other channels (that, on the contrary, refers to explicit conscious attitudes of the consumers), did not seem the most effective solution on forming individual attitudes and understanding choice behaviors. This is a strong as well as challenging assumption that can open large advances in robustness to the attitude construct. On the other hand, stressing too much on the self may also fortify the consumers' convictions of following a healthy diet in a self-sufficient way that might distance the consumers themselves from certain healthy recommendations. Since research stipulates that both explicit and implicit attitudes predict dietary behavior, it sounds wise to respectively apply both approaches – the food labeling (that will be discussed next) and the SR task to the mutual objective of the attitude construct validation represented in Figure 10.1 (McEachan et al. 2011; Prestwich et al. 2011). In this respect, it might be interesting to set up attitude models that are more consciously oriented and models more unconsciously oriented, or a mixture of both. More research on this novel ground is still needed.

10.4.2 The Role of Food Labeling and Nutrition and Health Claims in Food Choice

Placing food labels on products is the most common strategy applied by the food industry and encouraged by policy makers to provide mandatory information and/or more specific characteristics of the food product. Mandatory labels provide information on the name of the food, ingredients, the presence of ingredient(s) that may cause allergies or intolerances, the net quantity of the food, shelf life, storage conditions, the name of producer, the country of origin, the instructions for use if relevant, and a nutrition declaration (Berryman 2015). The traditional nutrition information in table form (Figure 10.2) can be supplemented by further information on nutrients and vitamins or simplified information that can appear on the front-of-pack, helping consumers to choose products when shopping. Although the food label approach sounds easy and workable, it has been found that many consumers do not pay much attention to

Nutrient	Per 100 g/ml	Per serving
Energy (kJ/kcal)		
Protein (g)		
Carbohydrates (g)		
of which sugars (g)		
Fat (g)		
of which saturates (g)		
Salt (g)		
Fiber		

→ Optionally

Figure 10.2 Typical nutritional label of a cereal-based product.

the information provided on the food labels (Guiné et al. 2016), having attenuated attitudes toward food labeling. Moreover, it has been found that consumers pay less attention to salt and fiber content listed on the label, than to the content of sugar and fat (Grunert et al. 2010).

Nutrition claims inform the consumers about the general benefits of the product, whereas the health claims (of so-called functional foods) inform the consumers of the prevention of a specific disease, thus allowing the consumers to gain more specific health information. Even if the consumers' food health literary is not high, these nutrition claims might help them with acceptance and understanding of some nutritional and health properties, therefore improving their nutrient intakes by making healthier food choices (Kim et al. 2016). Nutrition and health claims for products are controlled by EU legislation (Regulation (EC) No. 1924/2006, which states that claims should be based on scientific research and approved by the EU Commission. This legislation offered a legal framework for food producers to communicate with consumers and emphasize beneficial nutritional and health properties of a product (De Morpurgo and Botana 2016).

Cereal and cereal-based products abundantly contain ingredients with positive nutrition and health benefits. Cereal bran (5–15% of the grain, depending on its anatomy), contains the majority of non-digestible cell wall polymers (dietary fiber, DF) consisting mainly of cellulose, β-glucans, and pentosans (i.e. arabinoxylans). The bran layer is a source of different micronutrients such as vitamins (especially B vitamins), minerals, and phytochemicals, especially different phenolic compounds, and contains 40–70% of the grain's minerals (Cao 2019). The endosperm (80–85% of the grain), contains mostly starch and has the lowest protein and lipid content, so is a poor source of vitamins and minerals. The germ (up to 12% of the grain) is rich in the B-group vitamins, proteins, minerals such as potassium and phosphorous, unsaturated fats, antioxidants, and phytochemicals. Cereals are rich in glutamic acid, proline, leucine, and aspartic acid, but they are deficient in lysine. The amino acid content is mainly concentrated in the germ. Cereal grains are the most important dietary energy source worldwide, with wheat, rice, and maize providing over 50%, and with nutrient profiles that may allow claims for products with desirable nutritional characteristics (Rosell 2019). Although grains are nutritionally dense, the cereal processors have been challenged to substantially raise the level of fiber, vitamins, and minerals, both in white flour-based products and wholemeal-based products, to allow them to communicate the healthiness of bread and pasta to consumers, as well as to, for example, reduce the sodium salt content in bread (Van der Kamp 2008). Health claims for cereal-based products include normal bowel function, maintenance of cholesterol levels, and reduction of blood glucose rise after food consumption (Table 10.1).

Studies on health claims show that the basic product, the type of claim, and the functional ingredient used may affect consumers' perceptions of health benefits and their willingness to buy products with health claims. Foods with health claims, especially with risk reduction claims, imply that eating these products can influence a specific and relatively well-defined physiological function or related health factors. The interest in nutritionally healthy eating of cereal-based products is associated with the appeal of health messages, such as "contains whole grain" (Saba et al. 2010). On the other hand, some studies stipulate that consumers do not seem to have univocal views on food with such claims and thus their attitudes, understanding, and purchasing seem to be ambivalent (Hieke and Grunert 2018). It was shown that consumers do not always understand health and nutrition claims on foods. Moreover, they are

Table 10.1 Cereal-related health claims approved by EFSA (Annex 1, 2006).

Cereal-based material	Claimed health effect	Condition of claim use
Rye fiber	*Contributes to normal bowel function*	Applicable for food high in rye fiber as referred to in the claim HIGH FIBER[a]
Barley grain fiber	*Contributes to an increase in fecal bulk*	Applicable for food high in barley fiber as referred to in the claim HIGH FIBER[a]
Barley beta-glucans	*Reduction of blood cholesterol*	The beneficial effect is obtained with a daily intake of 3 g of barley beta-glucan. Applicable for food providing at least 1 g of barley beta-glucan per quantified portion.
Oat grain fiber	*Contributes to an increase in fecal bulk*	For food high in oat fiber as referred to in the claim HIGH FIBER[a]
Oat beta-glucans	*Reduction of blood cholesterol*	The beneficial effect is obtained with a daily intake of 3 g of oat beta-glucan. Applicable for food providing at least 1 g of oat beta-glucan per quantified portion.
Beta glucans from oats and barley	*Reduction of the blood glucose rise after meal*	The beneficial effect is obtained by consuming the beta-glucans from oats or barley as part of the meal. Applicable for food which contains at least 4 g of beta-glucans from oats or barley for each 30 g of available carbohydrates in a quantified portion as part of the meal.
Beta glucans	*Contributes to maintenance of normal blood cholesterol levels*	The beneficial effect is obtained with a daily intake of 3 g of beta-glucans from oats, oat bran, barley, barley bran, or their mixtures. Applicable for food which contains at least 1 g of beta-glucans from oats, oat bran, barley, barley bran, or their mixtures.
Wheat bran fiber	*Contributes to an acceleration of intestinal transit*	The beneficial effect is obtained with a daily intake of at least 10 g of wheat bran fiber. Applicable for food high in wheat bran fiber as referred to in the claim HIGH FIBER[a].
	Contributes to an increase in fecal bulk	Applicable for food high in wheat bran fiber as referred to in the claim HIGH FIBER[a].
Arabinoxylan from wheat endosperm	*Reduction of the blood glucose rise after meal*	Applicable for food which contains at least 8 g of arabinoxylan (AX)-rich fiber produced from wheat endosperm (at least 60% AX by weight) per 100 g of available carbohydrates in a quantified portion as part of the meal.
Resistant starch	*Contributes to maintenance of normal blood cholesterol levels*	Applicable for food in which digestible starch has been replaced by resistant starch so that the final content of resistant starch is at least 14% of total starch.

(*Source*: based on European Parliament and Council 2010; European Parliament and Council 2006)
[a] HIGH FIBER: Applicable for food containing at least 6 g of fiber per 100 g or at least 3 g of fiber per 100 kcal.

moderately doubtful about them, depending on the socio-cultural conditions and the overall trust in the food system and control (Klopčič et al. 2019).

In the case of grain products, consumers seem to react positively in terms of attitude toward both types of claims whenever they appear on these products, especially for staple grain products, even though the healthiness perception and likelihood of buying seem to be country dependent (Shepherd et al. 2012). This suggests that consumers' consensus on the healthfulness of food with claims has not been reached yet (Hieke and Grunert 2018). However, it is expected that the attitudes toward nutrition and specific health claims of those consumers who are more health motivated and/or affected by some disease, tend to be more positive and relevant (Dean et al. 2012). Furthermore, the perceived health benefits and sensory pleasantness have been found to be important for consumers' intentions to buy bread with risk-reduction claims (e.g. preventing heart coronary heart disease), even though they do not think themselves to be at the risk of contracting that disease (Vassallo et al. 2009). Even consumers with health issues still form their attitude toward grain foods with health claims on the basis of a pleasant taste (Siegrist et al. 2013; Verbeke 2006).

Therefore, the importance of information on nutrient content should be examined in order to find out which attribute is the most convincing for consumers when their purchase decisions are being made (Asioli et al. 2017). A Swedish-based survey on over a thousand cereal-product consumers found that three-quarters considered bread healthy when described using terms such as "coarse," "whole grain," "fiber rich," "sourdough," "crisp," "less sugar," "dark," "rye," "seeds," "a commercial brand," "homemade," and "kernels." These breads were perceived as having health benefits because they "contain fiber," are "good for the stomach," have good "satiation," and beneficial "glycemic properties" (Sandvik et al. 2018).

One of the drawbacks of food with health claims is that it may convince consumers to consider foods claimed to have potential health benefits as a medicine. This critical issue was widely studied within a European project (i.e. FP6 HEALTHGRAIN project) in the case of cereal-based products with health claims. This construct was initially defined as an attitude toward functional cereal foods "as tools to repair flaws in healthiness of the diet" (Dean et al. 2012). Hence, the latent construct was named "attitude towards using food as a medicine" (AFM) and measured by four items rated on a 7-point scale (1 = "strongly disagree;" 7 = "strongly agree"). Three of the four items were selected, adapting them from past works on "reward from using functional foods" (Urala and Lähteenmäki 2007): (i) "I can prevent diseases by regularly eating foods with health claims;" (ii) "Foods with health claims can repair the damage caused by an unhealthy diet;" and (iii) "Foods with health claims make it easier to follow a healthy lifestyle." On the other hand, the fourth item was completely thought out within the project to emphasize a possibly prevention of certain diseases with the eating of functional products, but with "help me" instead of "prevent:" "Eating foods with health claims will help me not to get some diseases." Psychometric properties of this AFM were cross-culturally explored across four European countries and the results highlighted that cultural differences mainly exist with the first two items, with a country-wise consensus of invariance on the third and fourth items (Vassallo 2013). This outcome confirmed an attitude toward healthy food (like grain-based food) with health claims – it is not perceived as a medicine and a way to cure health issues, but more likely as a means to follow a healthy lifestyle. Essentially, by more confident utilization of functional, healthy grain-food products and by association of these products with conventional healthy foods rather than medicine, an attitude toward healthy grain foods with claims as an ordinary way of healthy eating behavior will be created (Urala and Lähteenmäki 2007). On the other hand, Hieke and Grunert (2018) emphasized consumers' lack of concern about health claims, being irrelevant to them not only due to health motives but also to the specific health benefit addressed by the claim. In line with this, familiarity and direct experience with a food product is also a very important aspect of consumers' perception of safety and health (Siipi 2013). This is especially true for cereal-based products perceived as natural foods due to their origin. Therefore, the consumers may want accurate information (e.g. nutritional and health claims) about the potential benefits of the grain consumption. Labels on healthy grain food need to be as clear and effective as possible, as well as simple and direct. A good example might be the clean labels that will be discussed next.

10.5 Clean-Label Trend in Grain Products

Consumers are increasingly interested in both health and sustainability impacts on their life (Aschemann-Witzel et al. 2019), specifically in their diet. They are driven by factors such as modern health worries and concerns over processing, perceived risk and skepticism for certain ingredients, processing techniques, and also lack of trust in regulations. They demand foods which are more natural and organic (Janssen 2018), are less processed and *free-from* ingredients, and more plant-based food products (Hemmerling et al. 2016) (Figure 10.3).

There is no commonly accepted definition of a "clean-label" product (Asioli et al. 2017), but they are typically understood as products which consumers prefer due to the absence of negatively perceived ingredients. These absent ingredients can be allergenic ingredients (Venkataratnam et al. 2018), additives (Massini et al. 2016), processed ingredients, or those perceived as unfamiliar and chemical-sounding. Therefore, clean labels will have ingredients perceived as natural, not artificial and harmless, well-known to the consumers, and perceived as "kitchen cupboard ingredients" (Asioli et al. 2017). Food producers have responded by altering their ingredient lists in order to move closer to the idea of "clean label." This trend also triggers consumers to turn to products such as certified organic foods (Gilsenan et al. 2012). Naturalness in food is sought because of associations with more traditional and "authentic" processing, leading to assumptions about favorable health effects (Amos et al. 2014). In order to enable the accurate measurements of the degree of food naturalness, Sanchez-Siles et al. (2019) proposed the concept of the Food Naturalness Index (FNI), which integrated and was built on insights from consumer research and legal and technical perspectives. The proposed FNI was comprised of four component measures, such as farming practices, *free-from* additives, *free-from* unexpected ingredients, and degree of processing, which all included 10 relevant food naturalness attributes that can be evaluated from information given on the product label. The proposed concept is of high relevance for the reformulation of the existing and the development of new products, as well as understanding, tracking, and communicating food naturalness attributes in the market.

However, there is limited research on consumer categorization of food ingredients, even though food producers have assumptions about how consumers interpret the ingredient lists. The diversity of ingredients, food categories, trends and motives, consumers'

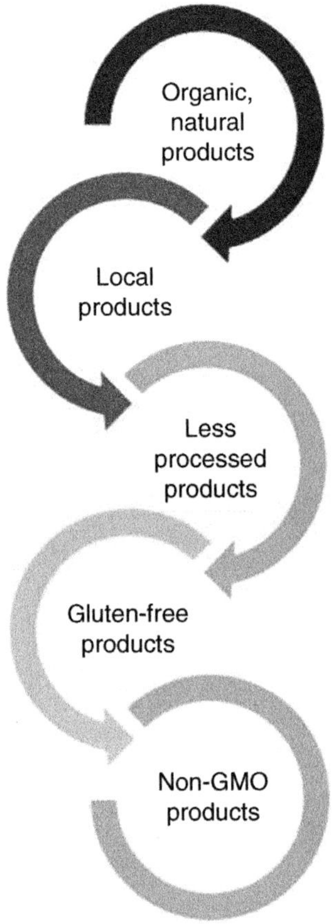

Figure 10.3 Trends in the cereal market and consumption.

categorization behavior, and ingredient perception all impact the market entry of new ingredients or the launch of new clean-label foods (Aschemann-Witzel et al. 2019).

Protein ingredients have been suggested as a potential "clean-label" ingredient and are investigated in relation to consumer trends toward both health and sustainability characteristics of food (Lazzarini et al. 2016).

Some commercial bread recipes include ingredients such as added vitamins, colors, and flavors, together with other ingredients. The raising awareness toward the clean labels in bread-baking leads to the use of only main ingredients, while the other ingredients such as additives are eliminated. For example, a bleaching agent like azodicarbonamide is banned in the EU, whereas this ingredient is still used as a whitening agent in cereal flours and conditioning dough in bread-baking in the United States, as approved by the U.S. Food and Drug Administration.

Nowadays, the clean label has been adapted by various food industries, including the cereal processing industry. With the increasing market shift toward clean-label food products, further specific regulations or legislation may be required (Busken 2015). It is quoted that about 28% of the global food and beverage companies have started using one or more clean label claims, indicating that their products as natural, organic, contain no additives/preservatives, and are non-GMO (contain no ingredients derived from genetically modified organisms). On the basis of a survey conducted by Health Focus International, St. Petersburg, Florida, which involved 22 countries, it was found that 39% of respondents considered eating clean more important over time. The main motives that drive eating clean were the avoidance of chemicals in food (39% respondents) and artificial ingredients (34% repondents) (Gelski 2019).

10.6 Healthy Grain Products on the Market

10.6.1 Whole Grain Products

There is no legal definition of whole-grain products and foods at the European level. The American Association of Cereal Chemists states that whole grain "consist of the intact, ground, cracked, or flaked caryopsis, whose starchy endosperm, germ and bran are present in the same relative proportions as they exist in the intact caryopsis (seed)." All whole-grain products have higher levels of DF and bioactive compounds than their refined equivalents. Examples include whole wheat, oatmeal, whole-grain cornmeal, brown rice, whole-grain barley, whole rye, and buckwheat (Jones et al. 2017).

Whole grains can be eaten in cooked form (after boiling), for example brown rice (wild, red, black), oatmeal, and corn (maize). However, whole grains are normally further processed into a variety of edible products (e.g. whole-grain flour), altering the grain's physical form and may also affect the nutritional value of the grain (Jones et al. 2017). Frequent consumption of whole-grain foods is associated with reduction in heart disease and type 2 diabetes. The benefits of whole-grain foods are not only due to fiber, but also to other biologically active compounds and to synergistic effects of DF and micronutrients. The need to promote a diet rich in whole grains is an important task in nutrition education. There is growing interest in the possible applications of cereals or cereal constituents in functional food formulations to provide health benefits over basic nutrition (Smulders et al. 2018).

10.6.2 Low Glycemic Index Products

The Glycemic Index (GI) is defined as the blood's response curve to 50 g available carbohydrate of a test food expressed as a percentage of the response to 50 g of carbohydrate from a standard food taken by the same subject. Table 10.2 gives an overview of the GI of a range of cereal-based products (Bornet et al. 2007).

The amount of starch depends on the grain and flour types, and consists of varying ratios of amylose and amylopectin which affect the digestibility. The starch can become more resistant to hydrolysis during boiling or baking and thus impacting the glycemic index. Apart from techniques for cereal processing that result in starch that is slowly digested, fibers, polyols, and some sugars (e.g. fructose and tagatose) are also considered as low-GI ingredients (Patil 2008). In addition, coarse, whole grain, fiber, sourdough, and rye were attributes that were perceived by consumers to be good for the stomach and bowel, as well as having good satiation and glycemic properties. Several of these health effects, important to consumers, cannot be communicated on the packages, as there are no officially approved health claims regarding the satiation or glycemic properties of bread. Consumers have to use their own cues to evaluate these health effects, and this includes nutritional labeling, product color, and terms like

Table 10.2 Glycemic index (GI) of various cereal-based foods (reference food is glucose).

Type of product	GI	Type of product	GI
Cereal grains		***Bread***	
Basmati rice	58	Dark rye bread	76
Parboiled rice	47	Rye crispbread	64
Pearl barley	25	Wholemeal bread	69
Buckwheat	54	Sourdough bread	57
Bulgur wheat	48	White bread	70–94.6
Brown rice	55	***Pasta***	
White rice	64	Instant noodles	48
Couscous	65	Macaroni	47
Sweet corn	53	Fettucine, egg	40
Breakfast cereals		Spaghetti	38
Breakfast cereals various	42–102.8	***Cookies and crackers***	
Porridge	46	Cream cracker	65
Corn flakes	84	Digestive cookies	55
		Water cracker	71

Low GI <55, intermediate GI: 55 to 70 and high GI: >70. (*Source*: adapted from Bornet et al. 2007; Yaman et al. 2019)

"rye bread," "whole-grain bread," "fiber," "seeds," or "sourdough bread." Other sensory properties of the bread, such as density, chewy texture, and a sour flavor, could also imply health benefits. According to EU food law, information on food packages should not be misleading. When designing packages and labels, food companies should consider how consumers may perceive the attributes communicated. Further research is needed on how bread is identified from a health perspective and how it can be simplified for consumers through labeling or other means, especially focusing on health attributes that consumers find important, such as satiety and glycemic properties (Bornet et al. 2007).

10.6.3 Fortified Grain-Based Products

Iron, zinc, iodine, vitamin A, and vitamin D are the most chronic and worldwide micronutrient deficiencies and the prevalence of folate and B vitamin deficiencies are also significant (Akhtar et al. 2011; Jan et al. 2019). Cereals and cereal-based products are the best vehicles for fortification in most developing countries, because 95% of the population consume cereals as a dietary staple. Mineral and vitamin deficiencies can be overcome with the use of fortification programs. Globally, whole-wheat flour serves as a dietary staple for millions of people, especially in Asian countries. The Food Fortification Initiative declared that 250 metric tons of milled wheat flour, 26 metric tons of milled maize flour, and 171 metric tons of milled rice were fortified in 2016 (Cardoso et al. 2019). The amount of flour fortification varies from 97% in the Americas, 44% in the Mediterranean area, 31% in Africa, 21% in Southeast Asia, 6% in Europe, and only 4% in the Western Pacific (Marks et al. 2018).

10.6.4 Supplemented Cereal-Based Products

Several studies have been carried out to improve the nutritive value of cereal-based products, supplementing wheat flour with protein-rich non-wheat cereals and legume flours. Composite flours, obtained by enriching the wheat flour with legume flours at 10–25% or more, depending on the type of protein source, have produced successful results due to the high levels of protein in legume flours which, when mixed with wheat flour, have the advantage of improving the nutritional value because of the better composition of amino acids. Soy flour, which is more economical, is the leading protein source. Several studies have shown that the fortification of wheat flour with soy flour increases the content and the protein quality of bread and other bakery products. Quinoa flour contains protein and tryptophan in quantities similar to those of wheat and spelt, but markedly higher than those of other cereals and legumes. Therefore, the use of quinoa could be promoted to enrich the nutritional value of bakery products (Comai et al. 2011).

The amino acid composition is often used to define the nutritional quality of a protein. Tryptophan is the least represented amino acid in the protein of cereals, which are an essential part of daily nutrition. Among cereals, spelt flour, which is rich in the protein tryptophan, contains both free and protein-bound tryptophan in considerably higher amounts compared to those of all cereals and quinoa. Legumes are good sources of proteins, with tryptophan (free and protein-bound) being markedly higher in soybean than in lupine flour, with the highest levels in chickpea flour. These sources of vegetable protein used in the fortification of wheat flour should be encouraged (Millar et al. 2017). Pulse flour offers a sustainable source of plant protein for innovation in protein-enriched cereal-based foods. Fava-bean (*Vicia faba*), green-pea, and yellow-pea (*Pisum sativum*) flour supplementation were shown to enhance the nutritional value of cereal-lupine-based foods (Millar et al. 2019).

10.6.5 Gluten-Free Products

Gluten-free (GF) products are primarily intended for celiac disease patients, as well as for non-celiac gluten sensitivity patients. Moreover, many consumers nowadays turn to a GF diet, considering it healthier, positive for their overall well-being, and beneficial for weight reduction, but without sound scientific evidence for the latter. These trends caused GF market growth characterized by the high prices of GF products, which are also of poor sensory and nutritional properties, despite the considerable advances in research and development of GF products. Therefore, the development of better-tasting and healthier GF products is still a challenging task for food technologists. Many GF products are poor in vitamins, minerals, phytochemicals, protein, and DF and of high glycemic response, all of which are important for a well-balanced and healthy diet (Fratelli et al. 2018; Scherf et al. 2018). Generally, two types of GF products can be distinguished: those obtained from GF raw materials (i.e. amaranth, quinoa, buckwheat) and products rendered GF during processing (Koehler et al. 2014). GF raw materials are nontoxic cereals (e.g. corn, rice, sorghum, and millet) and pseudocereals (e.g. amaranth, buckwheat, and quinoa) (Cardoso et al. 2019; Koehler et al. 2014). Traditional basic GF raw materials, such as refined rice, cornflours, and pure starches lack B vitamins, iron, calcium, and fiber, whilst pseudocereals have a high content of proteins of high quality, fiber, and minerals such as calcium and iron (Koehler et al. 2014).

To produce GF products that are palatable and sensory-acceptable, different flours, starches (e.g. from rice, corn, potato, cassava), proteins (e.g. from milk, egg, soy), and hydrocolloids (e.g. hydroxypropylmethyl-cellulose, carrageenan, xanthan gum) are employed. However, in order to increase the nutritional profile of GF products, they are enriched by the addition of psyllium husk, cellulose, or fiber from oilseed, fruit, and vegetable by-products (Fratelli et al. 2018; Pojić et al. 2015; Šarić et al. 2016).

Several attempts have been made to detoxify gluten proteins in raw materials and products by enhancing their hydrolysis via microbial enzymes or native enzymes of cereals. The use of selected *lactobacilli*, able to extensively hydrolase prolamin proteins, has been proposed in a complex fermentation process of wheat sourdough (30%) mixed with non-gluten containing flours. The resulting bread was comparable to common wheat sourdough bread, and clinical tests showed that this bread was tolerated by gluten-intolerant patients (Catzeddu 2019).

10.6.6 Reduced Salt and Sugar Products

Cardiovascular disease, including hypertension, is the leading cause of preventable death worldwide and excessive consumption of sodium chloride (NaCl) is linked to hypertension. The global dietary NaCl intake has increased extensively and limitations on sodium (Na) consumption were recommended by the World Health Organization (WHO) and other relevant international health agencies (< 2 g Na/day). Bread and other cereal products contribute about 30% to the daily intake of sodium in the Western human diet. Salt reduction in bread-baking is not easy, as salt has its role in gluten development and the control of the rate of yeast fermentation affecting the dough machinability and dough-handling properties, and overall appearance and structural properties of the final bakery products. European residents consume, on average, 59 kg of bread per year, with consumption being generally stable. Bread and bakery products were the most important sources of salt in most European countries, except for the Czech Republic, Poland, and Romania, where salt added during cooking is the most significant source of salt, and Norway and Spain, where meat products were reported to be the more significant sources of salt (Gębski et al. 2019).

To reach the reduction goals, long-term strategies and reformulation of recipes are required. A number of different techniques have been proposed to reduce the sodium chloride content and include salt replacers or taste enhancers. One of the promising strategies to reduce salt has been the addition of sourdough to bakery goods. Sourdough can counteract some of the negative impacts salt reduction has on bread (e.g. flavor and shelf life), thus improving the overall quality (Silow et al. 2016).

The World Health Organization also published guidelines on daily sugar intake, in order to prevent and control health issues associated with high sugar consumption. A reduced consumption of free sugars to less than 10% of the daily calorie intake was recommended (WHO 2015). To reduce high sugar consumption, various governments introduced a sugar tax on sugar-rich food and/or beverages. The taxation varies between countries and has influenced sugar consumption differently (Sahin et al. 2019). The reduction of sugar in sweet bakery products is challenging, however, since sugar provides not only sweetness and flavor, but also product bulkiness, water retention, and increased shelf life (Luo et al. 2019). It is essential to investigate the interactions of sugar with the main ingredients in baked goods and to understand the role of sugar in different products. Sugar can be reduced either by sugar replacement with a sweet bulking agent, such as polyols or sugar substitution by a combination of non-sweet bulking agents and high-intensive sweeteners (Clemens et al. 2016).

10.6.7 Fiber-Rich Products and Fiber Consumption

The increase of non-communicable diseases is attributable to the global change in lifestyle, including reduced physical activity and diets rich in salt, sugar, and fat, and with relatively low DF intake observed in many countries. While some consumers are more interested in a healthy lifestyle and healthy food choices, there are many barriers preventing a change in the dietary patterns of the general population. DF is the edible part of plants, including polysaccharides and lignin which are resistant to human digestive enzymes, and therefore resistant to digestion and absorption in the human small intestine. DF also includes non-starch polysaccharides (NSP) such as celluloses, some hemicelluloses, gums and pectins, resistant dextrins, and resistant starches. Increasing fiber intake is a crucial objective for public health worldwide to combat "diet-related" non-communicable diseases. Dietary fiber intake reduces the risk of chronic heart disease and diabetes, and a reduction of total and low-density lipoprotein cholesterol levels. Cereal products with a high DF level have a low glycemic index. Dietary fiber is one of the primary substrates for growth of the microflora in the large bowel, which increases stool bulk and improves laxation and reduces the risk of duodenal ulcers (Stojceska 2019).

Grain-based products are the largest source of fiber, providing 32–33% of fiber intake in the USA and Spain, and 48–49% in Ireland, the Netherlands, and Sweden. In some countries, bread was the major source of fiber (11–30% of total fiber), with much smaller contributions from breakfast cereals (5–8%), biscuits and pastries (3–11%), and pasta (1–4%) (Kranz et al. 2017). Bread is one of the products that should be taken into consideration for fiber supplementation, because it is an important component of diets in Europe, with average consumption of 59kg per year. Increasing fiber content in plain bread may be effective in ensuring that the population receives adequate amounts of fiber (Baixauli et al. 2008). These modifications can also have an impact on the acceptance of the reformulated products due to the changes in flavor and texture, and other attributes of food and consumer's expectations should be considered (Almeida et al. 2013).

Traditionally, bread is considered a nutritious food rich in carbohydrates, protein, DF, and vitamins and essential components of the daily diet. To provide more variety in functional breads, different sources of DF have often been used in recipes, such as wheat bran, barley, oat, rye, and rice brans (Rakha et al. 2010). Fiber-supplemented breads show a pronounced decrease in quality parameters, and there is a significant effect on mixing and viscoelastic properties and fermentation behavior during bread preparation. Dietary fiber addition increases water absorption, decreases loaf volume, and affects dough rheology and shelf life (Katina et al. 2006).

10.6.8 Sourdough Products

Consumer interest in the health aspects of food continues to increase, and traditional food is often perceived as having beneficial nutritional properties that could improve health and well-being. Sourdough fermentation has health-promoting properties and numerous beneficial effects have been reported with this bread due to its increased content of bioactive peptides, free amino acids, and γ-aminobutyric acid. Additionally, lactic acid bacteria (LAB) used during sourdough fermentation of mainly wholemeal flour can synthesize angiotensin I-converting enzyme (ACE) inhibitory peptides with an antihypertensive effect (Rizzello et al. 2008). Bread, especially whole-wheat, contains a significant amount of phytate and other anti-nutritional factors (e.g. condensed tannins, raffinose, and trypsin inhibitors), which interferes with the absorption of essential minerals such as calcium, zinc, and iron, and makes digestion difficult. Phytate can be degraded by sourdough fermenting microorganisms, due to the activity of phytase and lower pH, as well as other anti-nutritional factors (Catzeddu 2019; Montemurro et al. 2019).

The consumption of organic acids is important in reducing the postprandial glycemic response in human blood. High starch bread is usually rapidly digested and absorbed (exhibiting high GI), leading to hyperglycemia in people suffering from insulin-resistance syndrome. The organic acids produced in sourdough are responsible for a reduction of the glycemic index due to a delay in gastric emptying (Catzeddu 2019).

Sourdough products are mainly produced from rye and wheat flour, but also from other starch-containing resources, where recent trends of sourdough fermentation of non-wheat cereals has a high potential to improve and diversify the sensory qualities of bakery products in general and GF bakery products in particular (Brandt 2019; Dentice Maidana et al. 2020).

10.6.9 Cereal-Based Products with Bioactive Benefits

Various bioactive constituents associated with health promotion and disease prevention properties are present in whole grains, and include carotenoids, anthocyanins, and phenolic compounds. The anthocyanins found in blue corn are cyanidin and malvidin, whereas pelargonidin, cyanidin, and malvidin are found in red corn. The xanthophylls are the source of the yellow color in corn, and lutein and zeaxanthin are the major carotenoids in corn and have been associated with anti-tumor-promoting activity and prevention of age-related macular degeneration. Bioactivities of purple, blue, and red pigmented corn have been associated with the presence of anthocyanins, which have antimutagenic and radical scavenging activities (de la Parra et al. 2007).

An antioxidant named pronyl-L-lysine, produced in bread crust due to the Maillard reaction during baking, acts as a monofunctional inducer of glutathione S-transferase, which serves as an antioxidant chemopreventive activity *in vitro*. The amount of this antioxidant was higher in sourdough bread than in bread obtained by yeast fermentation and dependent on the pH value (Lindenmeier and Hofmann 2004).

10.6.10 Cereal-Based Beverages

The recent emergence of non-alcoholic functional beverages in the market is one of the most dynamic market trends, closely related to the increasing demands on plant-based dairy substitutes, with an annual growth rate of almost 20% in the United States (Srikaeo 2020). Although their sensory characteristics are commonly described as sour, sweet, cereal-like, and malty, and not being positively regarded by consumers (Dongmo et al. 2016), their popularity is increasing. This is due to the fact that they are suitable vehicles for delivery of nutrients and functional ingredients to the consumers, which include vitamins, minerals, antioxidants, omega-3 fatty acids, plant extracts, sterols/stanols, DF, amino acids and biopeptides, prebiotics, and probiotics, etc. Although a number of cereal-based beverages represent traditional products mainly produced in an artisan way (e.g. kvass, boza, togwa, etc.), a second generation of cereal-based functional beverages is emerging, innovatively designed to meet specific consumer requirements linked to aging issues, better athletic performance, higher energy, better digestion, longer satiety, better cognitive ability, better hydration, weight management, improvement of cardiovascular health, and prevention of non-communicable diseases, etc. (Basinskiene and Cizeikiene 2020; Blandino et al. 2003; Srikaeo 2020). Moreover, it appears that cereal-based beverages are extremely useful products for people with different food restrictions (i.e. intolerances, allergies, veganism, etc.) (Bernardo et al. 2019). Cereal-based beverages can be of different types: fermented and non-fermented (Basinskiene and Cizeikiene 2020). Non-fermented cereal-based beverages can be in the form of cereal-based milk substitutes designed to resemble cow's milk, or in the form of roasted grain beverages to serve as an alternative to coffee and/or tea (Basinskiene and Cizeikiene 2020). Moreover, they can be in the form of instant powder beverages, produced by spray drying, agglomeration, extrusion, or by a combination of these processes, which provide convenient preparation and high microbiological stability (Bernardo et al. 2019). The versatility of formulations of cereal-based beverages is interminable: cereal-based kefir-like riboflavin-enriched beverages from oat, maize, and barley flours (Yépez et al. 2019), yogurt-like beverages made of a mixture of rice, barley, emmer and oat, soy flour, and concentrated red grape must (Coda et al. 2012), or yogurt-like product based on fermented maize enriched in carotenoids and phytosterols (Gies et al. 2019), etc.

10.7 Conclusion and Future Perspectives

The problem of having meaningful and positive attitudes toward consuming *healthy* grains pass through a consumer's cognitive process that involves an evaluation on what is good, pleasant, and beneficial for their health. This process encompasses both external information and direct experience with healthy grains consumption, whereby the consumers express their views consciously and unconsciously. Considering this, the more the nutritional and health properties of grains are clearly communicated, in terms of helping nutritional deficiencies, and preventing non-healing health issues, the more the attitude toward these products will be robust and close to the actual consumption. Thus, it would be easier for the consumers' understanding to know how to fit *healthy* grains into their daily diets so their attitude toward this is clear. Otherwise, this construct, that is also sensitive to cultural differences, will be tenuous, and even more heterogeneous for understanding healthy grains food choices. In this respect and in conclusion, the most urgent future challenge to have a definitive acceptance of *healthy* grain products in a daily diet would be to turn their consumption into a common habitual food behavior above and beyond cultural barriers. Robust attitudes are good drivers and the consequences of right and uniformed food information campaigns with a common objective will make the consumers feel safe and certain while consuming healthy grains. This self-confidence will be the key to the consumers' habitual consumption and routine selection of these health foods.

Acknowledgment

M. Pojić would like to acknowledge the financial support of the Ministry of Education, Science and Technological Development of the Republic of Serbia (No. 451-03-68/2020-14/200222).

References

Ajzen, I. (2019). Theory of Planned Behavior. https://people.umass.edu/aizen/tpb.html (accessed 2 April 2020)

Akhtar, S., Anjum, F.M., and Anjum, M.A. (2011). Micronutrient fortification of wheat flour: recent development and strategies. *Food Research International* 44 (3): 652–659.

Almeida, E.L., Chang, Y.K., and Steel, C.J. (2013). Dietary fibre sources in bread: influence on technological quality. *LWT - Food Science and Technology* 50 (2): 545–553.

Amos, C., Pentina, I., Hawkins, T.G., and Davis, N. (2014). "Natural" labeling and consumers' sentimental pastoral notion. *Journal of Product and Brand Management* 23 (4/5): 268–281.

Aschemann-Witzel, J., Varela, P., and Peschel, A.O. (2019). Consumers' categorization of food ingredients: do consumers perceive them as 'clean label' producers expect? An exploration with projective mapping. *Food Quality and Preference* 71: 117–128.

Asioli, D., Aschemann-Witzel, J., Caputo, V. et al. (2017). Making sense of the "clean label" trends: a review of consumer food choice behavior and discussion of industry implications. *Food Research International* 99: 58–71.

Baixauli, R., Salvador, A., Hough, G., and Fiszman, S.M. (2008). How information about fibre (traditional and resistant starch). Influences consumer acceptance of muffins. *Food Quality and Preference* 19 (7): 628–635.

Basinskiene, L. and Cizeikiene, D. (2020). Cereal-based nonalcoholic beverages. In: *Trends in Non-Alcoholic Beverages* (ed. C.M. Galanakis), 63–99. Academic Press.

Bernardo, C.O., Ascheri, J.L.R., Carvalho, C.W.P. et al. (2019). Impact of extruded sorghum genotypes on the rehydration and sensory properties of soluble beverages and the Brazilian consumers' perception of sorghum and cereal beverage using word association. *Journal of Cereal Science* 89: 102793.

Berryman, P. (2015). The EU food information for consumers regulation. In: *Advances in Food and Beverage Labelling* (ed. e.P. Berryman), 3–13. Elsevier.

de Beukelaar, M.F.A., Zeinstra, G.G., Mes, J.J., and Fischer, A.R.H. (2019). Duckweed as human food. The influence of meal context and information on duckweed acceptability of Dutch consumers. *Food Quality and Preference* 71: 76–86.

Blandino, A., Al-Aseeri, M.E., Pandiella, S.S. et al. (2003). Cereal-based fermented foods and beverages. *Food Research International* 36 (6): 527–543.

Bollen, K.A. and Davis, W.R. (2009). Causal indicator models: identification, estimation, and testing. *Structural Equation Modeling: A Multidisciplinary Journal* 16 (3): 498–522.

Bornet, F.R.J., Jardy-Gennetier, A.-E., Jacquet, N., and Stowell, J. (2007). Glycaemic response to foods: impact on satiety and long-term weight regulation. *Appetite* 49 (3): 535–553.

Brandt, M.J. (2019). Industrial production of sourdoughs for the baking branch – an overview. *International Journal of Food Microbiology* 302: 3–7.

Busken, D.F. (2015). Cleaning it up – what is a clean label ingredient? *Cereal Foods World* 60 (2): 112–113.

Cao, R. (2019). Technologies for improving the nutritional quality of cereals. In: *Bioactive Factors and Processing Technology for Cereal Foods* (eds. J. Wang, B. Sun and R. Tsao), 19–31. Springer.

Cardoso, R.V.C., Fernandes, Â., Gonzaléz-Paramás, A.M. et al. (2019). Flour fortification for nutritional and health improvement: a review. *Food Research International* 125: 108576.

Catzeddu, P. (2019). Sourdough breads. In: *Flour and Breads and Their Fortification in Health and Disease Prevention*, 2e (eds. V.R. Preedy and R.R. Watson), 177–188. Academic Press.

Clemens, R.A., Jones, J.M., Kern, M. et al. (2016). Functionality of sugars in foods and health. *Comprehensive Reviews in Food Science and Food Safety* 15 (3): 433–470.

Coda, R., Lanera, A., Trani, A. et al. (2012). Yogurt-like beverages made of a mixture of cereals, soy and grape must: microbiology, texture, nutritional and sensory properties. *International Journal of Food Microbiology* 155 (3): 120–127.

Comai, S., Bertazzo, A., Costa, C.V.L., and Allegri, G. (2011). Quinoa: protein and nonprotein tryptophan in comparison with other cereal and legume flours and bread. In: *Flour and Breads and their Fortification in Health and Disease Prevention*, 2e (eds. V.R. Preedy and R.R. Watson), 113–125. Academic Press.

Cooper, D., Kable, M., Marco, M. et al. (2017). The effects of moderate whole grain consumption on fasting glucose and lipids, gastrointestinal symptoms, and microbiota. *Nutrients* 9 (2): 173.

Costa, V. and Johnson, A. (2019). Global food and nutrition trends: driving positive momentum for grains. *Cereal Foods World* 64 (3) doi: 10.1094/CFW-64-3-0032 https://www.cerealsgrains.org/publications/cfw/2019/May-June/Pages/CFW-64-3-0032.aspx (accessed 2 April 2020).

De Morpurgo, M. and Botana, P.C. (2016). The nutrition and health claims regulation applies to commercial communications addressed to health professionals. *European Journal of Risk Regulation* 7 (3): 634–641.

De Ridder, D., De Vet, E., Stok, M. et al. (2013). Obesity, overconsumption and self-regulation failure: the unsung role of eating appropriateness standards. *Health Psychology Review* 7 (2): 146–165.

Dean, M., Lampila, P., Shepherd, R. et al. (2012). Perceived relevance and foods with health-related claims. *Food Quality and Preference* 24 (1): 129–135.

Demartini, E., De Marchi, E., Cavaliere, A. et al. (2019). Changing attitudes towards healthy food via self-association or nutritional information: what works best? *Appetite* 132: 166–174.

Dentice Maidana, S., Aristimuño Ficoseco, C., Bassi, D. et al. (2020). Biodiversity and technological-functional potential of lactic acid bacteria isolated from spontaneously fermented chia sourdough. *International Journal of Food Microbiology* 316: 108425.

Dongmo, S.N., Procopio, S., Sacher, B., and Becker, T. (2016). Flavor of lactic acid fermented malt based beverages: current status and perspectives. *Trends in Food Science & Technology* 54: 37–51.

FAO (Food and Agriculture Organization of the United Nations) (2016). Plates, pyramids, planet. Developments in national and sustainable dietary guidelines: A state of play assessment. http://www.fao.org/3/I5640E/i5640e.pdf (accessed 2 April 2020)

European Parliament and Council (2006). Regulation (EC) No. 1924/2006 of The European Parliament and of The Council of 20 December 2006 on nutrition and health claims made on foods. *Official Journal of the European Union* L404: 9–26.

European Parliament and Council (2010). Annex I: (EC) No: 1924/2006 of 30 Dec 2006 lastly amended by Regulation (EC) No. 1116/2010. *Official Journal of the European Union* L317: 1–2.

Fishbein, M. and Ajzen, I. (eds.) (1980). *Understanding Attitudes and Predicting Social Behavior*. Englewood Cliffs, NJ: Prentice Hall.

Foschia, M., Peressini, D., Sensidoni, A., and Brennan, C.S. (2013). The effects of dietary fibre addition on the quality of common cereal products. *Journal of Cereal Science* 58 (2): 216–227.

Fratelli, C., Muniz, D.G., Santos, F.G., and Capriles, V.D. (2018). Modelling the effects of psyllium and water in gluten-free bread: an approach to improve the bread quality and glycemic response. *Journal of Functional Foods* 42: 339–345.

Gawronski, B., Hofmann, W., and Wilbur, C.J. (2006). Are "implicit" attitudes unconscious? *Consciousness and Cognition* 15 (3): 485–499.

Gębski, J., Jezewska-Zychowicz, M., Szlachciuk, J., and Kosicka-Gębska, M. (2019). Impact of nutritional claims on consumer preferences for bread with varied fiber and salt content. *Food Quality and Preference* 76: 91–99.

Gelski, J. (2019). 'No preservatives' stands out among clean label claims. https://www.foodbusinessnews.net/articles/14178-no-preservatives-stands-out-among-clean-label-claims (accessed 20 April 2020)

Gies, M., Descalzo, A.M., Servent, A., and Dhuique-Mayer, C. (2019). Incorporation and stability of carotenoids in a functional fermented maize yogurt-like product containing phytosterols. *LWT - Food Science and Technology* 111: 105–110.

Gilsenan, C., Burke, R.M., and Barry-Ryan, C. (2012). Do organic cherry vine tomatoes taste better than conventional cherry vine tomatoes? A sensory and instrumental comparative study from Ireland. *Journal of Culinary Science & Technology* 10 (2): 154–167.

Grunert, K.G., Fernández-Celemín, L., Wills, J. et al. (2010). Use and understanding of nutrition information on food labels in six European countries. *Journal of Public Health* 18 (3): 261–277.

Hemmerling, S., Asioli, D., and Spiller, A. (2016). Core organic taste: preferences for naturalness-related sensory attributes of organic food among European consumers. *Journal of Food Products Marketing* 22 (7): 824–850.

Hieke, S. and Grunert, K.G. (2018). Consumers and health claims. In: *Nutrients and Food Ingredients with Authorised EU Health Claims*, vol. 3 (ed. M.J. Sadler), 19–32. Woodhead Publishing.

Jan, Y., Malik, M., Yaseen, M. et al. (2019). Vitamin D fortification of foods in India: present and past scenario. *The Journal of Steroid Biochemistry and Molecular Biology* 193: 105417.

Janssen, M. (2018). Determinants of organic food purchases: evidence from household panel data. *Food Quality and Preference* 68: 19–28.

Jones, A.R., Mann, K.D., Kuznesof, S.A. et al. (2017). The whole grain content of foods consumed in the UK. *Food Chemistry* 214: 453–459.

Katina, K., Heiniö, R.-L., Autio, K., and Poutanen, K. (2006). Optimization of sourdough process for improved sensory profile and texture of wheat bread. *LWT - Food Science and Technology* 39 (10): 1189–1202.

Kim, H.-S., Oh, C., and No, J.-K. (2016). Can nutrition label recognition or usage affect nutrition intake according to age? *Nutrition* 32 (1): 56–60.

Klopčič, M., Slokan, P., and Erjavec, K. (2019). Consumer preference for nutrition and health claims: a multi-methodological approach. *Food Quality and Preference* 82: 103863.

Koehler, P., Wieser, H., and Konitzer, K. (2014). Gluten-free products. In: *Celiac Disease and Gluten: Multidisciplinary Challenges and Opportunities* (eds. P. Koehler, H. Wieser and K. Konitzer), 173–223. Academic Press.

Kranz, S., Dodd, K., Juan, W. et al. (2017). Whole grains contribute only a small proportion of dietary fiber to the US diet. *Nutrients* 9 (2): 153.

Lazzarini, G.A., Zimmermann, J., Visschers, V.H.M., and Siegrist, M. (2016). Does environmental friendliness equal healthiness? Swiss consumers' perception of protein products. *Appetite* 105: 663–673.

Lindenmeier, M. and Hofmann, T. (2004). Influence of baking conditions and precursor supplementation on the amounts of the antioxidant pronyl-L-lysine in bakery products. *Journal of Agricultural and Food Chemistry* 52 (2): 350–354.

Luo, X., Arcot, J., Gill, T. et al. (2019). A review of food reformulation of baked products to reduce added sugar intake. *Trends in Food Science & Technology* 86: 412–425.

Marks, K.J., Luthringer, C.L., Ruth, L.J. et al. (2018). Review of grain fortification legislation, standards, and monitoring documents. *Global Health: Science and Practice* 6 (2): 356–371.

Massini, L., Rico, D., Martin-Díana, A.B., and Barry-Ryan, C. (2016). Study of antioxidant properties of fractionated apple peel phenolics using a multiple-assay approach. *International Food Research Journal* 23 (5): 1996.

Mattavelli, S., Richetin, J., Gallucci, M., and Perugini, M. (2017). The self-referencing task: theoretical overview and empirical evidence. *Journal of Experimental Social Psychology* 71: 68–82.

McEachan, R.R.C., Conner, M., Taylor, N.J., and Lawton, R.J. (2011). Prospective prediction of health-related behaviours with the theory of planned behaviour: a meta-analysis. *Health Psychology Review* 5 (2): 97–144.

McGill, C. and Devareddy, L. (2015). Ten-year trends in fiber and whole grain intakes and food sources for the United States population: National Health and Nutrition Examination Survey 2001–2010. *Nutrients* 7 (2): 1119–1130.

Millar, K.A., Barry-Ryan, C., Burke, R. et al. (2017). Effect of pulse flours on the physiochemical characteristics and sensory acceptance of baked crackers. *International Journal of Food Science & Technology* 52 (5): 1155–1163.

Millar, K.A., Gallagher, E., Burke, R. et al. (2019). Proximate composition and anti-nutritional factors of fava-bean (*Vicia faba*), green-pea and yellow-pea (*Pisum sativum*) flour. *Journal of Food Composition and Analysis* 82: 103233.

Montemurro, M., Pontonio, E., Gobbetti, M., and Rizzello, C.G. (2019). Investigation of the nutritional, functional and technological effects of the sourdough fermentation of sprouted flours. *International Journal of Food Microbiology* 302: 47–58.

de la Parra, C., Serna Saldivar, S.O., and Liu, R.H. (2007). Effect of processing on the phytochemical profiles and antioxidant activity of corn for production of masa, tortillas, and tortilla chips. *Journal of Agricultural and Food Chemistry* 55 (10): 4177–4183.

Patil, S.K. (2008). Improving the use of dietary fiber and other functional ingredients to lower the glycemic index of cereal products. In: *Technology of Functional Cereal Products* (ed. B.R. Hamaker), 495–517. Woodhead Publishing.

PF Guiné, R., Duarte, J., Ferreira, M. et al. (2016). Attitudes towards dietary fibre on a multicultural basis: a fibre study framework. *Current Nutrition & Food Science* 12 (2): 132–141.

Pojić, M., Dapčević Hadnađev, T., Hadnađev, M. et al. (2015). Bread supplementation with hemp seed cake: a by-product of hemp oil processing. *Journal of Food Quality* 38 (6): 431–440.

Prestwich, A., Hurling, R., and Baker, S. (2011). Implicit shopping: attitudinal determinants of the purchasing of healthy and unhealthy foods. *Psychology & Health* 26 (7): 875–885.

Rakha, A., Åman, P., and Andersson, R. (2010). Characterisation of dietary fibre components in rye products. *Food Chemistry* 119 (3): 859–867.

de Ridder, D., Kroese, F., Evers, C. et al. (2017). Healthy diet: health impact, prevalence, correlates, and interventions. *Psychology & Health* 32 (8): 907–941.

Rizzello, C.G., Cassone, A., Di Cagno, R., and Gobbetti, M. (2008). Synthesis of angiotensin I-converting enzyme (ACE)-inhibitory peptides and γ-aminobutyric acid (GABA) during sourdough fermentation by selected lactic acid bacteria. *Journal of Agricultural and Food Chemistry* 56 (16): 6936–6943.

Rosell, C.M. (2019). Trends in science of doughs and bread quality. In: *Flour and Breads and their Fortification in Health and Disease Prevention*, 2e (eds. V.R. Preedy and R.R. Watson), 333–343. Academic Press.

Saba, A., Vassallo, M., Shepherd, R. et al. (2010). Country-wise differences in perception of health-related messages in cereal-based food products. *Food Quality and Preference* 21 (4): 385–393.

Sahin, A.W., Zannini, E., Coffey, A., and Arendt, E.K. (2019). Sugar reduction in bakery products: current strategies and sourdough technology as a potential novel approach. *Food Research International* 126: 108583.

Sanchez-Siles, L.M., Michel, F., Román, S. et al. (2019). The food naturalness index (FNI): an integrative tool to measure the degree of food naturalness. *Trends in Food Science & Technology* 91: 681–690.

Sandvik, P., Nydahl, M., Kihlberg, I., and Marklinder, I. (2018). Consumers' health-related perceptions of bread – implications for labeling and health communication. *Appetite* 121: 285–293.

Šarić, B., Mišan, A., Mandić, A. et al. (2016). Valorisation of raspberry and blueberry pomace through the formulation of value-added gluten-free cookies. *Journal of Food Science and Technology* 53 (2): 1140–1150.

Schaffer-Lequart, C., Lehmann, U., Ross, A.B. et al. (2017). Whole grain in manufactured foods: current use, challenges and the way forward. *Critical Reviews in Food Science and Nutrition* 57 (8): 1562–1568.

Scherf, K.A., Wieser, H., and Koehler, P. (2018). Novel approaches for enzymatic gluten degradation to create high-quality gluten-free products. *Food Research International* 110: 62–72.

Shepherd, R., Dean, M., Lampila, P. et al. (2012). Communicating the benefits of wholegrain and functional grain products to European consumers. *Trends in Food Science & Technology* 25 (2): 63–69.

Siegrist, M., Hartmann, C., and Keller, C. (2013). Antecedents of food neophobia and its association with eating behavior and food choices. *Food Quality and Preference* 30 (2): 293–298.

Siipi, H. (2013). Is natural food healthy? *Journal of Agricultural and Environmental Ethics* 26 (4): 797–812.

Silow, C., Axel, C., Zannini, E., and Arendt, E.K. (2016). Current status of salt reduction in bread and bakery products – a review. *Journal of Cereal Science* 72: 135–145.

Smulders, M.J.M., van de Wiel, C.C.M., van den Broeck, H.C. et al. (2018). Oats in healthy gluten-free and regular diets: a perspective. *Food Research International* 110: 3–10.

Srikaeo, K. (2020). Biotechnological tools in the production of functional cereal-based beverages. In: *Biotechnological Progress and Beverage Consumption* (eds. A.M. Grumezescu and A.M. Holban), 149–193. Academic Press.

Stojceska, V. (2019). Brewer's spent grain from by-product to health: a rich source of functional ingredients. In: *Flour and Breads and their Fortification in Health and Disease Prevention*, 2e (eds. V.R. Preedy and R.R. Watson), 189–198. Academic Press.

Tuorila, H. and Hartmann, C. (2020). Consumer responses to novel and unfamiliar foods. *Current Opinion in Food Science* 33: 1–8.

Urala, N. and Lähteenmäki, L. (2007). Consumers' changing attitudes towards functional foods. *Food Quality and Preference* 18 (1): 1–12.

Van der Kamp, J.W. (2008). The EU nutrition and health claim regulation – impact on cereal R&D. In: *Consumer Driven Cereal Innovation: Where Science Meets Industry* (eds. P. Weegels, C. Courtin, A. Arrachid, et al.), 41–44. American Association of Cereal Chemists International.

Van der Kamp, J.W. and Lupton, J. (2013). Definitions, regulations and health claims associated with dietary fibre and wholegrain foods. *Fibre-Rich and Wholegrain Foods: Improving Quality*: 3–24.

Vassallo, M. (2013). In defence of modeling simultaneity for a correct approximation of cultural aspects: implications for food consumers studies with latent variables. PhD Thesis, University of Bologna.

Vassallo, M., Saba, A., Arvola, A. et al. (2009). Willingness to use functional breads. Applying the health belief model across four European countries. *Appetite* 52 (2): 452–460.

Vassallo, M., Scalvedi, M.L., and Saba, A. (2016). Investigating psychosocial determinants in influencing sustainable food consumption in Italy. *International Journal of Consumer Studies* 40 (4): 422–434.

Velardo, S. (2015). The nuances of health literacy, nutrition literacy, and food literacy. *Journal of Nutrition Education and Behavior* 47 (4): 385–389.

Venkataratnam, H., Cahill, O., and Barry-Ryan, C. (2018). Food allergens in prepacked foods. https://www.safefood.eu/Publications/Research-reports/Food-allergens-in-prepacked-foods.aspx (accessed 2 April 2020)

Verbeke, W. (2006). Functional foods: consumer willingness to compromise on taste for health? *Food Quality and Preference* 17 (1–2): 126–131.

WHO (2015). *Guideline: Sugars Intake for Adults and Children*. World Health Organization https://www.who.int/publications-detail/9789241549028 (accessed 2 April 2020).

WHO (2018). Healthy Diet. News communication. https://www.who.int/news-room/fact-sheets/detail/healthy-diet (accessed 2 April 2020).

Yaman, M., Sargın, H.S., and Mızrak, Ö.F. (2019). Free sugar content, in vitro starch digestibility and predicted glycemic index of ready-to-eat breakfast cereals commonly consumed in Turkey: an evaluation of nutritional quality. *International Journal of Biological Macromolecules* 135: 1082–1087.

Yépez, A., Russo, P., Spano, G. et al. (2019). *In situ* riboflavin fortification of different kefir-like cereal-based beverages using selected Andean LAB strains. *Food Microbiology* 77: 61–68.

11 Assessing the Environmental Impact of Processed Healthy Grains

Nicholas M. Holden and Mingjia Yan

UCD School of Biosystems and Food Engineering, University College Dublin, Dublin, Ireland

11.1 Introduction

The context of this chapter is the processing of grains to supply safe and healthy food. Society and individuals are beginning to ask where food comes from, and what was the environmental impact of producing it. At a global scale, these questions are addressed by recent United Nations (2015) resolutions concerning sustainable food systems and operationalized through the Sustainable Development Goals (SDGs). Three of the SDGs are of relevance here, Goal 2 – Zero Hunger ("end hunger, achieve food security and improved nutrition and promote sustainable agriculture"), Goal 12 – Responsible Consumption and Production ("ensure sustainable consumption and production patterns"), and Goal 13 –Climate Action ("take urgent action to combat climate change and its impacts"). These goals reflect the desire to have a sustainable food system, i.e. one that conserves land, water, plant and animal genetic resources, is environmentally non-degrading, technically appropriate, economically viable, and socially acceptable (FAO 1998). Sustainable grain-based foods should be assessed by considering the *environmental*, including non-renewable resource depletion and polluting emissions; the *social*, including welfare, labor, rights, and equity; and the *economic*, including maintaining the financial flows necessary for a working economy and a viable business. This means that for grain-based foods to be sustainable, it must be possible to keep supplying them without adverse penalty for future generations, thus processing grain for food supply should not reduce future generations' ability to access resources and have a healthy living environment, should not disrupt social structures and networks, and should ensure long-term financial viability for individuals, communities, and countries. Obviously, it is difficult to relate a relatively simple food processing concept to such important impacts, yet it is the sum of all such

Innovative Processing Technologies for Healthy Grains, First Edition.
Edited by Milica Pojić and Uma Tiwari.

small impacts that determine future sustainability. A common, often tacit, assumption when considering sustainability is that the environmental concerns dominate. While this is not a tenable assumption, this chapter is going to focus on explaining a methodological approach that can be used for quantifying and understanding the contribution of processed healthy grains to global environmental impact. It will focus on the dominant method used for such analysis, Life Cycle Assessment (LCA) (ISO 14040 2006), and provide non-experts with a brief guide to understand how to best use the methods to answer questions about environmental impact. As with most environmental assessment methods, LCA is substantially less broad in scope than that required to address sustainability, which encompasses social, economic, productive, and environmental functions of a system (Basset-Mens et al. 2009). LCA can be extended to consider some social functions (UNEP-SETAC 2009) and economic costing (in the form of Life Cycle Costing) (Swarr et al. 2011), and recently these have been integrated to balance all aspects of sustainability (Chen and Holden 2018). The focus on environmental aspects of healthy grain processing is justified because in the strong sustainability model, the social and economic must work within the limits of the planetary boundary (Rockström et al. 2009), thus the environmental aspects are preeminent.

11.1.1 The Role of LCA in Grain Processing

Determining the environmental impact of processed *healthy* grains is not a trivial question to answer because for many people, particularly those living in more developed countries, food chains are long, complex, and globally interconnected, and publicly available data are scarce. Given that grains constitute only a (sometimes small) component of diet, and the processing of those grains is only a small fraction of the input of resources and time required to produce food, it is a conceptually challenging task to both ask and answer questions about the environmental impact of the processing stage. Actually, asking such a specific question might be of little value to anyone other than the processors themselves, so this chapter will take a more holistic view of the whole supply chain and the resources it requires to exist. There are several methodologies that can be used to evaluate the environmental impact of grain processing, summarized (using dairy production systems as a case study) by Yan et al. (2011). The differences between methods relate to focus, practice, or product, the purpose and the spatial scale at which they operate. While each method has its advantages and disadvantages, LCA is the only method that can be adapted to operate over a range of scales, to work retrospectively (historically) or prospectively (future scenarios), is truly holistic (if applied properly), and that is focused on the product rather than the practice. A focus on the processing stage itself would require a site-specific method, such as Environmental Impact Assessment (EIA) (Morgan 1998), whereas a study to understand the contribution of the processing stage to the impact of a processed food product might use LCA.

The main strength of LCA is the holistic nature of the analysis. Theoretically, the whole life cycle of a product should be included in the model used to estimate impacts. Grain processing only encompasses the range of activities that take the seeds of a cereal or legume and convert them into a more desirable food product. The processing stage is part of the larger system (Figure 11.1) that: (i) produces the commodity (farm production, agriculture); (ii) modifies it for storage and logistics (post-harvest technology); (iii) moves it to a central location for handling (logistics); (iv) converts, possibly combines with other ingredients and packages to create a new type of product (processing); (v) routes to shops (logistics); (vi) sells (retail); (vii) uses for human food (consumption); (viii) handles subsequent human waste; and (ix) handles management of co- and by-products as either wastes or raw materials for valorization (bioeconomy). This linear system can also include several loops (circular economy) to maximize resource use efficiency. Similarly, grain processing can be part of an animal feed system with similar steps. There is further complexity in the system because in some life cycle stages food and feed are in competition, and at others feed is a sub-set of a human food system with an animal converting a by-product into meat, milk, or fiber. Since 2013, global grain production has been around 2.5×10^9 tonnes per year, with estimates of around 40% being used as feed and perhaps 60% as food.

It is also worth noting that the added value of LCA (and similar tools) for the agri-food industry, and businesses involved in food processing, lies with both process optimization and communications. Process optimization can start from hotspot analysis, which is the identification of places in the systems that are responsible for a large proportion of environmental impacts. A hotspot might well be associated with

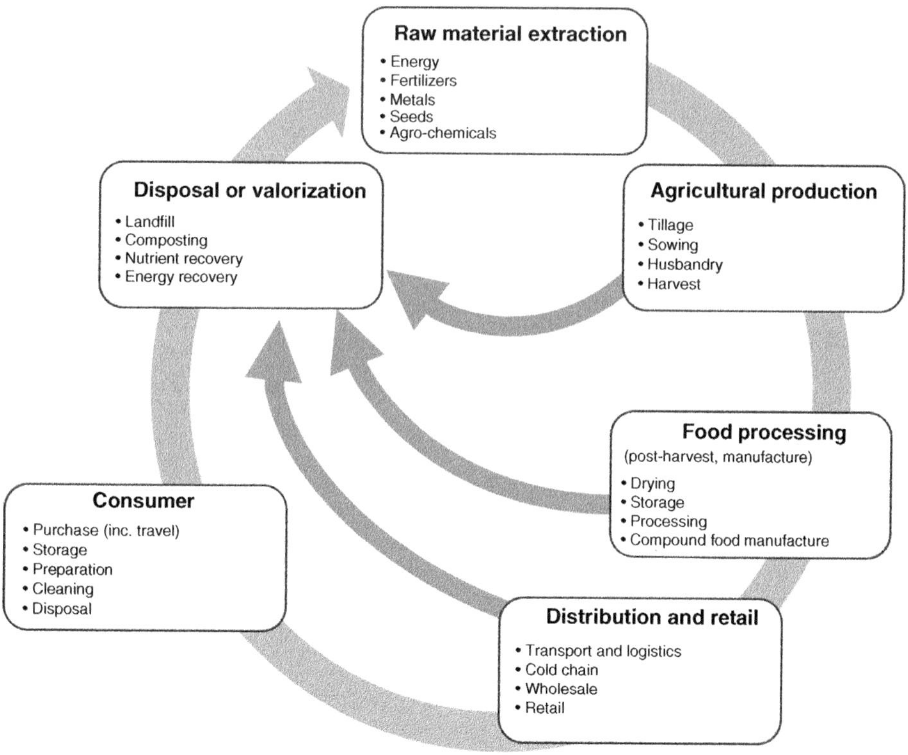

Figure 11.1 A generalized system for the production, processing, distribution, consumption, and end-of-life for grain-derived food products.

resource use inefficiency. A related purpose is to improve the environmental performance of all components, which for food products may be closely related to the processing stage, even if that stage appears to contribute relatively little to overall environmental impacts when compared to the agricultural production phase for some products (Yan et al. 2011). For grains, it is likely that the provision of resources such as fertilizer and energy may be more important, but this will depend on the impact category of interest (Cancino-Espinoza et al. 2018). Communications are expressed in terms of eco-labelling (Bougherara and Combris 2009; Gruère 2015), Product Environmental Footprinting (PEF) (Manfredi et al. 2015), and Environmental Product Declarations (EPD) (Schau and Fet 2008; ISO 14025 2006). Interestingly, it tends to be the companies responsible for food processing that drive this activity, rather than those responsible for producing raw materials or making complex consumer products. For instance, it is the dairy process and meat process industry that worked to develop PEF guidelines for dairy and meat products (European Commission 2015; European Commission 2016), rather than the companies that use the processed outputs as ingredients, even though a relatively small percentage of impact can be attributed to the processing stage (Yan et al. 2011).

In this chapter we outline how LCA works, and the general rules for using the method, using studies of grain-based food products to illustrate the method where possible.

11.2 Impact Assessment: Life Cycle Assessment

LCA formalizes life cycle thinking (Figure 11.1) into an accounting framework (inputs and outputs for each life cycle stage) that balances mass and energy flow throughout a system. The flows are called *valuable substances* if they remain in the *technosphere* and represent the work done by the system (e.g. seeds, oil, machinery, bread) or are *elementary flows* if they represent resource depletion from, or pollution of the *ecosphere* (Figure 11.2). The elementary flows are used to estimate environmental impacts and the valuable substances are used to calculate the work done by the system. Impact is then usually expressed per unit activity of the system, but can be expressed as an absolute total. The impacts are based on calculation of theoretical, or potential impact of the system using generalized impact models (Jolliet et al. 2016).

11.2.1 LCA Definition

The International Standards Organization (ISO) defines LCA as "... a technique for assessing the environmental aspects and potential impacts associated with a product, by compiling an inventory of relevant inputs and outputs of a product system; evaluating the potential environmental impacts associated with those inputs and outputs; interpreting the results of the inventory analysis and impact assessment phases in relation to the objectives of the study" (ISO 14040 2006), and defines a specific set of guidelines for conducting an LCA study (ISO 14040 2006; ISO 14044 2006; ISO 14044:2006/Amd.1:2017). Rebitzer et al. (2004) describe LCA as a "methodological framework for estimating and assessing the environmental impacts attributable to the life cycle of a product, such as climate change, stratospheric ozone depletion, tropospheric ozone (smog) creation, eutrophication, acidification, toxicological stress on human health and ecosystems, the depletion of resources, water use, land use,

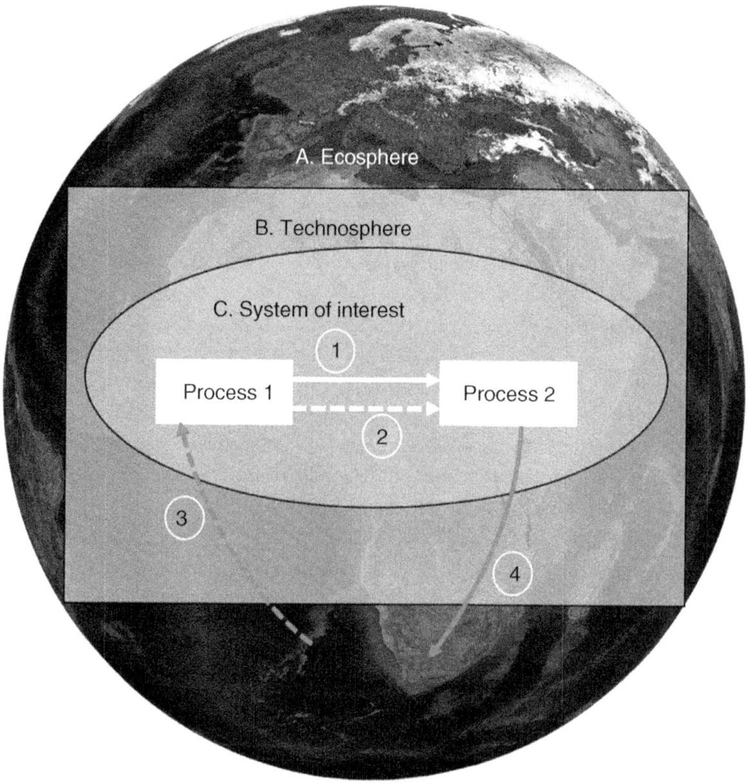

Figure 11.2 The interaction of human systems of interest (in this case grain utilization for food) and the planetary boundary. (A) The ecosphere represents the total resources available for use. (B) The technosphere is the human-constructed system to harness resources for society, usually valued in economic terms. (C) The system of interest in this case is food produced using novel healthy grains. Technosphere flows can be (1) mass flows or (2) energy flows. Elementary flows are those that (3) cause resource depletion and (4) cause pollution. Impacts associated with elementary flows are considered in terms of *areas of protection* including human health, natural environment, and man-made environment.

noise and others." Clearly these definitions are not compatible with assessing just the processing stage in the life cycle of a processed *healthy* grain, because they include the full life cycle of the system required to produce it and define the output of the system in terms of a usable product (e.g. the food available to eat, with processed grain as an ingredient). While LCA is normally focused on products, the methodology can be adapted to processes and services (Roy et al. 2009), and therefore has application to food processing rather than just food products.

11.2.2 The LCA Methods

While there have been many thousands of agri-food related LCA studies published over the years (see Yan et al. (2011); Roy et al. (2009); Hospido et al. (2010) for some examples spanning various sectors), most have not been complete LCAs including a full life cycle and a wide range of impact categories, and few considered impacts other than environmental. "Truncated" LCA methods have become widespread with a focus on single impacts such as Carbon Footprint (Clune et al. 2017; Carbon Trust 2018), Water Footprint (Mekonnen and Hoekstra 2011), Energy Footprint (Khan and Hanjra 2009), and the Ecological Footprints (not strictly an LCA method) (Wiedmann and Barrett 2010). System truncation is also common, particularly for studies from the *cradle-to-farm-gate* (Charles et al. 2006) or from *gate-to-gate* (Finnegan et al. 2018). The downside of "footprint" analysis and studies restricted to limited life cycle stages is the major strength of LCA, which is the ability to identify *burden shifting*. Burden shifting occurs when an intervention to reduce impact is successful for the target impact, but has the unintended consequence of increasing impact elsewhere in the system, either a different life cycle stage or a different type of impact. An example might be control of nitrogen becoming N_2O (a potent greenhouse gas) leading to an increase in NO_3 (causing greater eutrophication). It should be noted that LCA also requires consideration of the *value sphere*, which reflects the fact that where human systems interact with each other and the planet, there are rarely absolute truths, thus subjective choices are needed. The "footprint" approach is popular because simplified methods are available that are relatively easy to deploy and understand. For example, a Carbon Footprint can be calculated according to ISO standards (Casey and Holden 2005), Publicly Available Specification such as PAS2050 guidelines (Carbon Trust 2018), or using an online tool (Padgett et al. 2008; Sykes et al. 2017). The limited life cycle approach is popular because it makes data collection easier and allows the study to focus on the stage of most interest to the stakeholder, even if this is at the expense of proper understanding of impact. These simplifications can lead to a lack of comparability between results of different studies, even though most are based on an International Standard (ISO 14040 2006; ISO 14044 2006). Investing in a complete LCA is clearly advantageous because LCA can contribute to developing sustainable food processing systems by allowing us to understand how a process influences the impact of the whole system, how a local choice can have a global chain of consequences, and how different impacts interact.

11.2.3 Types of LCA

11.2.3.1 Attributional LCA The concept of LCA described by ISO 14040 (2006) is normally thought of as "attributional" LCA (ALCA). An ALCA is static, describing inputs (of resources) and outputs (of pollutants) that are attributed to a specified amount of product, process, or service (Rebitzer et al. 2004). This implies that ALCA is retrospective, i.e. the system must exist, and data must have been collected that represent the state of the system at a fixed and known time, and these are used to calculate the impact that can be attributed to the work of the system. A *retrospective, attributional* calculation, by far the most common for agricultural LCA, allows us to answer the question, what responsibility lies with the system actors (owners, consumers) for the impacts they have caused? (Who can we blame for what impact?) (paraphrased from Weidema 2003). It is possible to use a similar approach to estimate future impacts by imagining the future (a scenario) and estimating what the state of the system might be under that future scenario. It is then possible to make a *prospective, attributional* calculation to answer the question, what will happen in the future and who will be responsible for it? (paraphrased from

Weidema 2003). An example of such a calculation by Sharma et al. (2018) illustrates how this can be done. In the context of people starting to consume sustainable healthy grains, a prospective, attributional LCA might seem very appealing (Hospido et al. 2010). Theoretically it would allow us to compare a future with populations eating in a similar way to now, and with various future scenarios. Such information might allow us to make informed choices about personal consumption and policy needs for the agri-food sector. This ALCA method is however truncated because the introduction of new foods has other consequences. What happens to the commodities we no longer use for food? How does land use change? These are simple questions, but difficult to answer.

11.2.3.2 Consequential LCA Consequential LCA (CLCA) was developed to answer questions about how change in demand will influence environmental impact. The method is used to estimate how inputs (resources) and outputs (pollutants) change, with change in output of the product, process, or service (Ekvall and Weidema 2004; Rebitzer et al. 2004). A *retrospective, consequential* calculation allows us to explain past actions, i.e. why things are like they are and what would have happened if we had done things differently? (paraphrased from Weidema 2003). Perhaps of most interest for development of healthy processed grains is the *prospective, consequential* calculation that allows us to better understand the impact of our actions by predicting the future and allowing us to answer the question, what will happen if we do, or do not do this? (Weidema 2003). CLCA requires data that reflect the internal changes in a system in response to changes in the outputs it creates and market data to identify displaced products (Weidema et al. 1999). A market model is required to determine the consequences of a change in demand (Ekvall and Weidema 2004), and this can be made even more complex when considering multifunctional systems (Weidema et al. 2009; van Zanten et al. 2014). In the simplest case, assuming a direct functional substitution between something currently consumed (e.g. a conventional grain) and a new product (e.g. a novel grain) the convention is, if the market is increasing, the most competitive alternative is displaced and if the market is shrinking, the least competitive alternative is displaced (Weidema et al. 1999). To complete the CLCA model, only those data required to capture the change in demand for the new and displaced product need be included. This approach can be implemented as an *avoided burden* where the impact of the alternative is included in the system model (Sharma et al. 2018). There are arguments about whether prospective attributional or consequential methods are better for providing data on which to base decisions about the future (Finnveden et al. 2009). These revolve around the argument that the market modeling required for CLCA introduces uncertainty, but the static modeling for ALCA does not account for inevitable changes that will occur with changing demand. Industry-led initiatives such as PEF pilots (Manfredi et al. 2015) favor ALCA for its simplicity and certainty, but theoretically CLCA is emerging as the most appropriate approach to inform decision making (Lundie et al. 2007; Finnveden et al. 2009).

11.2.3.3 Economic Input–Output LCA A third type of LCA, which is conceptually very different from ALCA or CLCA is Economic Input–Output LCA (EIO-LCA). This is an extension of the economic method of Input–Output Analysis and its resulting Input–Output Tables (IOT) (Finnveden et al. 2009). An IOT states in monetary terms for each economic sector, how much is bought from other sectors for each unit produced, so when linked to resource use and environmental emissions it can be used to estimate the environmental burden of the supply chain required by a specific sector (Finnveden et al. 2009) without needing details of the specific processes involved. EIO-LCA has not been widely used to investigate individual products, processes, or services because data are aggregated at the sector level, and uncertainty arises due to disaggregation (Yan et al. 2013). To overcome this, a hybrid approach combining process-based LCA and EIO-LCA is becoming more common (Crawford 2008; Inaba et al. 2010). In this chapter, I-O data will not be considered further, and the discussion will focus on process-based data models.

11.3 LCA Study

An ISO standard LCA study comprises four mandatory stages: (i) Goal and Scope; (ii) Life Cycle Inventory (LCI); (iii) Life Cycle Impact Assessment (LCIA); and (iv) Interpretation, connected by a series of feedback loops (Figure 11.3). In the remainder of this chapter, key aspects of the methodology will be described, with a focus on application to grain-based food supply systems, grain processing in general, and healthy grains, where such studies have been published. The role of grain in animal feed and valorization of co- and by-products will not be considered in any detail.

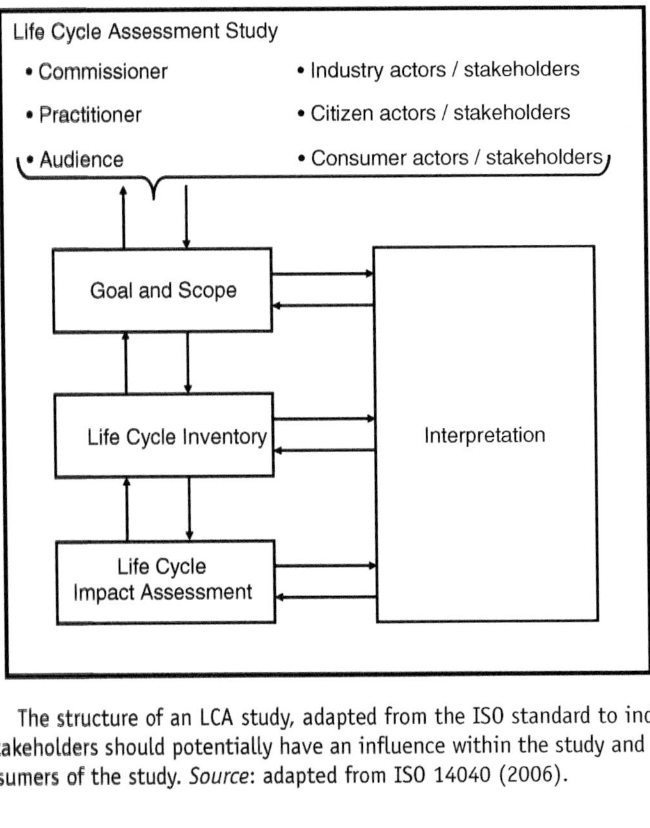

Figure 11.3 The structure of an LCA study, adapted from the ISO standard to indicate that all actors and stakeholders should potentially have an influence within the study and not merely as external consumers of the study. *Source*: adapted from ISO 14040 (2006).

Interestingly, a simple search for LCA and grain or cereal is dominated (the top 20 most cited papers) by studies about bioenergy and feed systems rather than food uses.

11.3.1 Goal and Scope

11.3.1.1 Goal The *goal* of an LCA study should define: (i) the intended application, i.e. how the study will be used; (ii) the reasons for carrying out the study, i.e. why it was commissioned; (iii) the intended audience, i.e. who it is aimed at; and (iv) whether the results are to be used to make comparative assertions about the product, process or service of interest. If the latter is intended, there are strict criteria about what methods can be used and a requirement for independent validation of the study. Relatively few academic LCA studies published in the literature explicitly define these four points, but they are mandatory for every LCA study that claims to meet the ISO standard requirements.

Applications that could be linked to food processing include identifying a product or process with the least environmental impact, combined with cost data, and this could be used for process or business optimization, hotspot analysis to identify those impacts that are within the control of the *commissioner*, and selection between new or alternative methods. Two applications are most likely to be relevant to stakeholders interested in processed healthy grains: the environmental impact of the product (retrospective, attributional LCA) and a comparison of a future diet including novel grains with the current dietary norm (prospective, CLCA). The application dictates the modeling approach that should be used and some of the technical decisions such as the system boundary and the functional unit. The reason for doing the study needs to be declared. Implicit in all studies is the relationship between a commissioner and the *practitioner*, who does the work. The reason provides a context for why the work was commissioned and why the application was considered important. Its primary purpose is to provide the reader with a context against which to judge the validity of the results. Typical reasons for commissioning an LCA study include generation of marketing materials, either based on stand-alone product merit or comparison with alternatives, strategic decision making for a company, and policy development for a sector. In the currently evolving food market, the most likely reason for commissioning an LCA study is perhaps to accrue competitive advantage. The specification of an intended audience is required to inform the reader of the intended stakeholder perspective and to ensure an appropriate use of language. An LCA aimed at consumers or company directors might require a completely different communication style to one aimed at designers, scientists, or production engineers with specialist knowledge and a different appreciation of the system under study. It is reasonable to believe that stakeholders in the healthy grain industry could wish to target any type of audience using LCA, ranging from company directors, to politicians through to consumers. The statement about comparative assertions is required to reveal potential bias, to help identify which stakeholders should be involved, and to flag that weighting is not permitted for impact assessment and mandatory use of independent peer-review of the study. As illustrated in Figure 11.3, during the process of completing an LCA the goal can change if necessary, perhaps due to data or technical limitations or because the original question becomes redundant with new understanding. It is essential for the final report that a coherent goal is stated, the methodology is consistent with the requirements of the goal as stated, and the history of change is transparently recorded and reported in the final LCA report.

11.3.1.2 Scope The *scope* of a study describes the technical implementation used for the LCA modeling in a *transparent* manner. It is worth noting that the ISO standard assumes that a formal report is prepared for each study, but when this is in the form of a journal paper, the description of goal and scope is often truncated to save space. This stage of the LCA should be fully recorded and documented regardless of how the work is reported, as these are the controls that guide the subsequent stages and ensure that both the practitioner and the reader understand the limits of the study. While the goal considers the "who" and "why" of a study, the scope addresses the "which," "what," "where," "when," and "how." The purpose of explicitly describing the limits of the study in the scope is to ensure both the author (while undertaking the study) and the end-user (afterwards) know exactly what the intentions, decisions, and limitations are. It is possible to adjust the scope as part of the study cycle (Figure 11.3), but the changes should be recorded. There are 14 items that should be defined *a priori* (i.e. derived from reason and existing fact and not from the study itself). Each will be considered in turn.

The *system to be studied* (1) should be described as concisely yet accurately and precisely as possible. This is normally accompanied by a system diagram that outlines the key life cycle stages and processes included in the model. More detail is provided about the *foreground* processes (also known as primary contributors, those most closely related to generating the output of the system, or the stage of greatest interest to the commissioner described using the best, most specific, typically empirically observed data). In the case of an LCA for grain processing, the foreground data could relate to cultivation and processing (Cancino-Espinoza et al. 2018) or perhaps be limited to the processing and post-processing into a food product (Andersson and Ohlsson 1999). The rest of the system, the *background* (also known as secondary contributors), is described in less detail using generic data (e.g. average data for production of energy carriers and fertilizers, assumed consumption patterns based on national statistics). Tertiary contributors, such as capital goods and office equipment, are not usually included. The specification of the system, how much material is needed and processed at each stage, and the capacity for output, should be understandable when reading the system description. The *function(s) of the system* (2) should then be explicitly stated to ensure the end-user knows exactly what the system does and why, for example "producing white bread" (Andersson and Ohlsson 1999) and "to deliver organic quinoa . . . to the main export destinations" (Cancino-Espinoza et al. 2018). This item is particularly important for multi-function systems because of the competing demands and pressures associated with each function. Many simple ideas prove to be quite complex once considered in light of function, for instance food packaging has multiple functions (e.g. protection, preservation, marketing, communication), therefore when comparing packaging options, it is necessary to ensure that all options offer similar functions. Likewise, the nutrition and function of grains is not necessarily interchangeable, so where comparisons are to be made (defined in the goal) it is essential that the comparison is for systems with exactly the same function in order for the study to be valid. Having clearly defined the system, it is necessary to choose a *functional unit* (FU) (3), which is a quantitative reference against which impact is expressed. The FU should be an accurate, quantitative description of the function(s) provided by the system based on having identified and quantified relevant properties and technical and functional performance criteria (Rice et al. 2018). Schau and Fet (2008) recommend a sophisticated approach to defining the FU that includes quality aspects. This is widely used in dairy LCA, but less common in grain LCA. An example FU used for studies involving grain and grain processing illustrates the range of possibilities that are available. For example, area (ha) used to produce the grain (Carranza-Gallego et al. 2018), mass (t or kg) of grain (Biswas et al. 2008), mass of processed grain (Meisterling et al. 2009), mass of processed food (e.g. "500 g packet of quinoa ready for retailing") (Cancino-Espinoza et al. 2018) and "annual production of ready-to-eat breakfast cereal products (388,000 tonnes)" (Jeswani et al. 2018), nutritionally adjusted mass of grain (e.g. "three weekly well-balanced diets, equivalent to one another for energetic and nutrient content") (Baroni et al. 2007), and per mass of nutrition provided (e.g. "per g protein") (Cancino-Espinoza et al. 2018). It is also necessary to define a *reference flow* (4), which describes the amount of material and resources necessary for the system to deliver the functional unit. In the case of comparisons, different systems delivering the same function may require different reference flows to achieve the function. The final part of the system description in the scope is to explicitly define the *system boundary* (5), indicating which processes and life cycle stages are considered in the model, and which are excluded. Ideally the system model should encompass all stages in the life cycle (*cradle-to-grave*, or according to more recent thinking *cradle-to-cradle*): raw material acquisition, manufacturing, use (including reuse and maintenance), and recycling/waste management (Figure 11.1). In practice it is common to ignore parts of the systems, such as cradle-to-retailer (Blengini and Busto 2009) or cradle-to-destination port (Pelletier and Tyedmers 2010), a focus on just processing (*gate-to-gate*) (Finnegan et al. 2018), or perhaps processing and use phases (*gate-to-grave*). There are many studies involving processing for valorization, particularly bioenergy, which set the system boundary to omit food functions (Parajuli et al. 2018), and there is an emerging focus on the end-of-life phase due to interest in the circular bioeconomy (Oldfield et al. 2016).

Having defined the system and its function(s), perhaps the most important decision for grain LCA practitioners is selection of *allocation method* (6). Allocation arises in LCA because most systems contain processes that produce more than one output (e.g. wheat produces grain and straw, a cow produces milk, meat, and skin for leather). According to the ISO standards, allocation should be avoided by splitting, which is often not possible, or by system expansion (van Zanten et al. 2014) where the model is expanded to account for the multiple outputs. However, the vast majority of agricultural LCA studies to date have used allocation of one kind or another. There is no accepted objective function that can be used to choose an allocation method, but economic (Carranza-Gallego et al. 2018; van Stappen et al. 2018), mass-energy (Hoffman et al. 2018), and mass (Van Stappen et al. 2018) predominate. It would also be possible to use nutrient content, such as protein. Rice et al. (2017) argue that one approach to choosing a method is to assess the quality of the data available for the calculation. The *data requirements* (7) and *data quality* (8) should also be defined before the data are collected. The type (e.g. site specific, national average), minimum standard (e.g. from surveys, national statistics, estimates and guesses), and rigor with which the data have been compiled and reviewed must reflect the intended goal of the study. A first-look, screening LCA for private use will have very different data standards to a study published associated with a named brand (Jeswani et al. 2018). It is also necessary to define the ideal spatial and temporal representation and technological specificity to meet the intended goal. This may be modified during the study depending on the quality of data available. Aspects of system description and data are often chosen using *assumptions* (9) and all assumptions must be documented. For example, Jeswani et al. (2018) assumed that breakfast cereals were consumed in the UK when defining their study because that it the largest market in Europe, which meant data for UK milk were used for the study. All assumptions should be documented to ensure transparency. The ISO standard does not define a minimum data specification, but it does define a minimum requirement for honest reporting of the quality of the data used.

There are many different *impact methods* (10) that can be used to model the impact of the system using the elementary flow data (the LCIA stage). The principle of the methods is to (a) assign an elementary flow to an impact category, (b) calculate an indicator of the damage that can be caused (the *midpoint*) and, if part of the method, (c) estimate the damage that will result (the *endpoint*). A good summary can be found in Baumann and Tillman (2004) and a more detailed description in Jolliet et al. (2016). Each impact method calculates results for one or more specific *impact category* (e.g. climate change, acidification, eutrophication, resource depletion, human toxicity, ionizing radiation, ozone layer depletion, particulate emissions, photochemical oxidation, land use, and ecotoxicity). There are now dozens of impact methods that can be used to calculate the consequence of a particular elementary flow and combine flows with similar impacts into a single result. Examples used for grain studies include ReCiPe (Cancino-Espinoza et al. 2018), the Intergovernmental Panel on Climate Change (IPCC) (climate change only) (Carranza-Gallego et al. 2018; Chaudhary et al. 2018), Ecoindicator 99 (Baroni et al. 2007), the Institute of Environmental Sciences (CML) (Charles et al. 2006; Taki et al. 2018), the Land Use Indicator Value Calculation Tool (LANCA) (Jeswani et al. 2018), EDIP97 (Nemecek et al. 2008), and the International Reference Life Cycle Data System (ILCD) (Van Stappen et al. 2018). The selection of a method depends on the goal, and the specific interests of the commissioner and target audience. For some impacts, results are similar among methods (e.g. climate change), but for others they can be very different (e.g. aquatic ecotoxicity) (Dreyer et al. 2003). The approach that will be taken to *interpretation* (11) will dictate what data are required and how they will be used. Typically, interpretation will include identifying *significant issues*, i.e. those parts of the system that contribute most to specific impacts and also some quality control including a *completeness check* (is the inventory complete, including elementary flows?), a *sensitivity analysis* (particularly important for parts of the model dependent on estimates or assumptions), and a *consistency* or *pedigree* check to confirm that all the data are similarly representative of the system. If the interpretation is going to be influenced by *value choices* (12), these should be declared, such as existing priorities due to legislation or role as a stakeholder, and any known *limitations* (13) should be clearly stated for the reader to judge. Finally, the *type of critical review* (14) should be declared. This may range from a single reviewer to a panel of experts and stakeholders, depending on the stated goal.

11.3.2 Life Cycle Inventory

The life cycle inventory (LCI) is the process of compiling the necessary data to quantify energy and material in the system. A formal system diagram/flow chart is constructed of the system, in the technosphere, that defines the relationship between each *unit process* (an activity that completes one task within the system) and *system process* (an activity that completes multiple tasks that cannot be disaggregated to unit processes) necessary for the system to function. Each process is linked by mass and/or energy flows (*activity data* for *valuable substances*), and each is linked to the ecosphere by polluting or resource depleting flows (*elementary flows*). Ideally, each process is defined in the simplest form possible (Figure 11.4) and these are combined to construct the complete system (Figure 11.5). The processes in the LCI should be described in terms of energy inputs, raw material inputs, chemical inputs, product, or service outputs and emissions to air, water, and waste (Baumann and Tillman 2004). Depending on the software used, the system can be constructed directly in software, or outlined separately and then implemented. An overview of available software and issues around its use can be found in Crioth (2012) and Jolliet et al. (2016). The flow chart is critical to the success of the LCA study. If focused on processing, most facilities will have extant design and process control diagrams that can be used. Where such information is unavailable, principles of engineering design can be used to work out what the most likely process combination would be, or the whole of a plant can be treated as a black-box, system process, without reference to internal detail. This will limit the interpretation possible so should only be used if compatible with the goal of the study. The diagram should clearly define the system boundary (as defined by the Goal and Scope) and each major process, while being as simple as necessary for the study.

When a company stakeholder is involved with a study, it will be relatively straightforward to capture technosphere data for foreground processes because there will be a bill-of-materials for most products and economic data and machinery specifications that can be used to describe energy consumption and throughputs. Data can also often be acquired from suppliers and customers, but capturing elementary flow that are polluting losses to the wider environment can be difficult. Some can be estimated from mandatory environmental monitoring associated with operational licensing, but typically these flows are estimated from other sources or using an appropriate calculation method. Obtaining downstream consumer and waste management data can be difficult, with studies having to rely on national consumer behavior survey data, which might not be

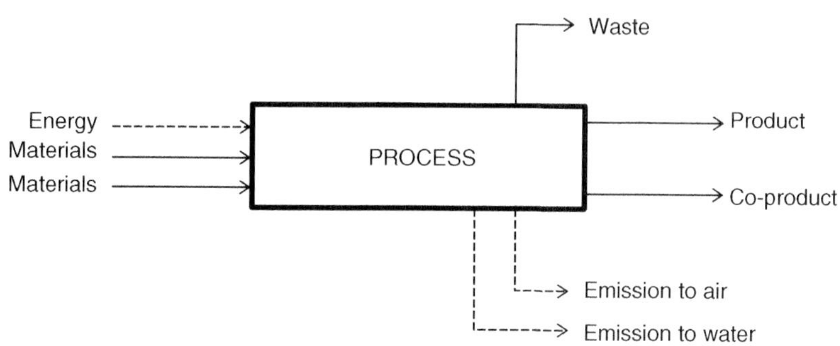

Figure 11.4 A simple representation of a process (ideally a unit process, but could be an open system process) with the types of data that need to be quantified for the life cycle inventory. The dashed arrows represent elementary flows and the solid arrows valuable substances (see Figure 11.2). Depending on how it is treated, "waste" can become a valuable substance if valorized, or can be returned to the ecosphere and beyond the reach of the technosphere. *Source*: after Holden and Yan (2016); Oldfield et al. (2016).

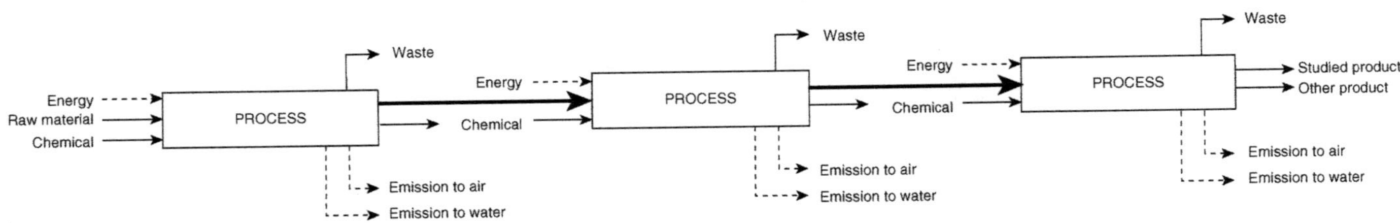

Figure 11.5 Conceptual structure of the LCI in terms of interconnected processes, each of which is described by *activity data* representing mass and energy flows into and out of the process.

reliable for a specific product such as a healthy grain. The effort required for collation of activity data depends on the goal of the study. If working at sectoral, or national level, industry statistics can be used to estimate average activity, but if working at processing site level, or specific product level, site-specific activity data will have to be collected. Foreground data should always be geographically and temporally specific (e.g. grain yield for the source field, farm, district, county, country, or region as required by the goal), but background data can be general (e.g. national average electricity grid mix for the country of interest). To give some idea of the types of data and sources used for grain LCA studies, a scan of the literature reveals a range of sources. Cancino-Espinoza et al. (2018) used data collected from 14 farms and agreed benchmarks for clusters of regional producers of quinoa, combined with background data from the Ecoinvent database, while Charles et al. (2006) used a range of theoretical calculations to assess wheat crop production, and Meisterling et al. (2009) used existing databases of agricultural and other statistics to assess wheat. Nemecek et al. (2008) relied on Ecoinvent to assess grain legumes in crop rotations, while Taki et al. (2018) used a combination of farmer interviews, the literature, and the Ecoinvent and Agri-Footprint databases to assess wheat production in Iran. Van Stappen et al. (2018) report using collated production statistics from Wallonia, Belgium. Jeswani et al. (2018) relied largely on previously published studies for source data to calculate the impact of breakfast cereals, and Baroni et al. (2007) used textbooks and scientific papers to collate data to study dietary patterns. For ALCA modeling, it is common to use average data that represent classes of processes or products rather than site-specific data, but this quick snapshot reveals both approaches have and continue to be used, depending on the goal of the study.

There is now a wide range of databases (Table 11.1) that can be used for scoping studies and to provide background data for studies specifically focused on the processing stage. Most studies can be completed using a combination of in-house data and commercial databases to achieve an LCI with suitable data quality.

11.3.3 Life Cycle Impact Assessment

The LCIA stage will not be discussed in great detail because for most studies, the implementation is automated in software. It is important to note that the impact calculation will only be as good as the completeness of the elementary flow data that are relevant for each impact category. While some studies stop at the LCI stage, with an estimate of total emissions and consumptions, this makes interpretation for both the expert and the general public difficult because it is not clear what the implications of a given emission might be. In the LCIA, the total emissions are processed to either midpoint or endpoint impact (Figure 11.6) and expressed per functional unit. Midpoint impacts are a "problem-oriented approach" because they model impacts on environmental mechanisms somewhere between the emission and the damage, while endpoint impacts are a "damage-oriented approach" because they estimate specific impacts of resource use or depletion, human health, and ecological consequences. The final value, reflecting the intensity of impacts per unit product, or *eco-efficiency*, is useful for comparing the relative impact of comparable products (Figure 11.7), but is not necessarily informative as it can lead to comparison of inherently unsustainable products and increased consumption (Anders and Hauschild 2012).

There are seven components in the LCIA stage: (i) selection and definition of impact categories, which is started, and potentially revised as part of the scope stage;

Table 11.1 Summary of some of the LCA databases containing process data related to grain crops.

Name	Source	Incudes
Agribalyse	Created by consortium of French partners. Available through software providers. See https://ecolab.ademe.fr/agribalyse for details	Durum wheat Soft wheat Rapeseed Faba bean Grain maize Barley Sunflower Triticale
Agri-footprint	Developed by Blonk Consultants (the Netherlands). Available through purchase of software. See www.agri-footprint.com for details	Rye flour milling Maize flour milling Maize starch (milling/drying) Oat grain peeled (milling) Wheat flour milling Wheat germ (milling) For 13 countries
AustLCI	Developed by the Australian Life Cycle Assessment Society. Available for download. See www.auslci.com.au for details	Wheat products Barley production Maize production Wheat production For Australia
Ecoinvent	Maintained by a not-for-profit organization in Switzerland. Available by access license or through purchase of software. See www.ecoinvent.org for details	Rye production Wheat production Barley production Rice production
ESU World Food	Data compiled by ESU-Services (Germany). Available for purchase. See http://esu-services.ch/data/fooddata for details	Cereals and cereal preparations For 1 country
exiobase	An input–output database managed by a number of partners. Free download available. See www.exiobase.eu for details	Cereal grains For 48 countries
Feed and Food	Available from thinkstep (Gernany and UK). Data available for purchase. See http://www.gabi-software.com/international/databases/gabi-databases/food-feed for details	Maize Wheat Soybean Rape/Canola Barely Triticale Lupine Sunflower Rice

Table 11.1 (*continued*)

Name	Source	Incudes
Feedprint	Developed by and available from Wageningen UR. Free to download. See http://webapplicaties.wur.nl/software/feedprintNL/index.asp for details	Barley Maize Millet Oats Rice Rye Sorghum Triticale Wheat Linseed Rape Soybean Sunflower Buckwheat Various legumes
Gabi	Part of a more general database offered by thinkstep with their LCA software. Available or purchase. See http://www.gabi-software.com/international/databases/gabi-databases for details	Linseed meal Linseed oil Wheat flour (milling) For 4 countries
idea	Japanese developed and maintained database. Available for purchase and integrated with software. See http://idea-lca.com/?lang=en for details	Wheat (various types) Barley (various types) Cereals (miscellaneous) Rice For Japan
PSILCA	A social LCA database developed by GreenDelta (Germany). Available for purchase. See psilca.net for details.	Breakfast cereal manufacturing Oat Sorghum Other cereals For 3 countries
US LCI	Held by the National Agriculture Library. Available for download. See https://uslci.lcacommons.gov/uslci/search for details.	Wheat production Rice production Maize production For USA

(ii) classification, where each elementary flow is assigned to at least one impact category (note some emissions are relevant to more than one impact category, in which case the method has to account for whether the emission can only have a single impact by being consumed during impact, or whether it applies to multiple impacts); (iii) characterization, where the polluting substance is expressed in terms of a standard impact, most people are aware of this in the context of converting methane and nitrous oxide to carbon dioxide equivalent units for expressing carbon footprint or climate change impact; (iv) normalization, where the impact is expressed as a proportion of a regional reference value making it possible to understand the relative importance of different impact categories; (v) grouping, where impacts are combined together, perhaps for transport, or national vs international; (vi) weighting, which is used to express

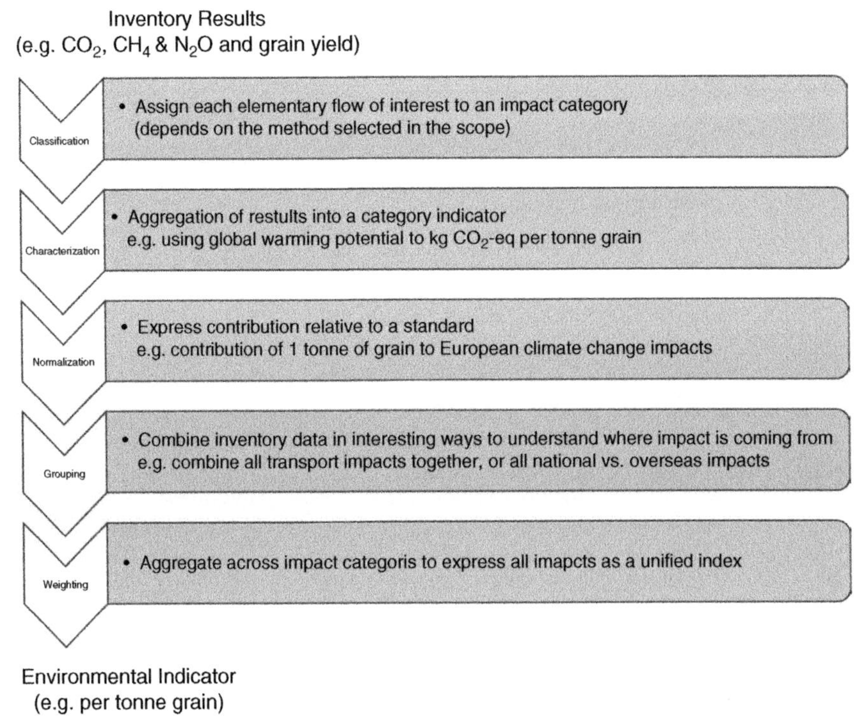

Figure 11.6 The generalized procedure for life cycle impact assessment (LCIA) illustrated using climate change associated with grain production.

the impact in terms of a perceived importance for stakeholders, usually defined by an expert panel or using monetary value; and (vii) evaluating and reporting, which overlaps into the interpretation stage. Normalization, grouping, and weighting are optional and should not be used for product comparison studies, but the others stages are mandatory for an ISO standard LCA study.

Typically, climate change impact is of most interest (Clune et al. 2017), but other impacts reported include ozone depletion, acidification, photochemical oxidant, particulate matter, metal depletion, fossil fuel depletion (Cancino-Espinoza et al. 2018), eutrophication (e.g. Andersson and Ohlsson 1999), land use (van Zanten et al. 2014), human toxicity, ecotoxicity, agricultural land occupation potential, water depletion (Van Stappen et al. 2018), abiotic depletion (Taki et al. 2018), and various water footprint indicators (Pfister et al. 2011). The selection of impact categories and methods should always reflect the stated goal, agreed in conjunction with the commissioner of the study.

11.3.4 Life Cycle Interpretation

The scope defined how the interpretation phase of the LCA is to be conducted. Significant issues may be those processes, stages, or classes of activity (e.g. energy generation, transport, domestic consumption) that make large contributions to the impact. These are readily found by graphing data, but formal methods such as *contribution*

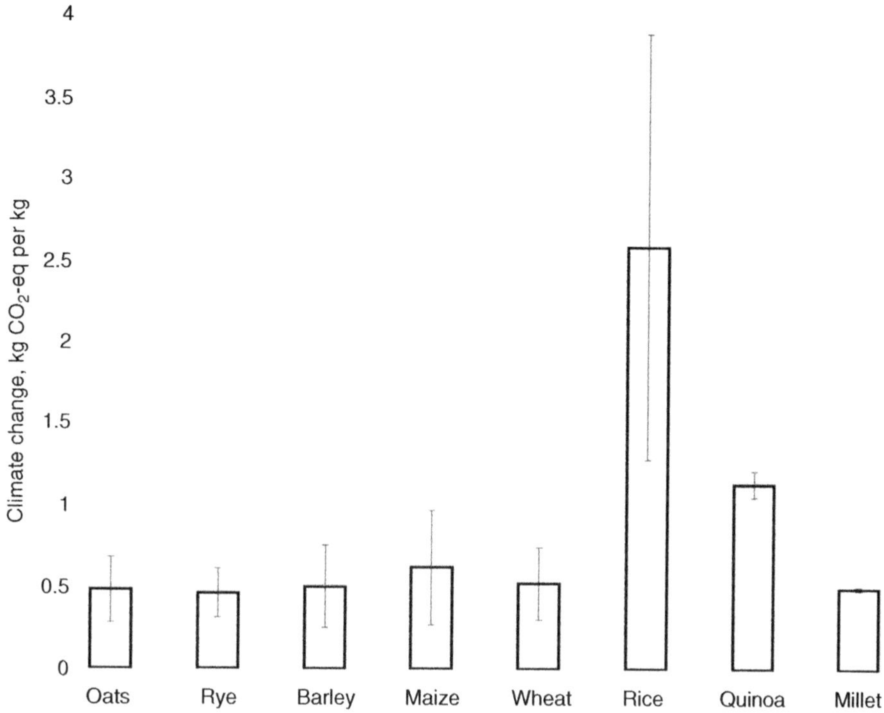

Figure 11.7 Example of LCIA data, by crop type, illustrated for the midpoint impact, climate change. *Source*: derived from data published by Audsley et al. (2009) as presented by Clune et al. (2017).

analysis (scoring relative impact, conveniently presented using a heat map), *dominance analysis* (using quantitative or statistical methods), and *anomaly analysis* (by comparison with published knowledge, experience, or expert knowledge) are available (Guinée 2002). Completeness analysis can require careful review not only of the system model, but also of the data used when implemented in LCA software (e.g. sphera Gabi, PRE Consult SimaPro, or GreenDelta OpenLCA and others). There is no guarantee that database entries are complete. Missing elementary flows can have an unforeseen influence on LCIA results unless spotted. The pedigree of the data can be judged using an agreed matrix, the most common of which was published by Weidema and Wesnaes (1996), which has been adapted more recently. Consistency checking should answer questions such as: (i) are the data quality and any differences within the system definition consistent with the goal and scope? (ii) have regional and temporal issues been dealt with in a consistent manner? (iii) has allocation and the system boundary been applied consistently? and (iv) have all elements of the impact assessment been applied consistently? *Sensitivity analysis* is typically used to evaluate whether estimates or assumptions are likely to have had an undue influence on the results, but it can also be used to evaluate decisions associated with allocation, cut-off (deliberate omission of data because it is thought, or known, to contribute very little to the overall impact of the system), and the boundary choice. While the best possible option should have been selected based on the scope of the study, it is always worthwhile understanding the impact of such decisions on the final conclusions. *Uncertainty analysis* is increasingly being used to interpret results in terms of the probability distribution of possible outcomes. The two most common approaches are to define possible scenarios or to use a quantitative method such as Monte Carlo analysis (Flysjö et al. 2011). The final stage of interpretation is to draw conclusions, and as appropriate make recommendations. Conclusions should focus on what the major impacts are, the relative magnitude of the different impacts analyzed, and a statement that the conclusions are consistent with any limitations of the methods and data that have been identified. If recommendations are made, they must reflect a logical and reasonable consequence of the conclusions, an explanation should be provided, and they must always relate to the intended application as defined by the goal and scope. If the study is comparative and for a public audience, then a full critical analysis must be performed.

11.4 LCA Studies on Cereal and Cereal-Based Products Processing

Looking at the studies published in journals, it is apparent that there has been a focus on bioenergy, and that food studies have concentrated on major cereal production from cradle-to-farm gate. A meta-analysis of 369 published LCA studies between 2000 and 2015 found that there were 51 LCA studies on wheat, 27 on rice, 13 on barley, 6 on maize, and 2 on quinoa (Clune et al. 2017). Rice tends to have higher climate change impact, mainly due to methane emissions from the flooded rice fields (Figure 11.7). Another study of worldwide crop production also found that rice has the greatest climate change impact of the 27 studied crops (Nemecek et al. 2012). There have been few studies that have gone beyond climate change, Jeswani et al. (2018) being an exception that focused on ecosystem services, and Achten and Van Acker (2016) that looked at non-renewable fossil energy, acidification freshwater eutrophication, marine eutrophication, and land occupation. While local and global estimates of the impact of common grain production systems have been published, and findings such as those of González et al. (2011) who ranked energy use and greenhouse gas (GHG) emissions in the order oat>rye>barley and wheat>corn>rice, Shiklomanov and Rodda (2003) who estimated global consumptive freshwater use, Foley et al. (2005) who estimated global land use, and Pfister et al. (2011) who estimated that global grain production needed 0.927 m^2yreq/kg (land use) and 182 Leq/kg (water use), are useful, these tell us little about novel grains, and less about the impact of processing. Millet and quinoa have been reported to have high impacts for both land and water due to very low yields and unfavorable climatic conditions, but buckwheat is reported to have significantly lower impacts than millet and quinoa due to higher production and yield (FAO 2016). Of the few LCA studies on cereal products beyond the farm gate for pasta, the trip to the supermarket and the cooking (consumer phase) were the largest contributor (47%) to the total energy consumption, followed by semolina production (23%) and pasta production (20%), while the wheat cultivation contributed less than 4% (Bevilacqua et al. 2007). A study on environmental impacts of 21 different types of traditional bread consumed across the European Union found that the cumulative energy demand ranged from 9.1 to 32.9 MJ/kg, and climate change ranged from 0.5 to 6.6 kg CO_2eq/kg (Notarnicola et al. 2017). Contribution of wheat and/or rye storage was small, while contribution of wheat/rye milling was around 10% across different breads. Jeswani et al. (2018) did not report contribution of different life cycle stages for breakfast cereal production, and Cancino-Espinoza et al. (2018) reported little data for the processing stage of quinoa, other than packaging, which represented about 1% of climate change impact.

Dietary choices have been found to have a significant impact on individual environmental impacts, which has implications for the development, promotion, and demand for healthy novel grains. A recent review of 14 journal articles assessing the GHG emissions and land use demand found that dietary change could result in an up to 50% potential reduction in GHG emissions and land use demand (Hallström et al. 2015). Furthermore, a study on the increasing preference for pasta and rice over potato in UK households found a transfer of burdens from the UK to Italy (dried pasta) and India (dried basmati rice) (Hess et al. 2016). Other potentially relevant considerations are the impacts processing has on chemical and nutritional properties of novel grains. Meharg et al. (2013) found milling reduces cadmium concentration in rice by 20–40% and Duarte et al. (2010) found processing could reduce mycotoxins in processed cereal products, probably by removal of surface layers by abrasive scouring or polishing and because mycotoxin tends to be concentrated in the bran. However, this inevitably leads to redistribution of Ochratoxin A in certain milling fractions. Autoclaving and extrusion have also been found to be effective (Duarte et al. 2010). It is important that empirical data become available to model the environmental impact of processed healthy grains before they emerge from the innovation pipeline into the marketplace, if they are to be associated with sustainability as well as nutritional messages.

11.5 Conclusion

As society and individuals are becoming more concerned about the sustainability of production and consumption, it is likely that the holistic assessment method, LCA, will become more important for those involved with development of novel grain production and processing technologies. As the method matures and more data become available, LCA is being applied to a wider range of grains and their products. It provides insights into how farming, processing, consumer behavior, and dietary choices together shape environmental impacts. The potential environmental and nutritional benefits of novel grains are yet to be assessed with sufficiently detailed data.

References

Achten, W.M.J. and Van Acker, K. (2016). EU-average impacts of wheat production: a meta-analysis of life cycle assessments. *Journal of Industrial Ecology* 20: 132–144.

Anders, B. and Hauschild, M.Z. (2012). Absolute versus relative environmental sustainability. *Journal of Industrial Ecology* 17: 321–332.

Andersson, K. and Ohlsson, T. (1999). Life cycle assessment of bread produced on different scales. *International Journal of Life Cycle Assessment* 4: 25–40.

Audsley, E., Brander, M., Chatterton, J. et al. (2009). *How Low Can we Go? An Assessment of Greenhouse Gas Emissions from the UK Food System and the Scope to Reduce them by 2050*. UK: World Wildlife Fund.

Baroni, L., Cenci, L., Tettamanti, M., and Berati, M. (2007). Evaluating the environmental impact of various dietary patterns combined with different food production systems. *European Journal of Clinical Nutrition* 61: 279–286.

Basset-Mens, C., Small, B., Paragahawewa, U.H. et al. (2009). Life cycle thinking and sustainable food production. *International Journal of Product Lifecycle Management* 4: 252–269.

Baumann, H. and Tillman, A.-M. (2004). *The Hitch Hiker's Guide to LCA*. Sweden: Studentlitteratur.

Bevilacqua, M., Braglia, M., Carmignani, G., and Zammori, F.A. (2007). Life cycle assessment of pasta production in Italy. *Journal of Food Quality* 30: 932–952.

Biswas, W.K., Barton, L., and Carter, D. (2008). Global warming potential of wheat production in Western Australia: a life cycle assessment. *Water Environment Journal* 22: 206–216.

Blengini, G.A. and Busto, M. (2009). The life cycle of rice: LCA of alternative agri-food chain management systems in Vercelli (Italy). *Journal of Environmental Management* 90: 1512–1522.

Bougherara, D. and Combris, P. (2009). Eco-labelled food products: what are consumers paying for? *European Review of Agricultural Economics* 36: 321–341.

Cancino-Espinoza, E., Vázquez-Rowe, I., and Quispe, I. (2018). Organic quinoa (*Chenopodium quinoa* L.). production in Peru: environmental hotspots and food security considerations using life cycle assessment. *Science of the Total Environment* 637-638: 221–232.

Carbon Trust (2018). Carbon footprinting guide: how to measure, calculate and communicate your carbon footprint. CTV043 v3. London: Carbon Trust. https://www.carbontrust.com/resources/guides/carbon-footprinting-and-reporting/carbon-footprinting (accessed 29 August 2018)

Carranza-Gallego, G., Guzmán, G.I., García-Ruíz, R. et al. (2018). Contribution of old wheat varieties to climate change mitigation under contrasting managements and rainfed Mediterranean conditions. *Journal of Cleaner Production* 195: 111–121.

Casey, J.W. and Holden, N.M. (2005). Analysis of greenhouse gas emissions from the average Irish milk production system. *Agricultural Systems* 86: 97–114.

Charles, R., Jolliet, O., Gaillard, G., and Pellet, D. (2006). Environmental analysis of intensity level in wheat crop production using life cycle assessment. *Agriculture, Ecosystems and Environment* 113: 216–225.

Chaudhary, A., Marinangeli, C.P.F., Tremorin, D., and Mathys, A. (2018). Nutritional combined greenhouse gas life cycle analysis for incorporating Canadian yellow pea into cereal-based food products. *Nutrients* 10 (4): 490.

Chen, W. and Holden, N.M. (2018). Tiered life cycle sustainability assessment applied to a grazing dairy farm. *Journal of Cleaner Production* 172: 1169–1179.

Clune, S., Crossin, E., and Verghese, K. (2017). Systematic review of greenhouse gas emissions for different fresh food categories. *Journal of Cleaner Production* 140: 766–783.

Crawford, R.H. (2008). Validation of a hybrid life-cycle inventory analysis method. *Journal of Environmental Management* 88: 496–506.

Crioth, A. (2012). Software for life cycle assessment. In: *Life Cycle Assessment Handbook, A Guide for Environmentally Sustainable Products* (ed. M.A. Curran), 143–158. Salem: Scrivener and Hoboken: Wiley.

Dreyer, L.C., Niemann, A.L., and Hauschild, M.Z. (2003). Comparison of three different LCIA methods: EDIP97, CML2001 and Eco-indicator 99: does it matter which one you choose? *International Journal of Life Cycle Assessment* 8 (4): 191–200.

Duarte, S.C., Pena, A., and Lino, C.M. (2010). A review on ochratoxin A occurrence and effects of processing of cereal and cereal derived food products. *Food Microbiology* 27: 187–198.

Ekvall, T. and Weidema, B.P. (2004). System boundaries and input data in consequential life cycle inventory analysis. *International Journal of Life Cycle Assessment* 9: 161–171.
European Commission (2015). PEFCR Pilot Fiche DAIRY. http://ec.europa.eu/environment/eussd/smgp/pdf/Fiche_dairy.pdf (accessed 29 August 2018).
European Commission (2016). PEFCR Pilot Fiche MEAT. http://ec.europa.eu/environment/eussd/smgp/pdf/Fiche_meat.pdf (accessed 29 August 2018).
FAO (Food and Agriculture Organization of the United Nations) (1998). *The State of Food and Agriculture 1998*. Rome: Food and Agriculture Organization of the United Nations http://www.fao.org/3/a-x9800e.pdf (accessed 29 August 2018).
FAO (2016). *Food and Agriculture Data*. Rome: Food and Agriculture Organization of the United Nations http://www.fao.org/faostat/en/#data/QC (accessed 29 August 2018).
Finnegan, W., Yan, M., Holden, N.M., and Goggins, J. (2018). A review of environmental life cycle assessment studies examining cheese production. *International Journal of Life Cycle Assessment* 23: 1773–1787.
Finnveden, G., Hauschild, M.Z., Ekvall, T. et al. (2009). Recent developments in life cycle assessment. *Journal of Environmental Management* 91: 1–21.
Flysjö, A., Cederberg, C., Henriksson, M., and Ledgard, S. (2011). How does co-product handling affect the carbon footprint of milk? Case study of milk production in New Zealand and Sweden. *International Journal of Life Cycle Assessment* 16: 420–430.
Foley, J.A., Defries, R., Asner, G.P. et al. (2005). Global consequences of land use. *Science* 309: 570–574.
González, A.D., Frostell, B., and Carlsson-Kanyama, A. (2011). Protein efficiency per unit energy and per unit greenhouse gas emissions: potential contribution of diet choices to climate change mitigation. *Food Policy* 36: 562–570.
Gruère, G.P. (2015). An analysis of the growth in environmental labelling and information schemes. *Journal of Consumer Policy* 38: 8.
Guinée, J. (2002). *Handbook on Life Cycle Assessment Operational Guide to the ISO Standards*. Dordrecht: Kluwer Academic Publishers.
Hallström, E., Carlsson-Kanyama, A., and Börjesson, P. (2015). Environmental impact of dietary change: a systematic review. *Journal of Cleaner Production* 91: 1–11.
Hess, T., Chatterton, J., Daccache, A., and Williams, A. (2016). The impact of changing food choices on the blue water scarcity footprint and greenhouse gas emissions of the British diet: the example of potato, pasta and rice. *Journal of Cleaner Production* 112: 4558–4568.
Hoffman, E., Cavigelli, M.A., Camargo, G. et al. (2018). Energy use and greenhouse gas emissions in organic and conventional grain crop production: accounting for nutrient inflows. *Agricultural Systems* 162: 89–96.
Holden, N.M. and Yan, M.-J. (2016). Life cycle assessment and sustainable food processing. In: *Sustainable Food Processing* (eds. B.K. Tiwari, T. Norton and N.M. Holden), 63–92. Oxford: Wiley-Blackwell.
Hospido, A., Davis, J., Berlin, J., and Sonesson, U. (2010). A review of methodological issues affecting LCA of novel food products. *International Journal of Life Cycle Assessment* 15: 44–52.
Inaba, R., Nansai, K., Fujii, M., and Hashimoto, S. (2010). Hybrid life-cycle assessment (LCA) of CO_2 emission with management alternatives for household food wastes in Japan. *Waste Management and Research* 28: 496–507.
ISO 14025 (2006). *Environmental labels and declarations – Type III environmental declarations – Principles and procedures*. Geneva: International Organization for Standardization.
ISO 14040 (2006). *Environmental management – Life cycle assessment – Principles and framework*. Geneva: International Organization for Standardization.

ISO 14044 (2006). *Environmental management – Life cycle assessment – Requirements and guidelines*. Geneva: International Organization for Standardization.

ISO 14044 (2017). *Environmental management – Life cycle assessment – Requirements and guidelines – Amendment 1*. Geneva: International Organization for Standardization.

Jeswani, H.K., Hellweg, S., and Azapagic, A. (2018). Accounting for land use, biodiversity and ecosystem services in life cycle assessment: impacts of breakfast cereals. *Science of the Total Environment* 645: 51–59.

Jolliet, O., Saade-Sbeih, M., Shaked, S. et al. (2016). *Environmental Life Cycle Assessment*. Boca Raton, London, New York: CRC Press Taylor and Francis Group.

Khan, S. and Hanjra, M.A. (2009). Footprints of water and energy inputs in food production – global perspectives. *Food Policy* 34: 130–140.

Lundie, S., Ciroth, A., and Huppes, G. (2007). Inventory methods in LCA: towards consistency and improvement – Final Report. *UNEP-SETAC Life Cycle Initiative*.

Manfredi, S., Allacker, K., Pelletier, N. et al. (2015). Comparing the European Commission product environmental footprint method with other environmental accounting methods. *International Journal of Life Cycle Assessment* 20: 389–404.

Meharg, A.A., Norton, G., Deacon, C. et al. (2013). Variation in rice cadmium related to human exposure. *Environmental Science and Technology* 47: 5613–5618.

Meisterling, K., Samaras, C., and Schweizer, V. (2009). Decisions to reduce greenhouse gases from agriculture and product transport: LCA case study of organic and conventional wheat. *Journal of Cleaner Production* 17: 222–230.

Mekonnen, M.M. and Hoekstra, A.Y. (2011). The green, blue and grey water footprint of crops and derived crop products. *Hydrology and Earth System Sciences* 15: 1577–1600.

Morgan, R.K. (1998). *Environmental Impact Assessment, A Methodological Approach*. Norwell, MA, USA: Springer US.

Nemecek, T., von Richthofen, J.-S., Dubois, G. et al. (2008). Environmental impacts of introducing grain legumes into European crop rotations. *European Journal of Agronomy* 28: 380–393.

Nemecek, T., Weiler, K., Plassmann, K. et al. (2012). Estimation of the variability in global warming potential of worldwide crop production using a modular extrapolation approach. *Journal of Cleaner Production* 31: 106–117.

Notarnicola, B., Tassielli, G., Renzulli, P.A., and Monforti, F. (2017). Energy flows and greenhouses gases of EU (European Union). National breads using an LCA (Life Cycle Assessment) approach. *Journal of Cleaner Production* 140: 455–469.

Oldfield, T.L., White, E., and Holden, N.M. (2016). An environmental analysis of options for utilising wasted food and food residue. *Journal of Environmental Management* 183: 826–835.

Padgett, J.P., Steinemann, A.C., Clarke, J.H., and Vandenbergh, M.P. (2008). A comparison of carbon calculators. *Environmental Impact Assessment Review* 28: 106–115.

Parajuli, R., Knudsen, M.T., Birkved, M. et al. (2018). Environmental impacts of producing bioethanol and biobased lactic acid from standalone and integrated biorefineries using a consequential and an attributional life cycle assessment approach. *Science of the Total Environment* 598: 497–512.

Pelletier, N. and Tyedmers, P. (2010). Life cycle assessment of frozen tilapia fillets from Indonesian lake-based and pond-based intensive aquaculture systems. *Journal of Industrial Ecology* 14: 467–481.

Pfister, S., Bayer, P., Koehler, A., and Hellweg, S. (2011). Environmental impacts of water use in global crop production: hotspots and trade-offs with land use. *Environmental Science and Technology* 45: 5761–5768.

Rebitzer, G., Ekvall, T., Frischknecht, R. et al. (2004). Life cycle assessment part 1: framework, goal and scope definition, inventory analysis, and applications. *Environment International* 30: 701–720.

Rice, P., O'Brien, D., Shalloo, L., and Holden, N.M. (2017). Evaluation of allocation methods for calculation of carbon footprint of grass-based dairy production. *Journal of Environmental Management* 202: 311–319.

Rice, P., O'Brien, D., Shalloo, L., and Holden, N.M. (2018). Defining a functional unit for dairy production LCA that reflects the transaction between the farmer and the dairy processor. *International Journal of Life Cycle Assessment* 24: 642–653.

Rockström, J., Steffen, W., Noone, K. et al. (2009). Planetary boundaries: exploring the safe operating space for humanity. *Ecology and Society* 14 (2): 32.

Roy, P., Nei, D., Orikasa, T. et al. (2009). A review of life cycle assessment (LCA) on some food products. *Journal of Food Engineering* 90: 1–10.

Schau, E.M. and Fet, A.M. (2008). LCA studies of food products as background for environmental product declarations. *International Journal of Life Cycle Assessment* 13: 255–264.

Sharma, P., Humphreys, J., and Holden, N.M. (2018). Environmental impacts of alternative agricultural uses of poorly drained farm land in Ireland. *Science of the Total Environment* 637-638: 120–131.

Shiklomanov, I.A. and Rodda, J.C. (2003). *World Water Resources at the Beginning of the 21st Century*. Cambridge: Cambridge University Press.

Swarr, T.E., Hunkeler, D., Klöpffer, W. et al. (2011). Environmental life-cycle costing: a code of practice. *International Journal of Life Cycle Assessment* 16: 389–391.

Sykes, A.J., Topp, C.F.E., Wilson, R.M. et al. (2017). A comparison of farm-level greenhouse gas calculators in their application on beef production systems. *Journal of Cleaner Production* 164: 398–409.

Taki, M., Soheili-Fard, F., Rohani, A. et al. (2018). Life cycle assessment to compare the environmental impacts of different wheat production systems. *Journal of Cleaner Production* 197: 195–207.

UNEP-SETAC (United Nations Environment Programme; Society of Environmental Toxicology and Chemistry) (2009). Guidelines for social life cycle assessment of products. United Nations Environment Programme.

United Nations (2015). Transforming our world: the 2030 Agenda for Sustainable Development, Resolution adopted by the General Assembly on 25 September 2015. https://www.un.org/ga/search/view_doc.asp?symbol=A/RES/70/1&Lang=E (accessed 29 August 2018).

Van Stappen, F., Mathot, M., Loriers, A. et al. (2018). Sensitive parameters in local agricultural life cycle assessments: the illustrative case of cereal production in Wallonia, Belgium. *International Journal of Life Cycle Assessment* 23: 225–250.

van Zanten, H.H.E., Mollenhorst, H., de Vries, J.W. et al. (2014). Assessing environmental consequences of using co-products in animal feed. *International Journal of Life Cycle Assessment* 19: 79–88.

Weidema, B.P. (2003). Market information in life cycle assessment. Environmental Project No 863 2003. Miljøprojekt. https://www.researchgate.net/profile/Bo_Weidema/publication/259645192_Market_Information_in_Life_Cycle_Assessment/links/0046352d1123201f38000000/Market-Information-in-Life-Cycle-Assessment.pdf (accessed 29 August 2018).

Weidema, B.P. and Wesnæs, M.S. (1996). Data quality management for life cycle inventories – an example of using data quality indicators. *Journal of Cleaner Production* 4: 167–174.

Weidema, B.P., Frees, N., and Nielsen, A.M. (1999). Marginal production technologies for life cycle inventories. *International Journal of Life Cycle Assessment* 4: 48–56.

Weidema, B.P., Ekvall, T., and Heijungs, R. (2009). Guidelines for applications of deepened and broadened LCA. 2.-0 LCA consultants, Aalborg, Denmark. https://pdfs.semanticscholar.org/8be5/9252f6790328a6360d506df522de78bbce4c.pdf (accessed 29 August 2018).

Wiedmann, T. and Barrett, J. (2010). A review of ecological footprint indicator-perceptions and methods. *Sustainability* 2: 1645–1693.

Yan, M.-J., Humphreys, J., and Holden, N.M. (2011). An evaluation of life cycle assessment of European milk production. *Journal of Environmental Management* 92: 372–379.

Yan, M.-J., Humphreys, J., and Holden, N.M. (2013). Evaluation of process and input-output-based life-cycle assessment of Irish milk production. *Journal of Agricultural Science* 151: 701–713.

Index

Innovative Processing Technologies for Healthy Grains, First Edition.
Edited by Milica Pojić and Uma Tiwari.

Printed and bound by CPI Group (UK) Ltd, Croydon, CR0 4YY
17/09/2026
14964268-0002